MAN

RICHARD J. HARRISON

Professor of Anatomy
University of Cambridge

WILLIAM MONTAGNA

Oregon Regional Primate Research Center
Professor of Experimental Biology
University of Oregon Medical School

MAN

second edition

APPLETON-CENTURY-CROFTS
NEW YORK • EDUCATIONAL DIVISION
MEREDITH CORPORATION

73 74 75 76 77 / 10 9 8 7 6 5 4 3 2 1

Library of Congress Card Number: 72-11660

This book was printed on recycled paper.

PRINTED IN THE UNITED STATES OF AMERICA

390-41648-7

CONTENTS

Preface — *vii*

1 What Kind of Creature? — 1

2 His Place in Nature — 31

3 Man's Many Kinds — 69

4 Man's Curiosity about His Anatomy — 87

5 Man's Large Brain — 105

6 Locomotion — 135

7 Man's Skin — 167

8 Characteristics of Teeth — 199

9 Man's Internal Environment — 215

10 Man's Reproductive Patterns — 233

11 Conception, Development, Pregnancy, Parturition, and Lactation — 261

12 His Sexual Behavior — 307

13 Means of Communication — 333

14 Other Sensory Systems — 357

15 Some Disadvantages of Being Man — 379

16 **The Ages of Man** 393

17 **Reflections on His Nature** 415

References 429

For Further Reading 431

Index 441

Preface

The first edition of this book was written when most other books on the biology of man were being written either by professional anthropologists exclusively for other anthropologists or by journalists and sensationalists. In the meantime, man has become more and more deeply interested in himself, and human biology is now admitted to be a subject increasingly worthy of study. Man's widespread anthropogenic and anthropologic concern for himself in serious biologic terms is relatively recent. Hence, the time is always ripe for us to look at man again. When we do so now, five years after our first assay, we find that much has changed in man's attitudes toward himself.

Someone once said that scientific books are written to be rewritten. Certainly this is true of books about man. Much of the first edition has been rearranged, rewritten, and brought up to date. Very little remains as it was. Sometimes we have had to confess ruefully to our inability to come up with satisfactory answers to some of the most pressing questions that today plague the scientist and the layman, mostly because anthropologists and biologists at large still have reached no mutual agreement. Among the most thorny of these questions are those about the origin of man and the systematics of living primates. Almost any stand one chooses to take on these topics is likely to be violently attacked. Concerning the classification of primates, for example, few agree whether the Chacma baboon should be called *Papio comatus, P. porcarius, P. ursinus,* or some other name; whether the mandrill should belong to the genus *Papio* or *Mandrillus;* or whether the Celebes ape should be placed in the genus *Macaca* (which, incidentally, we favor), *Cynopithecus,* or *Cyanomacaca.* Though such arguments may seem trivial, and we agree that they are, they are matters of life or death to systematists. In any event, we contend that it is the systematists themselves who, sooner or later and on firmer grounds than before, must resolve these problems before we mere mortals can come up with firm statements in our books.

In the chapter on human races, a topic fraught with controversy, we have unequivocally cast our lot with S. L. Washburn, who refuses to dignify the debate by assigning special names to groups of human beings who through isolation and the development of unique cultural patterns have emerged, at least physically, somewhat different from the others.

Certain books and articles have greatly influenced us in our reappraisal. Not the least of these are the many writings of Sherwood Washburn and two unique books: *Background for Man,* selected writings by Phyllis Dolhinow and Vincent Sarich; and *Human Genetics,* edited by James F. Crow and James V. Neel.

The preparation of any book that attempts to bridge the enormous span of human knowledge comprised in the single term, man, cannot be done alone. In the parturient task of editing this second edition, we have freely called upon and gratefully accepted the assistance of many and have shamefully capitalized on the generosity of friends and colleagues who have helped us in an infinity of ways for no other reward than that of seeing a job well done. From the Oregon Regional Primate Research Center, we must acknowledge the help of most of our colleagues but particularly of Drs. R. Van Horn, J. Resko, and T. Grand. How can we thank properly the punctilious and imaginative Margaret Barss, without whose constant encouragement, good cheer, and herculean help the book might have plodded on for a long time? Mrs. Maxine Allen has given both devotion and immaculate diligence to the preparation of the manuscript. Mr. Joel Ito remains an incomparable artist, scientific and otherwise, and Harry Wohlsein has produced many splendid photographs. Much of the time and talent of Miss Elizabeth Macpherson has gone into the preparation of the illustrations. Emeritus Professor Frank Goldby from St. Mary's Hospital Medical School, London, has helped rewrite sections on the brain, and D. A. McBrearty has provided photographs of material in the Anatomy School, Cambridge.

R. J. H.
W. M.

Courtesy of D. J. Chivers

Courtesy of D. J. Chivers

Courtesy of Fox Photos

San Diego Zoo Photo

San Diego Zoo Photo

San Diego Zoo Photo

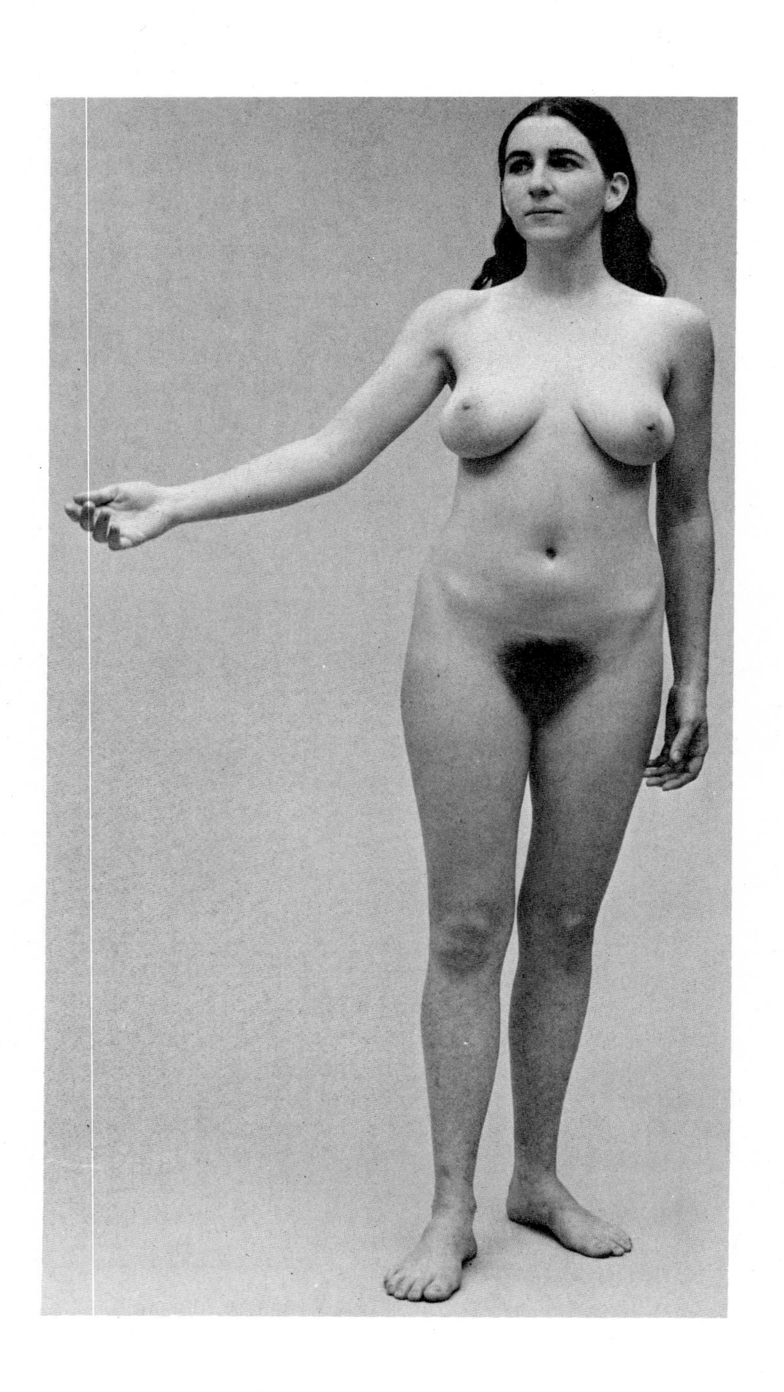

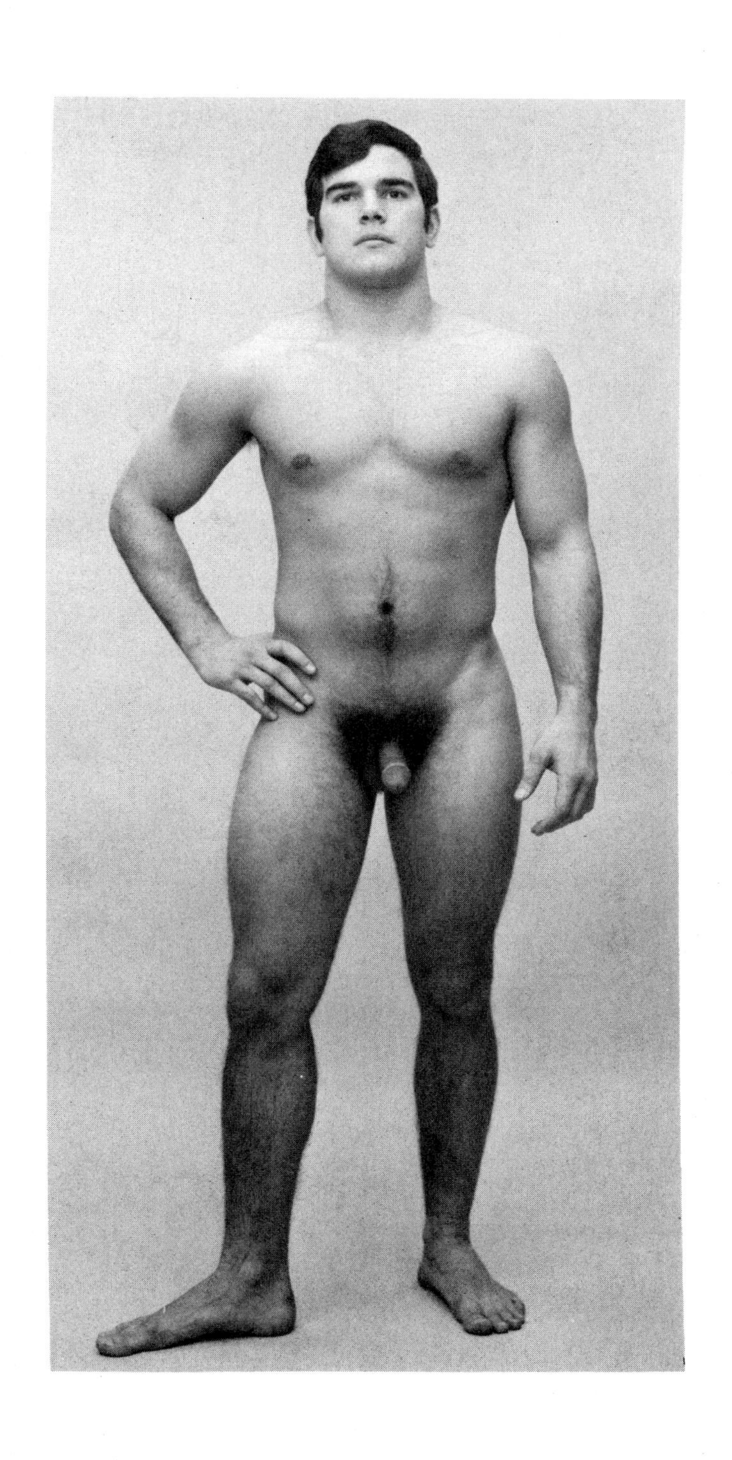

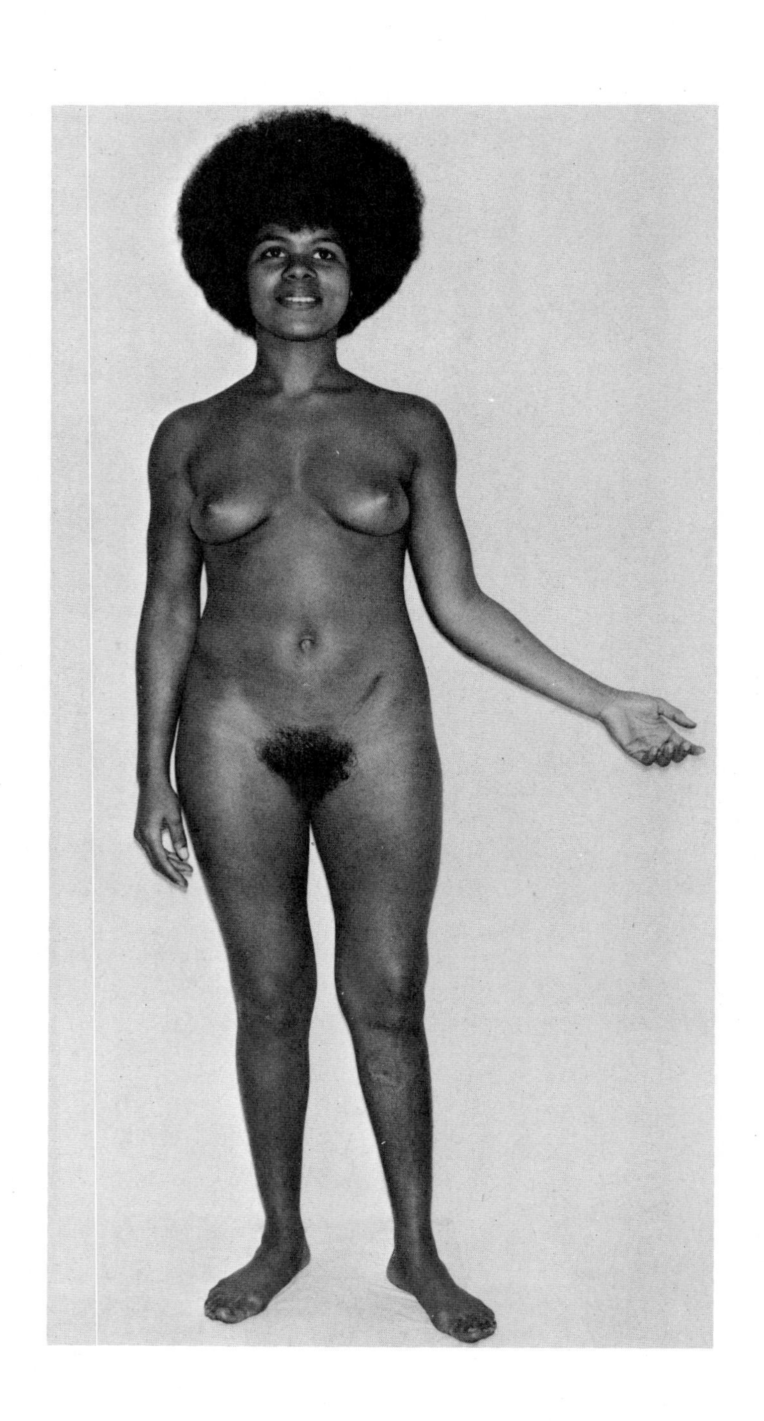

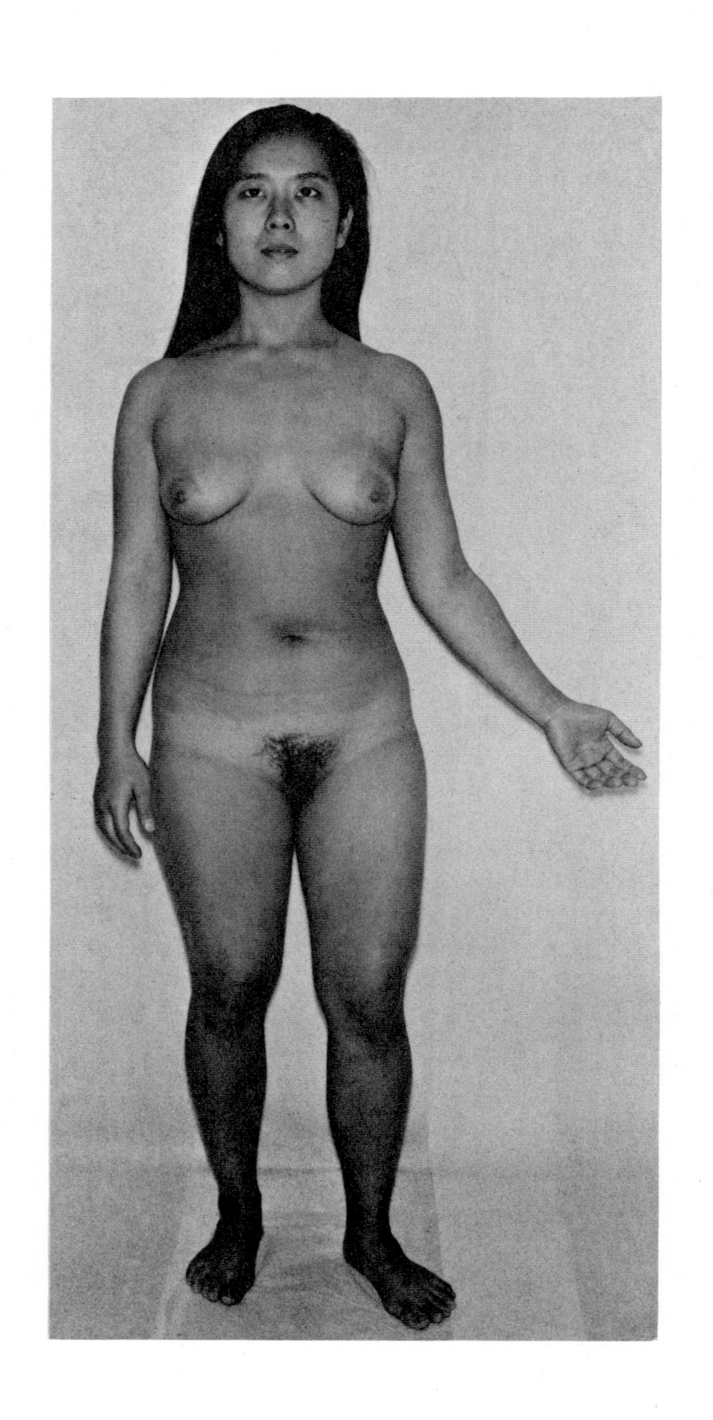

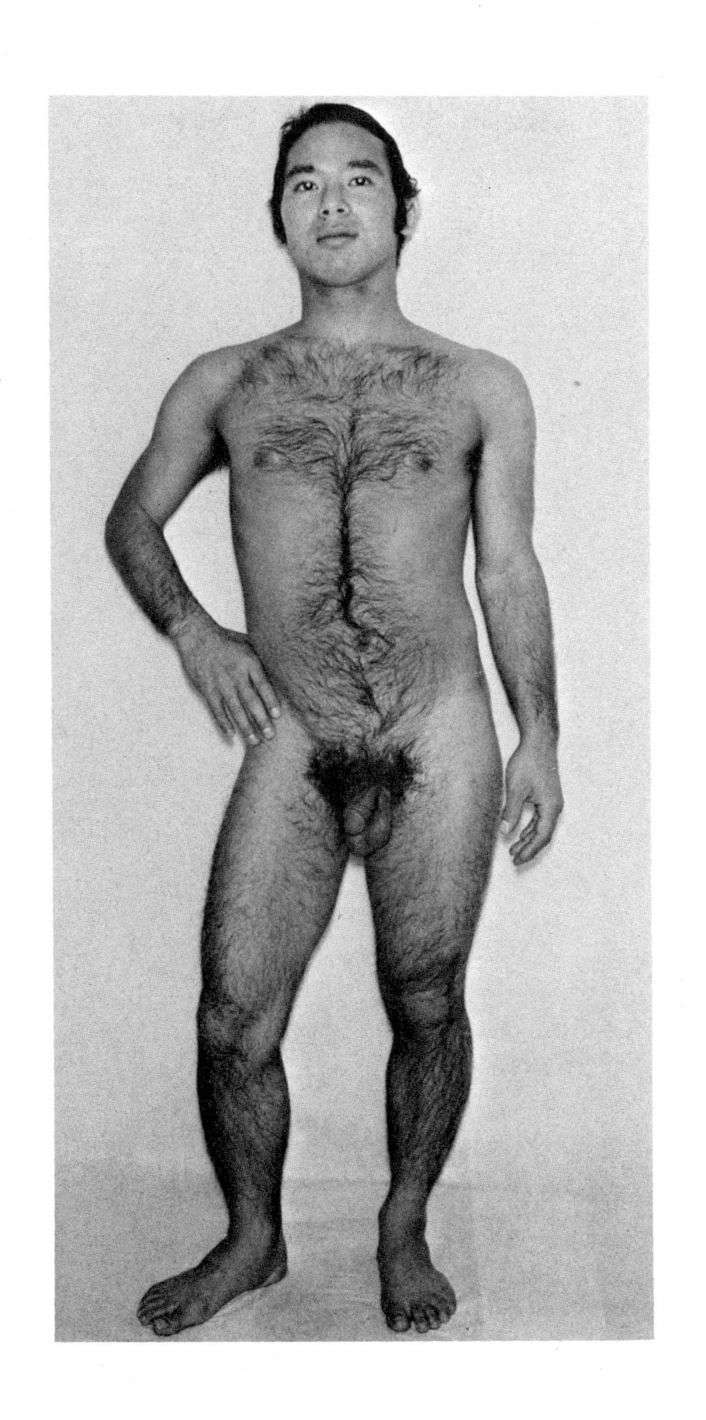

1

WHAT KIND OF CREATURE?

Am I satyr or man?
Pray tell me who can,
And settle my place in the scale;
A man in ape's shape,
An anthropoid ape,
Or a monkey deprived of a tail?

The above "prayer" appeared in *Punch* at the height of the nineteenth-century controversy on the origin of man, when Thomas Henry Huxley, then thirty-five years old, assumed the self-chosen role of Charles Darwin's advocate, became his bulldog, and dealt summarily with certain unwarranted criticisms. The immediate reaction of Victorian England to Darwin's theory of organic evolution was exemplified by Bishop Wilberforce at the 1860 meeting of the British Association for the Advancement of Science in Oxford. The Bishop demanded of Huxley whether it was through his grandfather or his grandmother that he claimed descent from a monkey. Huxley retorted that if he had to choose between an ape for grandfather or a man highly endowed by nature who employed his faculties for the "purpose of introducing ridicule in grave scientific discussion," then he unhesitatingly preferred the ape. An English anatomist, Richard Owen (1804–1892), was also opposed to the evolutionary theory, but his objection was based on the ill-chosen issue of whether or not apes and monkeys possess those parts of the brain that he claimed to be peculiar to man, i.e., the third lobe, the posterior horn of the lateral ventricle, and the hippocampus minor. Huxley answered that anyone who could not see the posterior lobe in an ape's brain was not likely to have much of an opinion on matters dealing with evolution. "If a man cannot see a church, it is preposterous to take his opinion about its altar-piece or painted window."

Huxley maintained that man's reverence for his own nobility should not be lessened by the fact that he is "in substance and structure, one with the brutes." He avowed that, in addition to his alliance with brute creation, man is endowed with rational speech, is able to accumulate and organize experience, and in his grosser nature reflects, here and there, "a ray from the infinite source of truth."

Let us ask once again what kind of creature is man. How should we begin to answer this question a century after Darwin and Huxley? Have the discoveries, during the intervening years, of fossil forms of ape-like man or of tool-using, manlike apes altered or confirmed the earlier theories? Can modern techniques of investigating man and other primates, either living or surviving only as incomplete remnants, settle these problems? Clearly, today's scientist has access to many more sources of information with which to answer these questions than had his predecessors. But even in antiquity, in fact ever since he has tried to find his place in the system of living beings, man has realized that he is an animal.

Aristotle (384–322 B.C.) and others after him envisaged the animal kingdom as a hierarchic system with man at the top as nature's perfect achievement. To determine man's affinity with the other animals, we must compare him with them. To classify an animal, we must name it and distinguish it from other animal types. We must also examine it, looking for relationships between it and the others. In essence, things must be sorted out in a formal order. With little regard for man's eminence as nature's perfect achievement, taxonomists, beginning in the eighteenth century with the Swedish botanist Linnaeus (1707–1778), using a number of subjective and necessarily arbitrary criteria, placed man fourth in the extant seventeen orders of eutherian (placental) mammals, among the Primates. The preceding three orders were the insectivores (Insectivora), flying lemurs (Dermoptera), and bats (Chiroptera). The most specialized of the orders are the hooved animals, the Artiodactyla and Perissodactyla, and the most aberrant are the whales (Cetacea).

The basic principle in classification is to group organisms according to the number and kinds of structural and functional characteristics they have in common. Each grade in a hierarchy corresponds to the characteristics possessed by all of the organisms on that level. Classification places the most specialized organisms higher than the relatively more generalized ones. Hence, the more features that are shared by the forms, the lower these forms are ranked in the hierarchy. Within any one rank such great diversities exist that secondary hierarchies become necessary. Among the order of Primates, for instance, little outward resemblance can be discerned between a bush baby (*Galago crassicaudatus,* Figure 2–6), a lowly African mammal classified as a prosimian, and man, the most developed mammal. Unfortunately, to define the primates precisely is difficult, and generalizations are nearly impossible to make. Consequently, the reasons for including certain species in the order are not always apparent.

Before going on we must pause briefly to comment on the biologist's meaning when he speaks of a structure as being *primitive* or *generalized,* and *specialized.* Much confusion has arisen from the carelessness with which these terms are used. Whereas the words *primitive* and *generalized* are clear enough when used to designate certain anatomical features, they become almost meaningless in other cases. *Primitive* denotes the possession of characters that existed in ancestral living forms whereas *specialized* implies a departure from an ancestral pattern—a particular adaptation for a specific function.

For example, some of the more ancient fossil skeletal remains of fishes show that their vertebrae were composed of two separate and distinct elements. Thus, when we find such a condition in extant vertebrates, as in the sturgeon and bowfin, we can say with some assurance that it is a primitive condition, since most vertebrates have vertebrae with the two elements completely fused. Today the dual origin of this bone is suggested only during development. Sturgeons and bowfins, then, have retained primitive or generalized vertebrae, but this alone does not warrant their classification as primitive.

Similarly, the ancient pattern of vertebrate appendages, which terminated in a hand or foot each with five digits and a specific number of supporting bones (Figure 1–1), has relevance in deciding whether the wing of a bird and the hand of a mole are primitive or specialized structures. During early development in the egg, the wings of most birds have vestiges of all five digits and their supporting metacarpal and carpal elements; however, in adult birds only the vestiges of three fingers and two carpals remain, and the fingers are reduced in size, fused, and modified to support feathers. The wing of a bird, then, is a specialized structure since it has markedly departed from the archetypical pentadactyl (five-digit) pattern of vertebrates. The hand of the mole has five fingers, which would seem to indicate the retention of a primitive or generalized anatomy. However, the total structure of this hand is so thoroughly modified and adapted for digging burrows that it is also specialized. Likewise, the foot of man, which has five digits, may not seem to be as highly specialized as that of the horse, which for practical purposes has only one digit; but man's foot has deviated almost as much as the horse's hoof from the ancestral pattern. The retention of any single ancestral trait, then, cannot be considered in itself to be ancient; one must always consider the specific changes in a pattern.

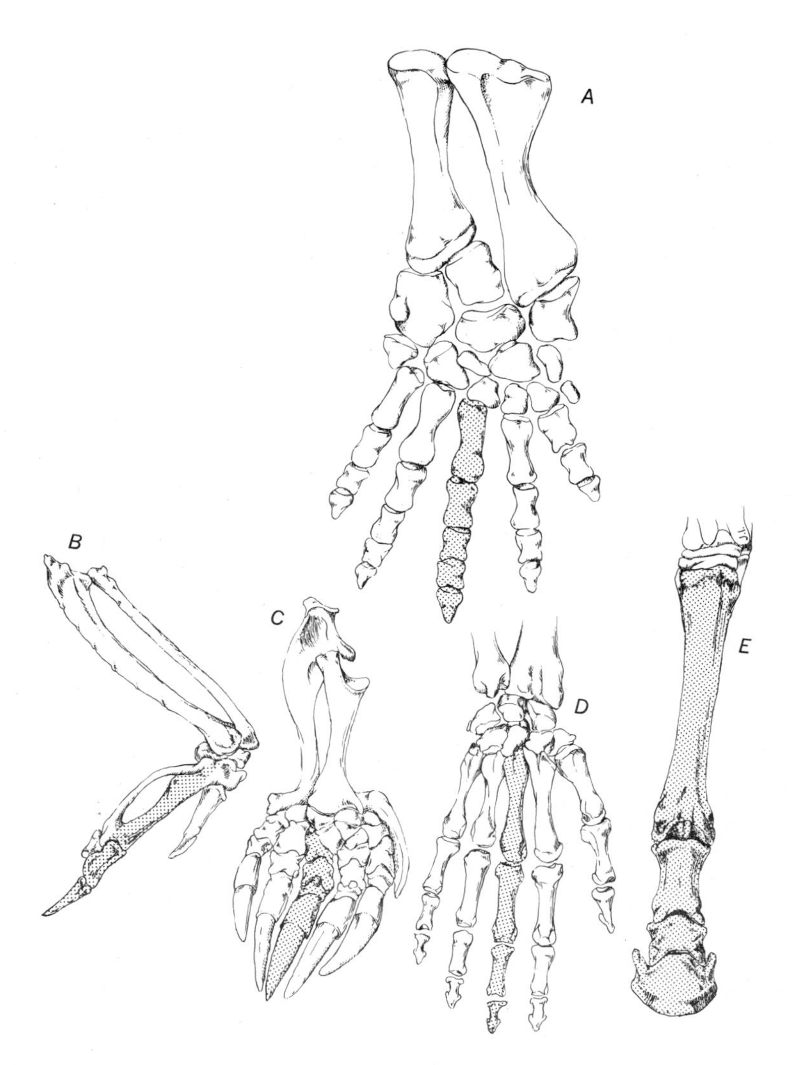

Figure 1–1. The bony elements of "primitive" and "specialized" "hands." The metacarpal and phalanges of the middle digit, number 3, are stippled. A: a primitive fossil mammalian hand with many bony elements. B: the highly specialized wing of a bird, where digits 1 and 5 have disappeared and 2 and 4 are greatly reduced; only digit 3 has two phalanges. The metacarpals of the three extant digits are fused with each other and with the distal row of carpals giving rise to the carpometacarpus. C: the many bones in the broad hand of a mole, specialized for digging laterally. D: the generalized hand of man. E: the highly specialized "hand" of the horse, consisting of digit number 3 (three phalanges) and its long metacarpal. Remains of the metacarpals of digits 2 and 4 are on the posterior side, out of sight. Digits 1 and 2 appear briefly during embryonic development and then disappear.

The biologist, therefore, should use these terms with caution, and the reader must interpret them relatively. The difficulty lies in attempting to assign a term of absolute connotation—primitive, generalized—to structures that exist within a specialized matrix. To speak of human structures as generalized or primitive is almost a deliberate attempt at vagueness. One should always ask, "Primitive in relation to what?"

Man, then, is a vertebrate animal in the class Mammalia and in the order Primates and along with monkeys and apes belongs to the suborder Anthropoidea. Because so many of his features resemble those of apes rather than of monkeys, he is classed with the apes in the superfamily Hominoidea. He is the tallest of the Hominoidea, and only the gorilla sometimes surpasses him in size. Superficially, he appears to have a relatively primitive structure, to be a seemingly hairless ape with a large brain, and to retain certain fetal characteristics in his adult anatomy.

Henceforth reference will be made in the text to various types of primates, so the reader should familiarize himself with the names of the primate families given in Table 1–1.

Man is bipedal, walking upright (orthograde) on his soles in a plantigrade fashion. This type of locomotion is associated with numerous adaptations in his skeleton, from the foot to the skull. Bipedalism is also observed in some nonhuman primates. When gibbons (a lesser ape)

Table 1–1. Primates

Prosimian families	*Examples*
Tupaiidae	Tree shrews
Daubentoniidae	Aye-ayes
Lorisidae	Lorises, bush babies
Lemuridae	Lemurs
Indridae	Indris
Tarsiidae	Tarsiers
Anthropoid families	*Examples*
Callithricidae	Marmosets, tamarins
Cebidae	New World monkeys
Cercopithecidae	Old World monkeys
Hylobatidae	Lesser apes, gibbons
Pongidae	Great apes
Hominidae	Men

The superfamily Hominoidea consists of the families Hylobatidae, Pongidae, and Hominidae.

walk, they do so mostly bipedally; and the Japanese macaques and Celebes black apes (two species of Old World monkey, probably both macaques) get along very well on two feet, particularly when they are clutching something in both hands. Since the upper limbs are not needed for locomotion, man can use them for prehension, particularly since the arms are strutted to the body by a collarbone (clavicle). Man's forearm is long and can be rotated with the arm by the side so that the palm faces forward (supinated) or backward (pronated). The large thumb (pollex) can be opposed to each of the other fingers to grasp either delicately or powerfully. The expanded pelvis, which helps to support the internal organs (viscera), has a capacious cavity associated in the female with the birth of a large-headed fetus. The legs are long and greatly strengthened; the hip in the upright position is extended, as is the knee joint; certain muscles involved in walking are enlarged. The great toe (hallux) is large, and the bones of the feet are arched. The skull possesses a large cranial cavity; the face is oriented forward in the upright position, and the head is poised above the vertebral column. Although man has a prominent chin, his jaws are weak and his teeth small.

The human population on earth is increasing rapidly, yet man usually bears only one young at a time, with twins occurring once in every 85 births and triplets, quadruplets, and quintuplets even more rarely. He has a gestation period of 250 to 285 days, and no evidence of a restricted breeding season exists. Both sexes reach puberty between 10 and 16 years, when in girls a 28-day reproductive cycle commences that is repeated for 30 to 35 years. The cycle has external manifestations in the phenomenon of a 3- to 5-day period of menstruation every month. Reproductive activity ceases in women at menopause, about the age of 50. Men have no definite sexual cycle nor a determinable cessation of sexual activity. Man is probably the longest-lived of all mammals.

A social, gregarious, and political animal, man lives a complex existence with a unique culture and an aesthetic sense. He displays various degrees of freedom from conduct enforced by primitively predetermined responses in behavior. He is educable and can develop complex mental processes. He can communicate in many ways, principally by articulate speech, and has a marked capacity for expressing his speech and thoughts in symbols.

Man's Size

A feature that is often overlooked is man's size. He is not so large that he has to spend all day finding and eating food to sustain his frame and energies, nor is he so small as to be at the mercy of predators and the climate. Certainly he takes a long time to reach his adult size: his body attains 5 percent of his mature weight during intrauterine life and 95 percent during a postnatal growth period of about 20 to 25 years.

Physical development is rapid during infancy. Infants increase close to 50 percent in length and nearly 300 percent in weight during the first year. Growth rates are halved in the second year and then continue more slowly and uniformly until puberty. Thereafter growth rates decrease, maximum adult height being reached before adult weight, which is greatly influenced by the disposition of fat. The size of any man or woman is determined by many factors—genetic, hormonal, nutritional—but all things considered, man's size would seem to be just right for survival. Pygmies and giants in the Congo and Nile regions have not flourished outside their environment. Tanner (1964), from his studies of the physique of Olympic athletes, concludes that with singular exceptions, such as the modern giants in basketball and the leviathans in American football, a man does not need to be particularly tall or short to be an outstanding athlete.

Compared with most animals, man is large. He seems small compared with the blue whale *(Balaenoptera musculus)*, the largest of which weighs up to 120 tons, but such bulk requires a marine environment to provide buoyancy. Dinosaurs and some extinct mammalian herbivores were the largest land animals, but their huge size was a specialization that probably led to their extinction. There are relatively few living terrestrial forms larger than man and fewer still that are biologically successful. Were the world to be suddenly deprived of gorillas, whales, and elephants, of the large hooved creatures such as giraffes, cattle, camels, and horses, and of such large carnivores as bears, walruses, and lions, man

Table 1–2. Average Heights and Weights for Boys and Girls

Age (years)		*Birth*	*1*	*2*	*5*	*13*	*15*	*20*
Height (inches)	Girls	19.9	26.0	29.0	39.6	59.7	62.5	64.0
	Boys	20.2	29.5	33.9	42.7	58.9	62.5	68.0
Weight (pounds)	Girls	8.0	21.25	27.25	41.25	93.0	111.0	*
	Boys	7.5	22.0	27.75	38.5	96.0	118.0	*

* *Weight varies so greatly at this age that an average would be meaningless.*

would qualify as the tallest and heaviest of the remaining mammals. The reader will quickly think of some nonmammalian exceptions; giant turtles and fishes, pythons and ostriches are taller or longer and perhaps heavier; but, one way or another, most of the extant animals are smaller than man.

Man's Shape

Inert and elderly, cadaveric man is not prepossessing in appearance nor anatomically striking (Figure 1–2). The sculptor and the painter see man in activity—vital, splendid, well proportioned, even godlike. Clothes, too, improve man; he becomes elegant, fashionable, eloquent of his breeding and taste. But naked on a slab, he appears much like a peglike or puffy doll, with an almost porcine trunk, unequal, gangling knobby limbs of pentadactyl pattern, a large, round, pedunculated head, and no tail. He lacks the striking external features of some animals: no covering coat of fur or whiskers, no tusks or muzzle, no elephantine trunk, no horns, claws, or hooves, no sexual skin of brilliant hue.

Man's external form, molded by his subcutaneous fat, has relatively smooth curves and contours, seen to best effect in babies and women. This blanket of fat helps to preserve body temperature, gives partial insulation from changes in external temperature, and affords protection from minor buffets. There is little subcutaneous fat in the eyelids and scrotal skin. The particular significance of this fat is seen in mammals like seals and whales, which spend much or all of their lives in the cold sea. The special thickness of fat in these animals indicates clearly its protective function; it also provides a source of reserve nourishment that can be called on when food is scarce or when the female is lactating.

In women, within a wide range of individual differences attributable to genetic factors, age, and parity, the breasts are large compared

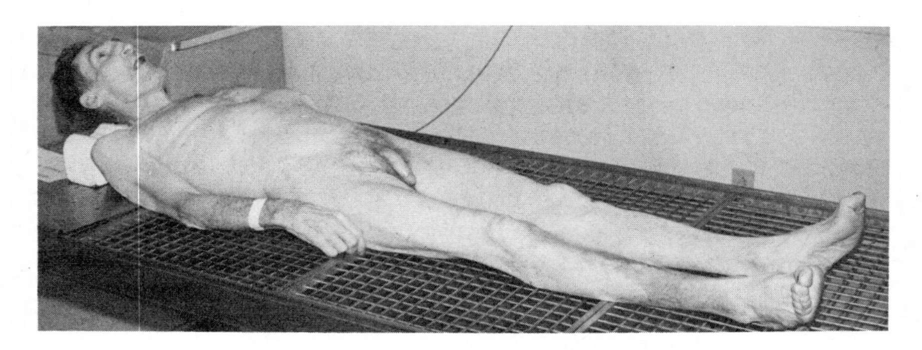

Figure 1–2. A dead man.

with those of other mammals. Unlike the breasts of most other mammals, which are abdominal, the two human breasts are pectoral. This indicates that one or possibly two offspring are born at a time. Most monkeys and apes also have two pectoral breasts. In South American monkeys and lemurs, the breasts are in the axilla, and some prosimians have three pairs of them. The curious aye-aye (*Daubentonia madagascariensis*, a prosimian) has a pair of nipples in the inguinal region. Only a few other mammals have pectoral breasts—bats, some aquatic forms, and elephants.

Man's head, the flower at the top of the human plant (as it impressed the German poet and anatomist Goethe) is that part where least tissue intervenes between bone and skin. This is particularly true of the cranial vault. Attempts to reconstruct a physiognomy from a skull have resulted in hopeless inaccuracies. An English anatomist, Frederick Wood Jones, wrote:

> It was my lot for seven years to share my room with the skeleton of Professor Collimore. As I worked at my desk his skull looked directly towards me and at the end of seven years I fancied that I knew very well what manner of face he had, and how his features were formed. It was only then that I managed to obtain a photograph of the living man, and it was at once apparent that the face was utterly different from the one I had reconstructed.

A Russian anatomist, M. Gerasimov, however, has claimed that remarkable, even recognizable, likenesses can be obtained by rebuilding artificial soft tissue onto a given skull, the average estimates of regional soft tissue thicknesses having been determined by repeated measurements of cadavers.

Attempts to confirm identity have also been made by superimposing an accurately enlarged photograph of the head on one of the skulls at similar magnification. This method has been used to confirm the identity of the skulls of famous persons long dead and to substantiate identity in legal matters. Yet, numerous snags can trip the unwary, and the method is perhaps really valuable only in demonstrating that a certain skull could not possibly be that of a particular individual.

A well-defined neck, which is related primarily to an ability to rotate the head, separates man's head from his trunk. The neck is not a human nor even a primate characteristic. Among the primates the primitive tarsier uses its neck to best advantage: its head can look east or south when the animal faces west or north, respectively. Man's neck is usually slender; in those people with scanty subcutaneous fat one can clearly see the contracted sternomastoid and other muscles beneath the neck skin. A head well balanced on the vertebral column does not require thick, heavy muscles to keep it upright as does the massive cranium of a gorilla. The midline prominence in the front of the neck, more marked in men and called the Adam's apple, is produced by the thyroid cartilage of the larynx. This pomum Adami, associated in mythology with original sin, can be felt in other primates, but is not usually visible.

The human head is composed of two parts (Figure 1–3). The larger, which contains the brain, is supported on the vertebral column by the strong skull base and is surrounded on the sides and top by the thinner bony vault (calvaria) covered by the scalp. The smaller part, the face, is given its prominent features by the underlying skeleton of jaws and facial bones, nasal cartilages, and facial muscles. The braincase, even some months before birth, is the larger, rounder, and more imposing component. At birth it is eight times larger than the facial component, but only one-fourth its potential adult size.

The human skull is so large that the cranial capacity of the more highly civilized races of man is about 1,500 cc. On the other hand, the largest recorded cranial capacity in a great ape, an adult male gorilla, is about 650 cc. Sir Arthur Keith, a Scottish anatomist who worked in London around the turn of the century, once suggested a dividing line in cranial capacities at about 750 cc; anything below this would indicate apehood, anything above might well indicate manhood.

Man's skull lacks the great overhanging frontal ridges, the large median bony crests on the crown, and bony flanges at the back of the head (occiput) that characterize the skulls of great apes (Figure 1–4). This buttressing of the simian skull is associated with the powerful jaw musculature and with the need to hold the heavy skull upright. A rounded skull also indicates that its contents are encased with the greatest economy.

The human scalp consists of skin, subcutaneous tissue, a thin fibromuscular sheet (the epicranius muscle and the galea aponeurotica), a layer of loose subaponeurotic tissues and, deepest of all, the fibrous pericranial layer of the outer skull. The first three layers are closely united and move slightly on the pericranial layer, which firmly adheres to the bones of the vault of the skull. When a person is scalped or when a flap of scalp is turned down during surgery, the outer three layers separate away through the lax tissue covering the pericranium. Monkeys can move their scalps more easily than men. However, Charles Darwin mentioned a family whose members could pitch books that had been placed on their heads by moving the scalp with the sheetlike epicranius muscle.

Man's Face

At birth the human jaws and face are small compared with the size of the braincase. They grow much larger in adulthood, but man never typically possesses a heavy, projecting face like the apes. Modern man is more

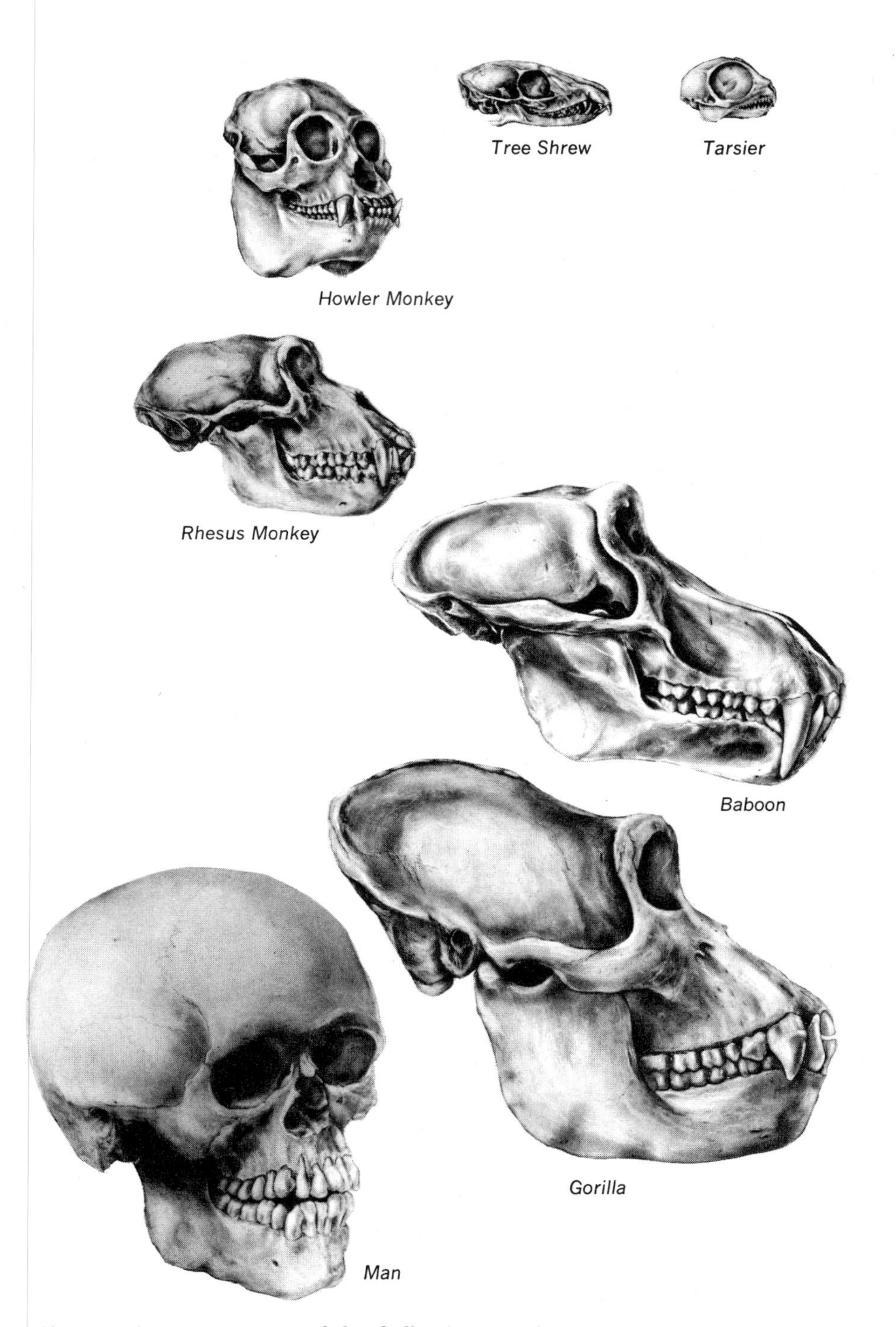

Figure 1–3. A comparison of the skulls of man and other primates, drawn to scale.

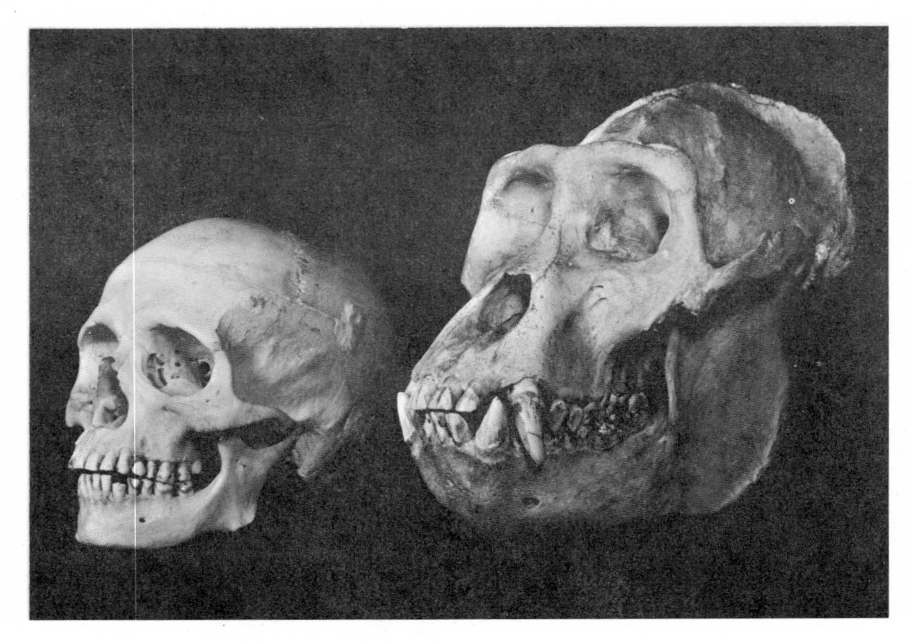

Figure 1–4. Skulls of a male gorilla and a man. Note the great overhanging brow ridges and the bony flanges at the back of the gorilla skull.

properly described as orthognathous than prognathous because of his receding jaws and flattened face. Great apes, fossil men, and primitive extant men all show progressive lessening of jaw prominence. This recession has occurred, moreover, without a reduction in the number of teeth (see chapter 8).

Man's high forehead, his noble brow, is an anatomical indication of marked development of the frontal lobes of his brain, and the transversely creased skin of his forehead increases his ability to express emotion. This persistence of a heightened forehead is considered one of the fetal characteristics that survive into adult life. The forehead is even more marked in the adult because from the age of ten certain air sinuses develop in the underlying frontal bone. These numerous air-containing spaces contribute considerably to the shape of man's adult face. Just before birth, they originate as hollow evaginations of mucous membrane from the nasal cavities. They soon invade or are enclosed by the developing bones of the young skull—the maxillary sphenoid, ethmoid, and frontal bones of each side—and the spaces formed become the paranasal air sinuses. Poorly developed at birth, the sinuses grow slowly until about the age of four, more rapidly during the period of teeth eruption, and cease at about twenty-one to twenty-five years of age. They may become enlarged again in old age as the bone surrounding them is resorbed.

Sinuses play an important part in hollowing out, modeling, and lightening the weight of facial bones. They act as resonating chambers and, because they are relatively larger in men than in women, are responsible for some of the voice differences between the sexes. Since small apertures connect the sinuses with the nasal cavities, nose infections can spread to them. Inflammation of the sinuses often causes headaches in the forehead region and aches between or behind the eyes or in the cheek, depending upon which sinuses are involved.

Other mammals also have sinuses. Notwithstanding its great size, the skull of elephants is surprisingly light because numerous air spaces make up much of its bulk. In quadrupedal mammals the sinuses drain more freely into the nose. Being upright, man has the disadvantage that some sinuses, such as the maxillary, do not drain into the nose from their most dependent part, and fluid and pus can, therefore, accumulate in them.

The nose of mammals is involved with breathing and with olfactory reception. The external opening is usually paired, even in fish, which have no internal opening into the mouth. The evolution of a palate has resulted in long nasal cavities or air passages opening into the back of the mouth (choanae). Coiled scrolls of bone (turbinals), covered with epithelium, subdivide the air passages in mammals, filtering, warming, and moistening incoming air. Olfactory reception occurs at the top and back of the airway. The turbinals reach their highest complexity in mammals with a keen sense of smell. Man possesses only three scrolls on the lateral wall of each nasal cavity.

Man's external nose shows some racial differences, but in general it is higher than it is broad. A cartilaginous skeleton connected to the bony bridge gives the nose some mobility. Since the skin adheres tightly to the cartilages, inflammation of one of the many skin glands can be unduly painful. The narrow opening is guarded by inner nasal hairs, not found in other mammals, and by a moustache. In other mammals, the nose may be variously modified: into a proboscis, like the elephant's trunk, which is a most delicate muscular specialization, or into a single or double blowhole, like that on the highest point of the head in whales. Man can move his nostrils only slightly whereas marine mammals and some others can close them tightly by means of muscles or complicated pocket valves.

The hairless, rosy lips (L. *labia*) of man are so freely movable that the word *labile* arises from this attribute. The human upper lip has a central vertical furrow, called a philtrum, limited on each side by a slight ridge and sometimes ending below in a slight prominence or tubercle. In certain lower primates, the philtrum is naked and the upper lip is tethered at the back; but in higher forms, including apes and man, it is much freer and so more mobile. Man's lips are essential in verbal communication and in eating and sucking. A child born with a cleft in its upper lip (harelip) has difficulty in sucking, especially when the harelip is associated with a cleft palate. The lower lip sometimes pouts outwards more than the upper, the so-called Hapsburg lip seen in members of the Hapsburg royal line and in other persons. A British zoologist, Sir Gavin de Beer, has suggested that "their present function of kissing may have been of considerable importance in the sexual selection which has undoubtedly accompanied the evolution of man." The highly differentiated muscles of facial expression, the so-called mimetic muscles that also move the lips, stamp man with an unmistakable sign of intellectual development. In the functional perfection of these muscles, man surpasses all other mammals, including the great apes. From a dissection of fetuses and adults of different races, one gains the impression that the evolution of facial musculature continues and has progressed farthest in Europiforms.

Studies of mammalian facial elements indicate that the development of facial musculature has always been accompanied by increasingly complex facial expressions as a means of social communication. Facial musculature is present in various degrees in all mammals, but only in the carnivores, ungulates, and especially the primates is a very elaborate system of social signals concentrated in the face. Some scientists believe that this development is probably correlated with the sense of vision. Because of the extraordinary development of their facial musculature, primates have a number of typical "compound expressions" that are related to the basic motivational state of the animal. Each of these compound expressions consists of combinations of movements and postures of the jaw, lips, corners of the mouth, tongue, eyebrows and scalp, eyes and eyelids, and ears. Even the uninitiated will have recognized that some monkeys grin, some grimace, others smack their lips, look complacent, fearful, or angered. These are special signals of communication. Those who have made detailed analyses of these facial expressions have come to recognize the specific pattern of each.

Man's tongue is relatively short and cannot be protruded very far. The lowest vertebrates have only a rudimentary tongue; certain fish have teeth-bearing bars in the floor of the mouth that bite against palatal teeth. In frogs, toads, and chameleons the tongue is specialized for capturing food. Life on land means that no water helps to manipulate food and ease its passage through the mouth. Hence, salivary glands are important to land animals, and the tongue becomes an essential manipulator of saliva-moistened food. The tongue also guides food between the teeth and takes part in swallowing movements after chewing. In the ungulates the tongue is long and muscular and is used to convey food to the mouth. Ant-eating mammals have a highly specialized tongue that can be extended great distances. In all mammals the tongue acts as a sensory organ, detecting objects in the food that are dangerous to swallow or damaging to the teeth, and discerning the temperature of food. The tongue also possesses special sense organs—taste buds—which react to sweet, salt, acid, and bitter substances in food and report their presence to the brain. Since many naturally occurring poisons are bitter or acid to the taste, the selection and retention of this gustatory attribute seems to be an admirable adaptation aiding in the survival of land vertebrates.

Man also uses his tongue to produce articulate speech. The complex intrinsic and extrinsic muscles of the tongue make it movable and malleable so that it helps to produce many different kinds of sound. It plays a part as well in the language of gesture—biting the tongue to indicate uncertainty or anguish, licking the lips in anticipation, and protruding the tongue in rudeness. The tongue cleans the teeth, and its movements ensure that conditions in the mouth return rapidly to normal after meals. The shape of the tongue shows some variation with an individual's build, e.g., it is long and pointed in tall, thin people.

Man possesses a distinct chin whereas apes and most fossil men do not. The two main components (rami) of the human lower jaw show some anterior convergence—when one grips it between thumb and forefinger, one discovers that the tips come closer together as the point of the chin is reached. The rami of the simian mandible are less convergent because the posterior teeth are arranged in parallel rows. Unlike that of man, the mandible of apes is reinforced at the inside of the chin by a shelf of bone called the simian shelf.

The external ear (pinna) of man is not particularly different from that of the higher primates (Figure 1–5). As in other primates, it is developed by the fusion and subsequent molding of six embryonic hillocks that arise around the outside of the ear passage or external auditory meatus. The ear is subject to remarkable variation in shape and size and is composed of a corrugated plate of cartilage covered by closely adherent skin. Rudimentary muscles are attached to the ears, and a few gifted people can actually move their ears. In the ungulates, carnivores, and some primates, these muscles are fashioned so that they bring the ear into an erect position, rotate it, and depress it. Tarsiers and galagos (prosimians) have large mobile ears, monkeys have relatively shorter ones, and those of most apes resemble man's. Interestingly enough, the reduction of the external ear in the higher primates took place at the same time that these animals acquired the habit of sitting up. In this posture the head can be so turned that sounds go directly to the ear.

An English sculptor, Thomas Woolner, showed Darwin a peculiarity in the external ear—a small blunt point that he had noticed on the inwardly folded margin of the rim of the pinna while working on a figure of Puck. Darwin regarded such points, which vary in their position on the rim (Figure 1–6), as the "vestiges of the tips of formerly erect and pointed ears." Now called the satyr point when at the top of the pinna,

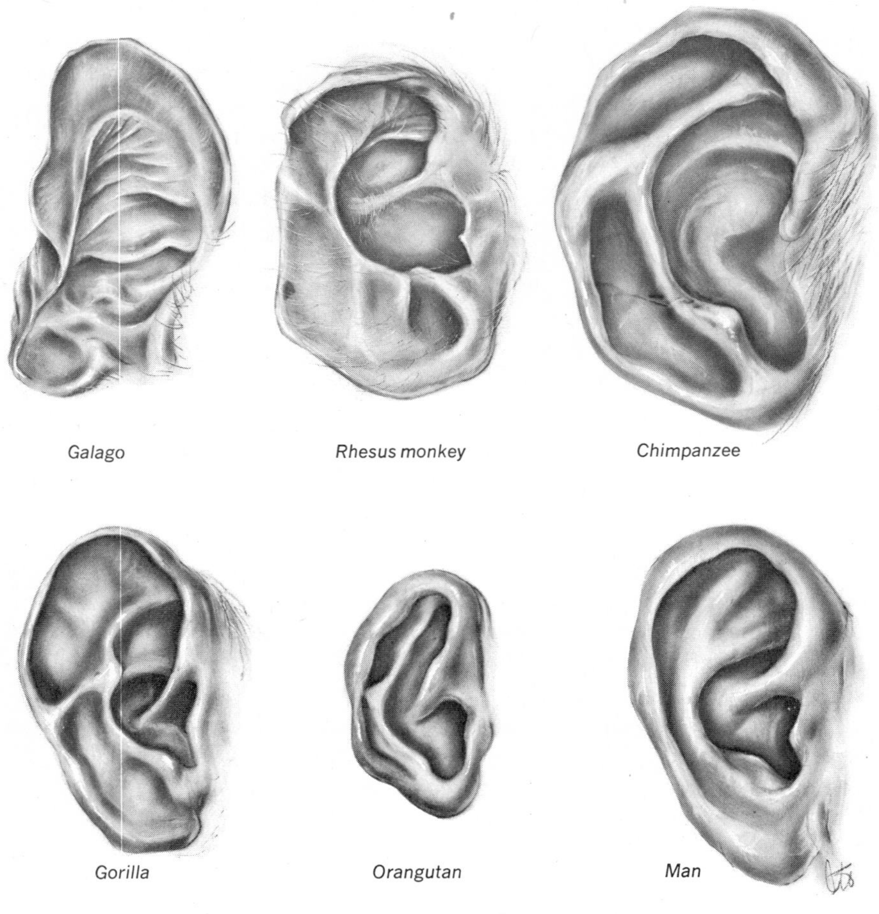

Figure 1–5. The various shapes of the external ears of some primates.

this occasionally occurs also in the gorilla. The bony external ear passage is short in man, its configuration and that of the neighboring jaw joint being different from those of other primates.

Man's Teeth

The marked recession of the jaws and the flattened, more upright face are associated with an upright posture and with the setting of the enlarged brain case in a better-balanced position on the vertebral column. Likewise, the reduction of the jaws and widening of the face are asso-

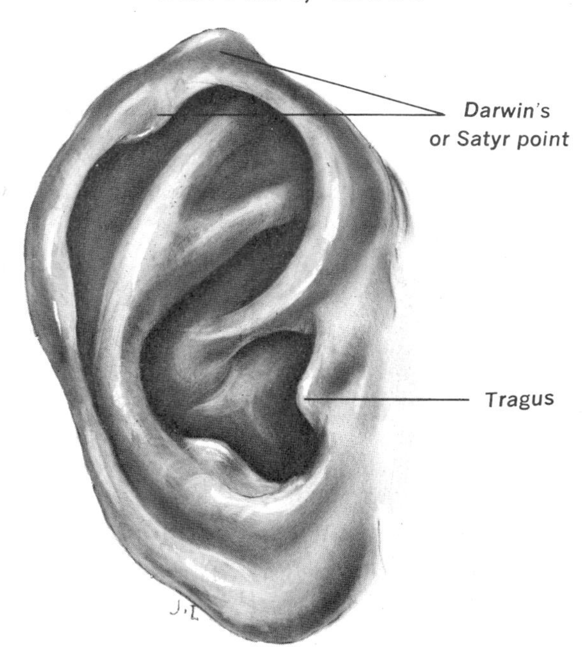

Figure 1–6. Human ear, showing the two locations of Darwin's or satyr points.

ciated with the rounded anterior part of the dental arcade and the shorter series of smaller teeth. It is as if the human jaws had been more elegantly constructed, built not on the strong, robust plan of the ape with its reinforcing simian shelf, but rather according to a plan in which all the masticatory features are much less marked. The absence of ridges for the attachment of masticatory muscles thus leaves the skull smooth and rounded.

In common with Old World monkeys and apes, man has ten deciduous and sixteen permanent teeth in each jaw. Some evidence of overcrowding of teeth in man's jaws exists, as if the recession of the jaws had not left enough room for the teeth. The relatively small teeth, arranged in a parabolic curve in the lower jaw and an elliptical one in the upper, all show certain human refinements. The incisors are more chisel-edged and somewhat spadelike (but not so spatulate as those of the apes) and their vertical position in the jaws is more pronounced in western Europeans. The canines, about the same size in men and women, are reduced in length so that the tip of the crown hardly projects beyond the level of the incisors and premolars. In the higher subhuman primates, canines are larger in the males. The premolars in the upper jaw have one root, sometimes two, while those in the lower jaw most often have a single root. The remarkable similarity of all the premolars often makes their exact identification difficult. They possess two cusps—an inner lingual and an outer buccal—which in the lower first premolar are connected by a ridge, a primitive characteristic found in fossil apes. The three molars in man are not aligned in parallel rows on each side of the two jaws as they are in apes, but rather the last molars are farther apart than the first and are smaller than the anterior two. The shape of the upper and lower molars differs considerably in detail; however, in general, the crowns are rhomboidal or cuboidal, and the cusp arrangement (four on each upper and an additional fifth on each lower) shows modification of a fairly primitive pattern.

During their evolution, primates have tended to develop a specialization of the incisors and canines and to retain relatively primitive molars. Most have eight incisors, four on each jaw. In some prosimians the extremely small upper incisors contrast with the lower ones, which are elongated and extend forward in a comblike pattern. In many of the higher primates the canines, usually large and pointed, attain a prodigious size in the males and are effective tools for tearing. They are also used in display; male baboons, for example, when alarmed, frightened, or angered, yawn elaborately, displaying their saberlike canines. Most higher primates have eight premolars, two on each side and four on each jaw, but all prosimians (and New World monkeys) have twelve. All primates, except marmosets (p. 202), have twelve molars, which, when compared with the archetypical dentition of mammals, have remained relatively primitive.

Because of their toughness, teeth or their fossils are often among the few relics of extinct animals. Hence, teeth have been extremely useful to anthropologists in piecing together the evolutionary history of vertebrates.

Man's Nakedness

Aside from the features briefly mentioned above, man's skin immediately distinguishes him from all other mammals and all other primates. Erroneously referred to as glabrous (without hair), man is in reality hairy, even though over the greater surface of his body the hair is rudimentary. As man's hair has become regionally vestigial, his skin has undergone adaptive changes that depart radically from the basic structural and functional patterns of the skin of most other mammals. When considering the uniqueness of man's skin, we must link all its specific attributes to his apparent nakedness.

Man has a sharper awareness of and concern with his skin than with any other tissues. After all, he can see it and feel it constantly. Skin is also one of his largest organs, functioning as a shield for the entire body against physical, chemical, and mechanical injuries, and as a preventive against the escape of body fluids and the entry of alien fluids. In mammals, skin gives rise to hairs, spines, nails, claws, hooves, scales, horns, and indirectly even to antlers. In birds, it produces feathers, scales, a horny beak, spurs, and claws. Various subcutaneous skin glands manufacture and secrete substances that may be watery or viscid, fatty or proteinous, colorless or pigmented, odorless or fetid.

Skin is man's major medium of sexual attraction. In order to enhance its attractiveness, man has subjected it to centuries of abuse. He has stretched, compressed, scraped, gouged, soaked, and burned it without, however, causing it irreparable harm. Despite constant application, it resists the ravages of powders, poultices, oils, and chemical and physical irritants. The regimen of daily baths dictated by a superlatively fastidious society can injure the skin by washing away natural emollients and essential substances from the surface, but man's skin has remained remarkably indifferent to these daily and other torments.

Over and above the size and shape of the body, skin hallmarks the identity of animals. For example, only mammals have hair: birds have feathers; reptiles and fish have scales and spines; and most amphibians have a genuinely naked, shiny skin. As mentioned earlier, skin and the underlying muscles mold the facial and body contours, and one person can be distinguished from another by his cutaneous characteristics. The human body, stripped of its skin, is much the same even in individuals who look different.

Man's nakedness, then, is merely apparent; he is really quite hairy. The only genuinely glabrous areas are the palms and soles, the rosy part of the lips, the lower surface of the wrists, and the glans penis. The number of hairs per unit area of the human body, both on the head and elsewhere, is greater than in most monkeys and great apes. Over most of man's body individual hairs are very small. At puberty stout hair appears in the axilla, on the face in men, and over the pubic region in both sexes. Hairiness in man at any given age varies not only in degree but also in texture, color, and curliness. In addition, the character of hair varies in the main races of man.

Notwithstanding the few exceptionally hairy individuals one sees on a beach, man's body is becoming progressively more naked, and one can predict that in time he will be totally glabrous. Since his capillary cover is waning, perhaps those existing ethnic groups that are the least hirsute may be the more advanced ones. Man's pelage, even that of hairy individuals, has limited effectiveness in protecting him from his environment. Reflect upon the absurdity of a practically naked animal with rich tufts of hair growing here and there over his body. Since, aside from scalp hair (see p. 188), these strange tussocks have little usefulness, they must be considered ornamental (Figures 1–7, 1–8). This development of ornamental hair is not unusual in mammals; many display peculiar

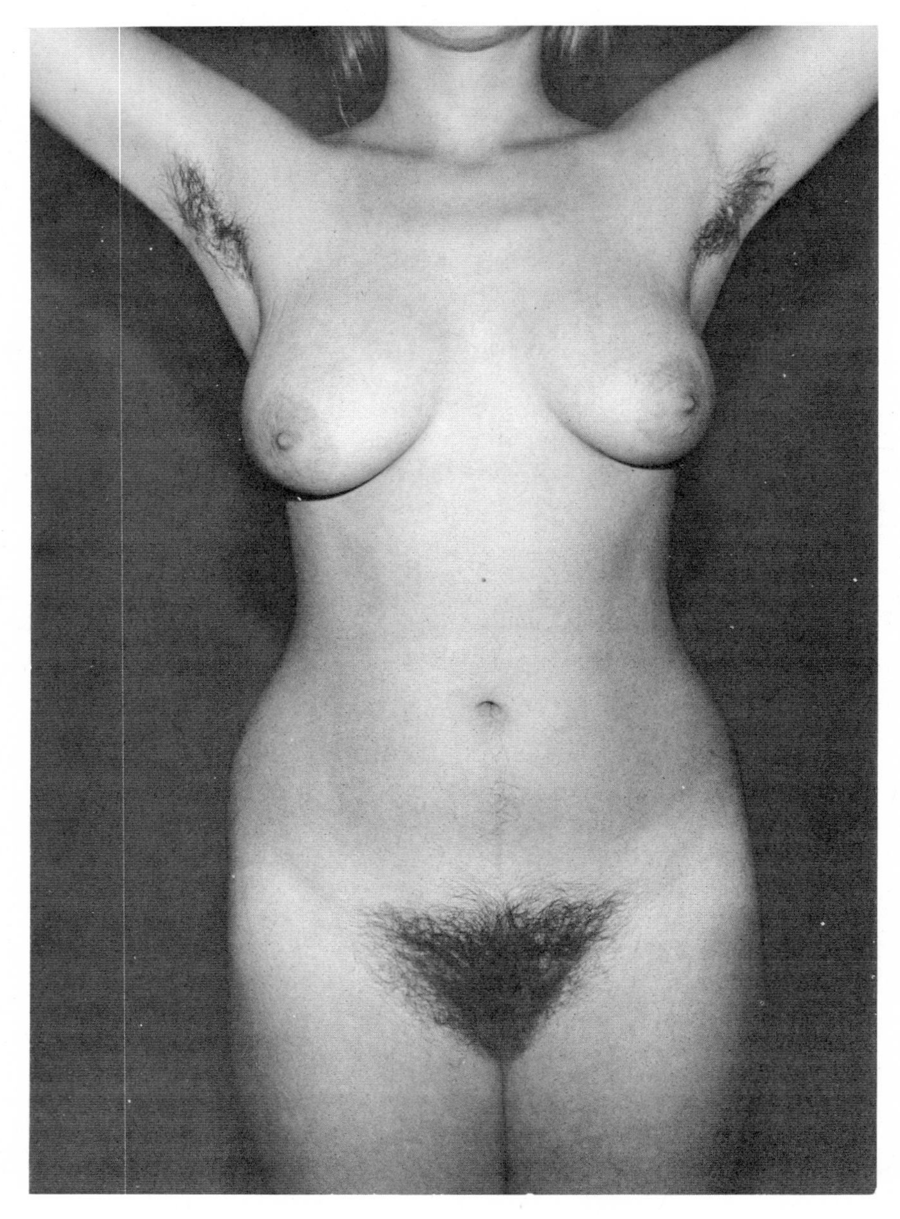

Figure 1–7. "Ornamental" hairs on the body of a woman.

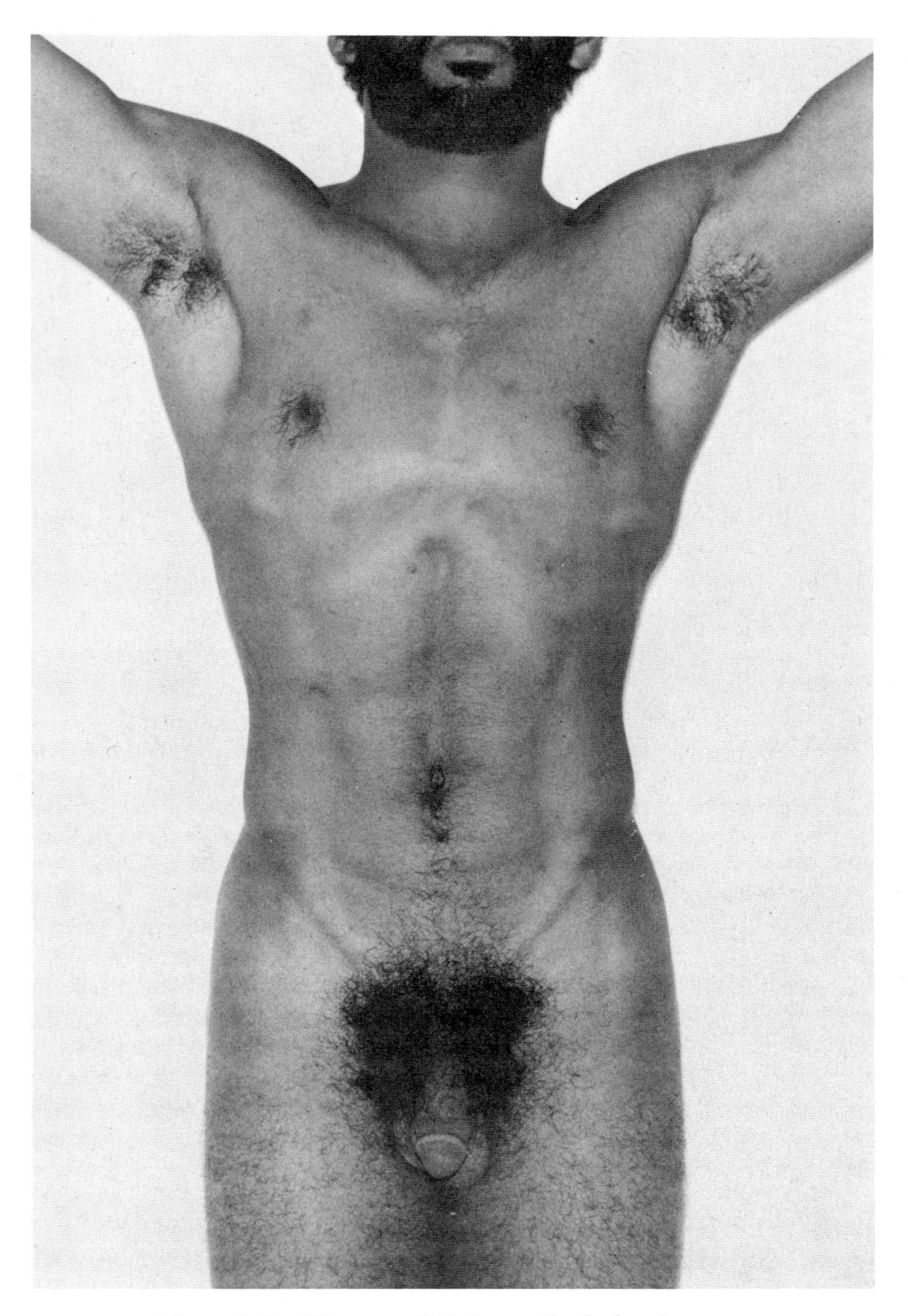

Figure 1–8. "Ornamental" hairs on the body of a man.

patterns of hair growth with no apparent functional or survival value. The coarse hairs on the axillae and around the genital areas are ornamental and peculiar to man. His scalp hairs and beard grow much too long to be useful. Left uncut these would grow so long as to be hazardous. What protection scalp hairs give the individual is negated either by their excessive length or, in some men, by baldness.

With the exception of the hairs on man's eyebrows, eyelids, and scalp, nearly all other hairs attain full bloom at sexual maturity. The axillary hair, the beard and moustache of men, the tufts of hairs on the tragus of adult men (so-called barbula hirci), and the hairs of the anogenital surfaces in both sexes are all dependent for normal growth upon the proper function of the gonads; they do not develop in individuals with gonadal insufficiency. Male pattern baldness (alopecia), which is controlled by male sex hormones, makes its appearance when all the other hairs dependent upon male hormones become coarser. This point deserves some consideration for it may foreshadow things to come. Aristotle, who, judging from extant effigies, was himself balding, made some perceptive observations on baldness.

> Of all animals, human beings are the ones which go bald most noticeably . . . no one goes bald before the time of sexual intercourse, and . . . those who are naturally prone to intercourse go bald. Women do not go bald because their nature is similar to that of children; both are incapable of producing seminal secretion. Eunuchs, too, do not go bald, because of their transition to the female state, and the hair that comes at a later stage they fail to grow at all, or if they already have it, they lose it, except for the pubic hair; similarly, women do not have the later hair, though they do grow the pubic hair. This deformity constitutes a change from the male state to the female.

Man, while regarding his scalp hair as an ornament, invests it with unnatural properties. Evidence of an exaggerated interest in hair can be found in the Bible, mythology, folklore, and the words of poets and artists, while modern man has what amounts to a hair fetish. Think of the many loving ways in which commercial advertisements refer to scalp hair—satiny, glowing, shimmering, breathing, living. Unfortunately for their vaunting claims, hair is as dead as rope. Carried away by a hair cult, men and women pamper the scalp hair at the same time that they savagely attack that which grows elsewhere. The quintessence of femininity demands a totally naked skin except for scalp, eyebrows, eyelashes, and mons veneris. All other hair is unwanted and the fastidious woman must scrupulously depilate the shins and the axillae. Yet, she may not shave her face but must remove any excess hair here with other devices.

By contrast, while most men, depending on style, shave their chins and jaws, it is considered unmanly to shave the axillae. Is it not ironical that man should choose to shave his beard, whose growth is dependent entirely upon his sex hormones, while at the same time he encourages the growth of his scalp hair, which becomes progressively thinner through the action of the same hormones? It is also amusing that man should regard the daily nuisance of shaving his chin as a ritual symbolic of his virility. Clearly, the man who sports a bald head, a rich inflorescence of moustache and beard, and hairy chest is showing the world that nature has been generous to him.

A large number of men undergo various degrees of baldness, which is more or less apparent. Male pattern baldness, however, is not peculiar to man. Among the nonhuman primates all adult male and female chimpanzees show some signs of it in the frontal area. Stump-tailed macaques *(Macaca speciosa)* become partially bald, and all uakaries *(Cacajao)* from South America experience complete postadolescent baldness.

The hairs around the body orifices have a limited function. Those inside the nostrils reduce the velocity of incoming air, trap dust particles, retard the ingress of insects or small objects, and prevent mucus from flowing too freely over the upper lip. These functions, however, are confined to man. The hairs in the external meatus and those around the anogenital orifices form a picket fence that keeps out insects and small objects. The eyebrows not only prevent sweat from flooding the eyes but also participate along with the rest of the face in communication. Similarly, the eyelashes protect the vulnerable and highly sensitive external surface of the eye (conjunctiva) and cornea.

Man characteristically possesses no tactile whiskers (vibrissae). All other mammals including the primates have at least vestiges of them. Vibrissae attain greatest sizes in nocturnal carnivores (Figure 1–9, 1–10) and in seals. Their function in these animals is evidently sensory. Although man has no vibrissae and his body hair is vestigial, the hair follicles in his skin are surrounded by sensory nerves that record any pressure upon the hair.

Man's skin envelops his whole body and is continuous at the openings on its surface with the linings of the passages of alimentary, respiratory, and urogenital systems. Modified somewhat in such regions as the conjunctiva and outer surface of the eardrum (tympanic membrane), it is a diffuse and highly active organ continually replacing its superficial layers. The skin of a man covers 2,500 square inches more or less, depending upon his size. Looking at a man, one sees only the dead outer covering of his skin, some of which is rubbed off every time he washes. In a postsunburn peel, even more dead surface cells flake off, occasionally leaving raw, painful areas exposed.

Genetic and racial factors, exposures to light or heat, and certain pathological conditions such as anemia and jaundice all influence the pigmentation of man's skin. According to skin color, mankind can be

Figure 1–9. The tactile hairs (whiskers) of a cat.

Figure 1–10. The tactile hairs (whiskers) of a seal.

divided into at least three main groups: white-skinned (Leukoderms), yellow-skinned (Xanthoderms), and black-skinned (Melanoderms). The many normal variations in the color of "white" races may be due partly to the inherited disposition of pigment cells in the skin and partly to the degree of activity of those cells when exposed to light. The thickness of the horny superficial layers of the skin also influences the color as does the amount of blood circulating in veins immediately beneath the skin. Thus we can account for the olive, tanned skin of Mediterraneans and the apparently whitish palms of Negroes.

The variety in skin texture (i.e., damp or dry, soft or hard, smooth or rough), better observed in living persons than in cadavers, is to a great extent controlled by the activity of the skin glands. In addition, the thickness and nature of the superficial horny layer, the structure of the subjacent tissue (bone, muscle, and thickness of subcutaneous fatty layer), and the profuseness of blood supply all modify skin texture. Finally, the number of sweat and sebaceous glands and the degree of their activity in different regions influence skin texture profoundly. Thus, since sweat glands and blood vessels are controlled by nerves, skin texture also reflects the temperament and temperature of its owner. At different times, the same skin area can be bluish white, cold and almost inert, cold and clammy, or warm, red, and dry.

Man's Skin

Although the skin of most mammals displays various degrees of topographic differences, in no other animal are the diversities of skin surfaces so striking as they are in man. These differences involve the total composition of the skin, not just the surface. Within the relatively small area of the face, for example, the skin that molds the various structures is thick, coarse, and hairy, as in the eyebrows, or thin and hairless, as in the lids. The skin is thin and velvety over the forehead and around the eyes, glabrous in the vermilion surface of the lips, and coarsely hairy in the bearded areas. Normally, the surface of the forehead, the facial disc, and particularly the nasal and perinasal areas is oily, unlike the relatively greaseless surface of the jowls and neck. Other contrasting features are found nearly everywhere on the body surface.

Aside from degrees of hairiness, skin, even the smoothest, is roughened or marked by the orifices of hair follicles and sweat glands (pores), by flexure lines, creases, folds, ridges, and furrows. The fixed creases or joint flexure lines are congenital and represent places where skin is more tightly bound down by anchoring fibers. They are quite apparent on the volar surfaces of fingers and palms. Although the configuration of these lines cannot foretell the future, much about an individual's past can be learned from a careful examination of his hands. Occupational marks are clues, so that even casual observers can learn to detect a seamstress, typist, writer, gardener, road-driller, or golfer. The acuity of observation possessed by a Scottish clinician, Dr. Joseph Bell, gave Conan Doyle the inspiration to create his immortal Sherlock Holmes.

Other creases and cleavage lines, not always so marked or so deep as joint flexure lines, are also present from birth. These are found where skin is particularly mobile and liable to be creased by movements—on the back of the hands or wrist or about the neck. Many of these creases become more deeply etched and new ones appear as a result of certain occupations or advancing years. Some creases become so deeply indented that they gain attachment and become anchored to underlying tissue, and even when the skin is relaxed, such creases are still visible. They appear at the outer angles of the eyelids as tell-tale crow's feet, at the sides of the mouth as "smile" wrinkles, and on the forehead as frown wrinkles. Over the entire surface of the body the skin is chiseled by fine inscriptions of intersecting lines that delineate geometric patterns characteristic of each area. These obviously allow the skin to expand in response to many subtle movements. They are also hallmarks of identification since they never form identical patterns in two individuals. Although fingerprints have become the usual means of identification because they have a higher relief and are more easily repeated, the patterns from other skin areas could also be used for identification. The skin of other primates lacks most of these marks except on the backs of their hands.

The furrowing found on the friction skin of the hands and feet is more pronounced than that just described. Numerous ridges and grooves run in parallel rows or in concentric whorls or loops to form specific patterns, which are fixed from birth, cannot be altered, and are entirely individual to each person. Even identical twins have distinctive patterns. So marked is the individuality of these patterns (dermatoglyphics) that they are used universally for personal identification. The impressions of these dermatoglyphics produce fingerprints which, when the water has evaporated from them, have enough organic and inorganic components left to be reproduced by one of several chemical methods.

Fingerprint patterns are not limited to man. Apes and monkeys, if they were to indulge in safe-cracking, would leave characteristic, recognizable prints. Comparison with the dermatoglyphics of other primates reveals a certain primitiveness in the human patterns. There is some evidence of divergent specialization in the dermatoglyphics of the great apes and man, but a similarity between those of man and of Old World monkeys prompted the American fingerprint experts, C. Midlo and H. Cummins, to suggest that "man stemmed from an ancestral stock more primitive than any recent ape, having dermatoglyphic traits more closely allied to those of the monkeys."

Nonhuman primates have many specialized modifications of skin not found in man. Only two will be mentioned here. Anyone who has visited a zoo must have noticed the naked sitting pads (ischial callosities) in the lower buttocks of Old World monkeys that are useful in protecting the bony ischial tuberosities underneath. In addition, they may have a sensory function comparable to that of the sole and palm, but this is speculation since no sensory nerves have been found there. All Old World monkeys, including gibbons, have highly developed callosities, which are absent in most apes and not found in Prosimii or in New World monkeys.

The macaque tribe, including the baboon, possesses what is known as sex skin. This is either localized only around the perineum or, as in the rhesus monkey, extends to nearly the entire back and thighs. This sex skin is highly vascularized and variously pigmented and passes through periods of rest, turgescence, and deturgescence. In the females it undergoes alterations in intensity of color during each sexual cycle but remains unaltered in the males. These changes in perineal skin and in other parts of the body in certain monkeys must be considered a secondary sexual adornment. Since women have no apparent sex skin, it would appear that, as Lord Zuckerman, Secretary of the Zoological Society of London, has speculated, man has descended from animals without this structure. Mention will be made of the areola around the nipples as a structure possibly analogous to the sex skin.

Man's Limbs

Man walks upright on two legs and thus calls attention to one of the most striking features of his anatomy—the lower limbs. His method of progression will be discussed later and only a few particularities will be introduced here. At birth a baby's legs and arms are almost the same length (Figure 1–11), but by the time he attains full size, his legs are longer than his arms. Apes have the reverse situation, the upper limbs being longer. Gibbons, which are brachiating animals, have the upper limbs so elongated that the hand lies below the knee when the animals stand erect. Man's pelvis, vertebral column, thigh, and leg bones (femur, tibia, and fibula) are all adapted in such a way as to increase ease of walking. His foot, however, is one of man's distinct characteristics. Robinson Crusoe, without any anatomical training, knew that it was a human footprint that he saw in the sand. The large big toe, unable to touch the tips of other toes (i.e., not opposable), the reduction of the other toes,

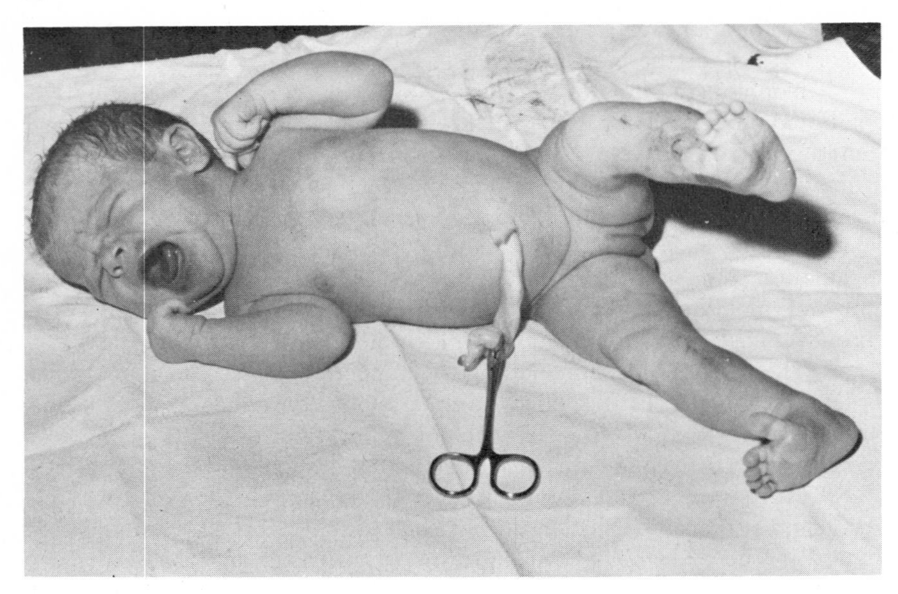

Figure 1–11. The length of a newborn baby's arms and legs is about equal.

the very small little toe, the enlarged ball of the foot, its arched form and broad heel are characteristically human. Perhaps the one generalized or primitive feature of the foot of man is the long mid-foot segment and the leg and foot musculature.

Despite the fact that the forelimbs have been freed from weight-bearing and locomotion, they possess few anatomical specializations. The relative shortness of the limb, a stout, powerful, gripping thumb capable of being opposed against the other fingers, the arm and hand musculature all point to a generalized structure. The presence of a clavicle (Figure 2–2), an outward deviation of the forearm on the arm (palm turned upward) to provide a "carrying angle," and the ability to rotate the forearm so that the palm can face upward or downward with a bent elbow (supine and prone) are features shared with many mammals besides primates. It is as if the special adaptations in his lower limbs for upright posture and bipedal gait freed his more primitive, mobile upper limbs to permit their generalized structure to become the expressive tools of his most highly developed brain.

A nineteenth-century English anatomist, F. O. Ward, gave a fitting description of the ordinary actions of the hand:

> Consider the swiftness of its movements in following a speaker with a pen; their variety in loosening a tangled knot; their nicety and precision in passing a thread through the eye of a needle. How steadily it guides

> the edge of the scalpel in a critical operation of surgery; with what singular truth it shapes the course of the schoolboy's marble, or adjusts his arrow to its mark!
>
> . . . Trained to the juggler's sleight, its joints become nimbler and more pliant. Its evolutions, in the practice of several mechanical arts, are swifter than the eye can follow, of unerring regularity, independent of the guidance of vision, and productive of the most surprising results. In the musician, the statuary, the painter, it becomes the minister of more subtle volitions, and a higher instinct; in them accordingly, it acquires still greater freedom and fluency of motion, a yet more exquisite refinement and fidelity of touch. In the orator it assumes a new character, and functions of an entirely different order. For him, it is a powerful organ of expression, an indispensable auxiliary to speech. . . .

Man's Trunk

The two main compartments of the human trunk are the thoracic (chest) and abdominal cavities, separated by a muscular partition, or diaphragm. Phylogenetically a dividing partition, it is seen in its simplest form in crocodiles, where though nonmuscular, it can be stretched by a muscle attached to the sternum. The mammalian diaphragm, tendinous in the center and muscular at its periphery, is innervated by the phrenic nerve. The origin of this nerve in the neck suggests that most of the diaphragm arose in the more cranial part of the embryo. Although the embryological history of the diaphragm is not completely clear, it is mainly derived from the septum transversum, an incomplete partition that originates in the head region and then migrates below the heart, taking its nerve supply with it. One is tempted to theorize that the diaphragm muscle in early vertebrates lay in the neck, later becoming associated with the descent of the lungs into the thorax. The diaphragm of man is not thick and, except in trained singers and athletes, seldom contracts forcibly in everyday adult life except when coughing and during the act of defecation. Only a small number of adults are even capable of active diaphragmatic breathing. The diaphragm of aquatic mammals is thick and muscular and obliquely placed. In pinnipeds (seals) the diaphragm has a curious continuation upward in the form of a sling, or cufflike sphincter, about the great vein (posterior vena cava) that returns the blood from the abdomen to the heart. This sphincter can occlude the vein and prevent blood from returning to the heart. This is no doubt related to the marked slowing of the heart rate that occurs when seals dive. By damming blood from the abdomen, the sphincter can prevent overloading of the right side of the heart.

The thorax is surrounded by the bony thoracic cage, composed of vertebrae, ribs, and the sternum. The evolution of ribs is not fully known, but phylogenetically, they show some interesting trends. In frogs and toads ribs are short; snakes have many ribs but have lost the sternum. Birds have articulated ribs attached to a very large sternum with a keel (except in some flightless birds) for the attachment of flight muscles. Most mammals have thirteen, fourteen, or more ribs on each side; man has twelve. Occasionally there may be a small rib on either side below the last one or just in front of the first one. The thorax of most mammals, including the majority of primates, is flattened from side to side and is deep and elongated in cursorial tetrapods. Compared with all other mammals, man is broad-chested. Apes show some flattening from front to back, which in man is even more marked. Thus, with the line of gravity centered near man's back (through the second sacral vertebra), balance in standing and walking is facilitated. The rate and volume of respiration vary more in man than in any other terrestrial mammal and are influenced by man's will and emotions.

An upright posture means that the abdominal wall does not hang below a horizontal backbone as in four-legged mammals but faces in a forward direction. Man is elongated between thorax and pelvis, a primitive feature that allows lateral flexure of the trunk. This means a relatively long abdominal cavity made even deeper by a capacious pelvis. Both the anterior abdominal wall and the pelvic floor are muscular and play important parts in respiration, defecation, and other functions. Both have a retentive function, which has both strong and weak aspects.

This brief survey of man's anatomical features will be expanded and considered in greater detail in subsequent chapters. From this introduction the reader can see that man is a primate, possessing a mixture of generalized, or apparently primitive, features and of curiously specialized ones that have evolutionary and thus adaptive significance.

2

His Place in Nature

After all, if he is an ape he is the only ape that is debating what kind of ape he is.
(G. W. Corner)

His Relation to Primates

"My vanity will not suffer me to rank mankind with Apes, Monkeys, Maucaucos (lemurs) . . ." wrote English zoologist Thomas Pennant in 1781. However, Pennant's vanity, like that of many others, has had to suffer this ignominy. Like it or not, man is now officially classified as a primate; what remains is to explain, if we can, this classification.

An understanding of the other members of the order Primates is a prerequisite if we are to appreciate our evolutionary history and our position in the animal kingdom. To say that man is a primate means that he belongs to a group of animals that have gradually evolved from a common ancestral stock. Except for some of the Prosimii (tree shrews, pottos, angwantibos, galagos, lorises, lemurs), the members of the order are easily recognizable, and even the uninitiated can tell at once whether an animal is a monkey, an ape, or a member of some other order. Yet, to define a primate today is almost as difficult as it was a hundred years ago when St. George Mivart gave his remarkable version:

> Unguiculate, claviculate placental mammals, with orbits encircled by bone; three kinds of teeth, at least at one time of life; brain always with a posterior lobe and calcarine fissure; the innermost digit of at least one pair of extremities opposable; hallux with a flat nail or none; a well developed caecum; penis pendulous; testes scrotal; always two pectoral mammae.

All things considered, not a bad, if somewhat compressed, definition.

When Sherwood Washburn, professor of anthropology at the University of California, Berkeley, was asked to define a primate and the characteristic features that should be included in or excluded from the order, he did so in both specific and general terms. Having called his readers' attention to the great antiquity and common ancestry shared by all mammals, Washburn spoke of the structural adaptation of primates to an arboreal life in which trees, particularly those of the tropical forests, offered vast resources of food. In this long process of adaptation, the primitive short fingers and toes of primates became elongated, nails replaced claws, and climbing was accomplished by grasping with hands and feet. This basic arboreal adaptation distinguished primates from the other orders of mammals and led to a series of characteristic adaptive trends. The relatively soft diet and the frequent use of the hands in feeding doubtlessly influenced the retention of a simple and reduced dentition. Moreover, this adaptation to an arboreal existence favored vision over smell; perhaps, as a result, the sense of smell decreased somewhat. Stereoscopic color vision in monkeys and apes evolved together with an increase in the size of the brain.

The way that primates thus adapted to an arboreal life, however, was not necessarily the only successful one. Squirrels, for example, are a mobile arboreal group, but they climb with claws, have very large incisor teeth and small brains, lack stereoscopic vision, and produce many young, which are left in a nest. The structural and behavioral adaptations of primates, therefore, represent the climax in a series of unique evolutionary events of a particular group. The similarities they share are due to a common ancestry, not to parallel or convergent evolution.

One reason why it is not easy to define primates simply is that although they made an ancient, successful adaptation to arboreal life, in the process they evolved into many forms. For example, most of them have a characteristically powerful great toe that can be widely abducted. Yet, the great toe of man cannot be used in this way. When the arboreal primate began to adapt to a bipedal life on the ground, his great toe lost its prehensile function and acquired new ones adapted to terrestrial support and bipedal walking. The apparent discrepancy between one primate and another disappears when both evolutionary history and structural adaptation are viewed together.

An additional problem in arriving at a satisfactory definition of man is that contemporary primates, whose ancestry was already distinct in the Eocene period (Table 2–1) some 50 million years ago, have attained numerous and varied forms. Some have changed very little from that period and can be regarded as living progeny whereas others have experienced a radical evolution. As Washburn indicates, it would be difficult to define a horse if representatives of three-toed and other long-extinct forms were still extant. Yet, except for purposes of definition, the study of primates is greatly enriched by the existence of the prosimians, particularly of true lemurs, that remarkable fauna preserved on the island of Madagascar.

Washburn cautions that classifications are meant not to confuse but to be useful in promoting an understanding of animal life. Consider, for example, the question of whether or not tree shrews (Tupaiidae) should be included in the order. These animals are either primates or nonprimates most similar and most closely related to primates. Since placental mammals were derived from insectivores at the end of the Age of Reptiles, the point at which any order of mammals is regarded as separate is somewhat arbitrary, and the decision must be based both on known fossil records and on an understanding of all extant forms. Primates became separated from other mammals near the beginning of the Age of Mammals and were distinguished by an enlarged brain, prosimianlike skull, and adaptation to arboreal life. If we accept the tree shrews in the order, then its definition can no longer include such features as grasping hands and feet, nails rather than claws, reduction in the number of teeth, shortening of the snout, enlargement of the brain, emphasis on vision, reduction in the sense of smell, reduction in the number of young and ways of caring for them. However, all other primates have so much in common that ad-

Table 2–1. Geological Periods of Quaternary and Tertiary Eras

Geological periods	*Started*
Holocene (wholly recent)	10,000 years ago
Pleistocene (most recent)	2–3 million years ago
Pliocene (more recent)	13 million years ago
Miocene (less recent)	27 million years ago
Oligocene (few recent)	38 million years ago
Eocene (dawn of recent)	56 million years ago
Paleocene (old of recent)	66 million years ago

mitting the tree shrews into the order causes some confusion about their evolution and adaptation. Detailed studies of the skull of tree shrews suggest that they are not particularly close to the primates, but immunochemical results are more favorable. However they are ultimately classified, tree shrews give us some hint about the kind of animal from which primates have evolved.

Washburn summarizes his discussion of the thorny question by stating that simple definitions based on structure are inevitably misleading because primates have been long evolving. Basically, their survival seems to have depended on their locomotory adaptation for climbing, together with such ramifications as tooth structure and digestion, special senses and the brain, reproduction, maternal care, and even social life.

Primates display a wide range of heterogeneous characteristics. Despite their common ancestry, dwarf galagos *(G. demidovii),* pygmy marmosets *(Cebuella pygmaea),* and mouse lemurs *(Microcebus murinus)* are no bigger than large mice, whereas man can exceed 300 lbs. and gorillas even 400 lbs. Although most members of the order are arboreal, man, gorillas and chimpanzees, baboons, macaques, and patas monkeys (*Erythrocebus patas* from sub-Saharan Africa) prefer a terrestrial life. Most of them survive on a vegetarian or omnivorous diet, but the tarsiers are carnivorous, and lorises and some galagos eat meat. Whereas most primates are diurnal, many of the prosimians are nocturnal, and some are crepuscular in habit. Most are alert and move quickly, but pottos and lorises are slow moving and apparently slow witted. Some are social, others solitary; some vociferous, others taciturn. No wonder it is difficult to formulate generalized definitions!

As mentioned before, *generalized* and *specialized* are often relative rather than absolute terms. Despite their many generalized structural features, primates also possess some highly specialized structural, physiological, and behavioral characteristics. Some of the primate specializations have also occurred in other mammals, perhaps for the same adaptive advantages (see the following chapters). Nevertheless, it is their apparently generalized biological attributes that serve most conveniently to distinguish primates. They lack such striking specializations as the trunk of elephants, the foot of the horse, the appendages of seals, or the teeth of rodents. Although they have probably evolved from a single precursor stock, some have retained certain ancestral characteristics that in others have become modified and specialized. Still others, perhaps man among them, while preserving some primitive ancestral features, have shown a constant tendency toward specialization.

One of the oddities of man is that he carries into adult life some fetal characteristics. Yet, apes, particularly gorillas, show the opposite tendency by developing adult features early in life (Figure 2–1). Man, apes, and monkeys resemble each other much more during early development than as adults. Some believe that, except for size, the anatomy of

Figure 2–1. A two-year-old lowland gorilla with well-developed adult features. Compare this with Figure 2–14.

adult man resembles that of a fetal chimpanzee more than that of an adult ape or monkey.

These considerations do not imply that man evolved from some primitive type of chimpanzee; they suggest only that adult man has certain features found in fetal apes. The theory of fetalization is primarily attributed to the Dutch anatomist and odontologist of the 1920s, L. Bolk, who produced evidence of differential degrees of fetalization in apes and in different races and families of living man. American slang grasps the anatomical point fully when it refers to a young woman as a "chick" or a "babe." The retention of fetal and early infantile characteristics is best seen in young women, most markedly in western Caucasians. The epitome of the beautiful woman is hairless, smooth skin; round-cheeked, dimpled face; high forehead; linear arched eyebrows; large blue eyes and long eyelashes; broad, uptilted nose; pouting full lips; soft, fine hair; a well-contoured figure with rounded buttocks. They are also the details of fetalization.

If we keep in mind the evolutionary trends and differential tendencies described above, and perhaps more important, if we do not get too enthusiastic about the minutiae of generalization or specialization, we can give a fair analysis of primate attributes. Anatomists in general have been too zealous in labeling the small differences in man's structure as hallmarks of his identity and as indications of his elevation to the primate peerage.

The retention of a generalized structure of limbs has enabled primates to move them freely, particularly the toes and fingers. This freedom of the digits is most marked in the thumbs and big toes, and in most forms at least one of these is opposable. Opposition means that a finger, say a thumb, can be rotated, or turned around, on its axis so that its palmar surface can be brought against that of any of the other fingers. This essential movement in grasping also involves other movements. All primates show some degree of true opposition even though not all human beings can truly oppose the thumb. The movement is one of great antiquity among mammals and probably indicates their ancestral adaptation to an arboreal existence. Certain arboreal marsupials can also oppose their digits. Perhaps this very useful characteristic of primates was retained during their long residence in trees. The freely opposable digits of primates have broadened, flattened nails rather than claws. This differentiation is not nearly so striking as that which took place in the hooved animals; however, it is fairly distinctive of primates. Among the prosimians, pottos, lorises, and galagos have retained one digit bearing a claw. The nails of lemurs are keeled and pointed like claws, and all prosimians have a claw on the second toe. Tree shrews have claws as does the peculiar "lemur" *Daubentonia* (also called aye-aye), except on the big toe, which bears a nail.

A well-strutted attachment of the upper limbs to the trunk by a clavicle is a distinct advantage for animals that swing on boughs by the arms. The word *clavicle* means "little key" (cf. *clavichord,* a stringed instrument plucked with keys) and implies that it "keys" the arm to the trunk. Swift-running four-legged animals have little use for such a structure; thus in cats and dogs, deer and horses it is much reduced. The retention of a clavicle is most useful for an arboreal life; still it is another of the more generalized and primitive anatomical features of primates (Figure 2–2). When man became bipedal, the clavicle assumed additional importance by providing a strut that allowed him to move his upper limbs in space. Together, his two clavicles provide him with a ready-made bony yoke to help carry weight. The importance of the clavicle is forcefully underscored when it is fractured: the patient has to support his now useless arm, which hangs down, by means of his intact clavicle and the associated shoulder muscles of the opposite side.

Being principally adapted for tree dwelling, primates must be able to grasp as well as to depend on the mobility and strength of the limbs

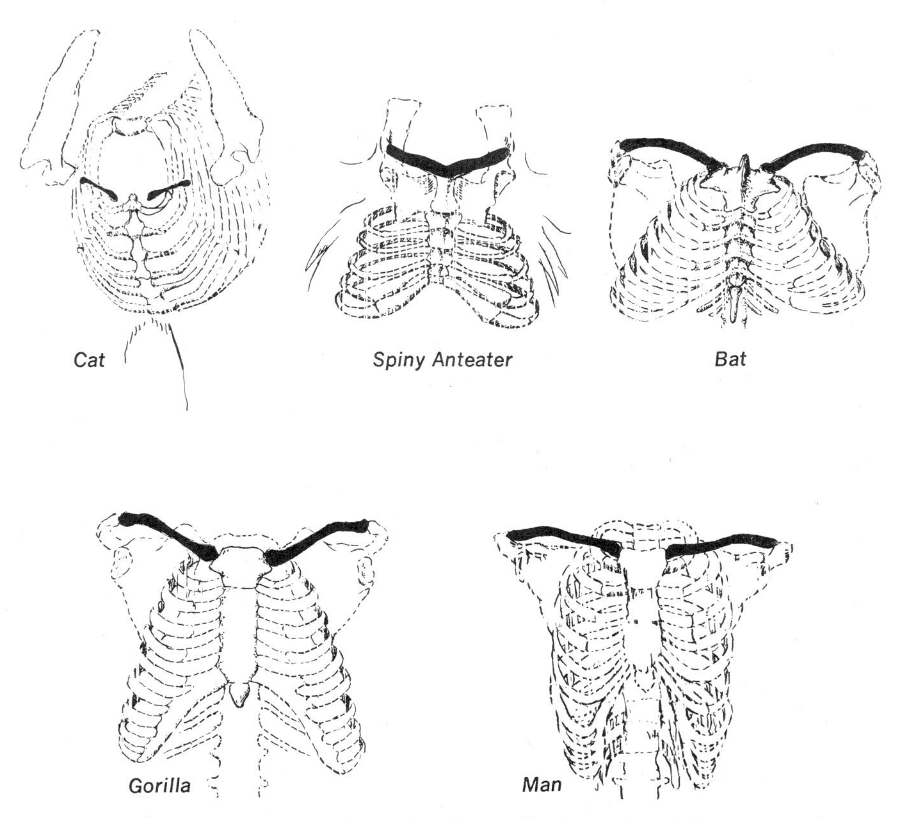

Figure 2–2. A comparison of the clavicles (collar bones) in different animals.

and girdles to which they are attached. Certain muscles in the shoulder region and trunk are markedly developed. Although terrestrial primates retain the prehensibility of their hands and feet, most animals that embark on a terrestrial life tend to lose their grasping power and all that is associated with it, and instead their limbs become adapted for running. Deer run on the tips of two-hooved fingers on each forelimb and two toes on each hind limb. Horses walk or gallop on the tip of one much modified digit of each limb.

Arboreal life not only involves a modification in limbs and musculature, but makes available items of diet that are unobtainable on the ground. We are so conditioned to man's dietary requirements of proteins that we forget that most primates feed largely on young shoots, leaves, fruits, and only occasionally on insects and other small animals. Such a diet does not require that their dentition develop grinding molars for pulping grass (as in ruminants), carnassial teeth for shearing meat (as in carnivores), heavy tusks for digging up roots (as in elephants), or sabre-like canines for ripping open a victim (as in the members of the cat family). For the most part, the dentition of living primates is generalized, that is, it has retained some features of numbers, cusp pattern, and functional categories possessed by primitive mammals. It consists of small cutting teeth (incisors), variously developed holding teeth (canines), and moderately developed premolar and molar teeth that can grind and crush. However, in some primates, as will be seen later, this fundamental plan is modified and specialized.

Primates also have some characteristic specialization of their special senses. All of them, but in different degrees, have undergone a reduction in their powers of smell and a corresponding regression in those parts of the nervous system concerned with olfaction. Mammals with a markedly developed sense of smell, such as carnivores, are called macrosmatic; those with no sense of smell, such as whales, anosmatic; and those with an appreciably reduced sense, such as most primates, microsmatic. This last condition is usually associated with a shortening of the muzzle or snout since there is no longer any need to smell or "nose out" food. Freedom of the upper limbs and a grasping pentadactyl limb relieved the primates from having to use their jaws to grasp and carry things. There are exceptions to this, for example, the baboons, which have a long doglike snout. Life in trees could be successful even with a reduced sense of smell, but without keen eyesight animals would probably miss their grasp, particularly during the twilight hours, and hurtle to the ground. Therefore, elaborate visual powers and an enlargement of that part of the brain that receives and interprets visual images are characteristic primate specializations.

The most extraordinary and important characteristic of primates is their large, well-developed brains. In most primates, particularly in man, the increase in brain size is relative to the size of the body and is mainly due to the increased growth of the cerebral cortex. We cannot, however, agree with Sir Arthur Keith that the difference between the brain of man and that of apes is only quantitative. (Let it be said, however, that Sir Arthur did add that one should not exaggerate the importance of this difference.) Human primates owe their success not so much to the increase in quantity but to the increasing complexity of their brain. The late Sir Wilfred Le Gros Clark, an expert primatologist, summed it up this way:

> The wile and cunning of the earlier primates have become the intelligence of the higher primates, and man himself has surpassed all other members of the animal kingdom in his capacity for mental activities of the most elaborate kind.

He also points out that it is difficult to explain why this precocious expansion of the brain "began earlier, occurred more rapidly, and proceeded further than in any [other] mammalian Order." It may, indeed, have been favored and speeded up by arboreal life, but this cannot be ascribed as the cause.

To sum up, primates are characterized by numerous generalized and almost primitive mammalian features, as well as by highly specialized ones. No single feature, however, either general or special, distinguishes them from other mammalian orders. Rather all these features combined, plus an expanded brain capable of giving added importance and functional skill to these features, make them unique. J. R. and P. H. Napier summed it up in their *Handbook of Living Primates* by saying that "overall judgment of affinity needs to be made not on the basis of the presence or absence of isolated characters but rather on total morphological pattern."

Man's Relation to Other Animals

Having placed man taxonomically with a group of animals recognized as primates, we must now backtrack to assign his place among other animals (Figure 2–3). Thus we shall have a better perspective of his position in nature, his evolutionary history, and his ascendancy on earth. But to do this, we must give some thought to classification. The schema below indicates the breakdown of the classification of man:

KINGDOM—Animalia
  PHYLUM—Vertebrata
    CLASS—Mammalia
      ORDER—Primates
        SUBORDER—Anthropoidea
          FAMILY—Hominidae
            GENUS—Homo
              SPECIES—sapiens

As this schema shows, classification is based on a hierarchic system in which animals, living or extinct, are incorporated progressively into ever larger, more abstract units. The most general division of this system, kingdom, separates living beings broadly into plants and animals, and the smallest category, species, is composed of individuals so alike that they are nearly indistinguishable from one another. Such a rigid hierarchy contains many uncertainties about the precise position of some animals; to fit them into the system required further subdivisions and additional categories, such as subphylum, superclass, subclass, infraclass, and so on. In this arbitrary system, even species has been further divided into subspecies and races. None of the steps is absolute, and only species is more or less

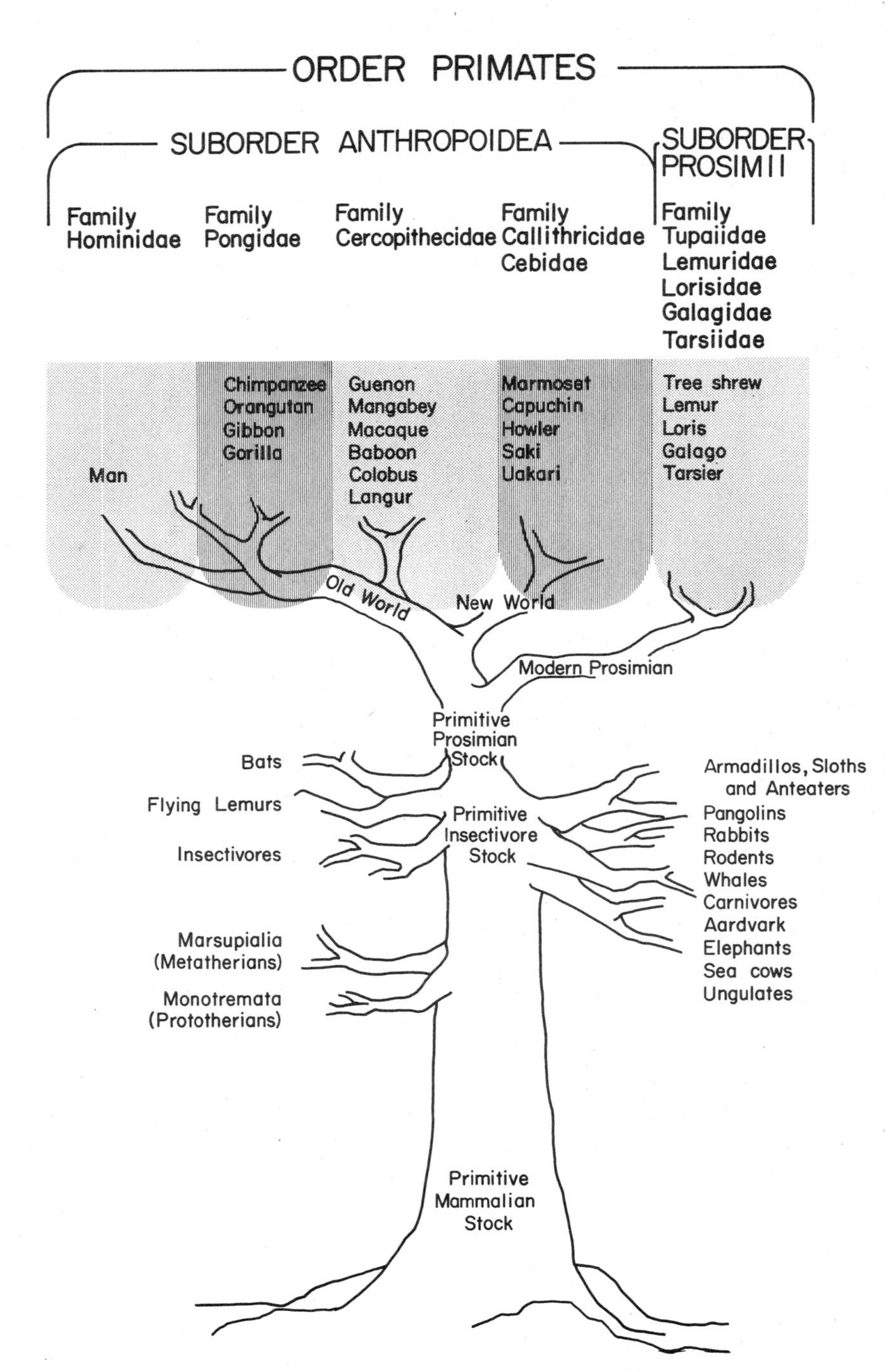

Figure 2–3. Man's place among the primates.

definable. According to George Gaylord Simpson, species is a group of living beings, the hereditary characters of any member of which can be passed on to the descendent of any other member. Unfortunately, even this definition has shortcomings, and authorities do not always agree about the exact identification of many animals. Classification has been the preoccupation of serious-minded biologists to whom the exact position of a plant or animal in a hierarchic system is of cardinal importance. These classifiers have separated themselves into two irreconcilable groups: lumpers and splitters. Each is so sanguine about its stand that one wonders whether any sanity will emerge from their deliberations and polemics.

Man is a vertebrate, that is, an animal with a backbone or vertebral column (L. *vertere,* to turn), so called because the individual bones of the vertebral column, vertebrae, can rotate upon each other. Being warm-blooded, he is a mammal, with a hairy skin, mammary glands to suckle the young, and embryonic existence within a maternal womb (uterus), and a placenta to nourish the fetus. Man's teeth, jaws, and respiratory system are typically mammalian, but numerous variations and adaptations of the basic structure have occurred in the different members of the class Mammalia. Linnaeus first used the term *Mammalia* to include all of those animals that give birth to young *(viviparous,* mostly mammals) rather than to eggs (*oviparous,* most other vertebrates).

The duckbill, or platypus *(Ornithorhynchus),* an interesting and evidently primitive Australian mammal of the order Monotremata, was thought to be viviparous until about a century ago, when, on September 2, 1884, a telegram from zoologist W. H. Caldwell in Australia to the meeting of the British Association for the Advancement of Science read, "Monotremes oviparous, ovum meroblastic." This terse piece of data indicated that the animal lays eggs, which, during the early stages of development, divide in a manner similar to that of birds and reptiles. But it did not change the duckbill's status as a mammal, and it did not mean, as was sometimes erroneously maintained, that *Ornithorhynchus* was a link between reptiles and mammals, though admittedly egg-laying is a primitive feature inherited from ancestral reptilian forms. The possession of hair, primitive mammary glands, a diaphragm, and many other anatomical features identifies the platypus as certainly a mammal. Its leathery "bill" is actually a muzzle, which in the living animal is soft, and the webs between its toes are larger than, but not too different from, those between man's fingers. The animal, therefore, is a surviving representative of an otherwise extinct subclass that branched off the evolutionary tree more than 100 million years ago. Along with it also survived the spiny anteater *(Tachyglossus)* and the long-billed anteater (Figure 2–4). These mammals are all included in the subclass Prototheria, or primitive animals.

The next group of surviving primitive mammals are the marsupials,

Figure 2–4. Echidna, or spiny anteater *(Tachyglossus aculeatus),* a monotreme from Australia.

which include opossums, wombats, kangaroos, and many others. The uniqueness of these animals lies in their methods of reproduction. In most of them the period of intrauterine life (gestation) is from twelve to twenty days, and the embryos establish only a simple connection with uterine maternal tissues. Their placenta, therefore, is most primitive, and the young are born only partially developed. These born embryos crawl into a special pouch within a ridge or fold of skin on the belly of the mother, inside of which they attach themselves to nipples. Thus they remain until fully developed when they can crawl out. These peculiarities of reproduction place the marsupials in the infraclass Metatheria.

Finally we come to the infraclass Eutheria (literally, good animals), or true mammals characterized by a relatively long intrauterine period of gestation, a well-developed placenta, well-structured female mammary glands, and the birth of more fully developed young. These animals are divided into seventeen extant orders (Table 2–2) and many extinct ones with only fossil representatives.

If the lineage of every animal were known, a precise natural hierarchy could be constructed, beginning with the apparently simplest and earliest forms and progressing to the later more complex ones. But the phylogenetic details of most animals have been lost or their records have not yet been uncovered, and only occasionally are enough fossil remains available to connect the past with the present. One should be cautioned that it is easier to be dogmatic about the distinctness of a group of animals

Table 2–2. The Living Mammals

Class MAMMALIA	*Examples*
Subclass PROTOTHERIA	
Order Monotremata	Spiny anteaters, duckbills
Subclass THERIA	
Infraclass Metatheria	
Order Marsupialia	Opossums, kangaroos
Infraclass EUTHERIA	
Order Insectivora	Hedgehogs, moles, shrews
Order Dermoptera	Flying lemurs, colugos
Order Chiroptera	Bats
Order Primates	(See Table 2–3)
Order Edentata	Sloths, armadillos
Order Pholidota	Pangolins
Order Lagomorpha	Rabbits
Order Rodentia	Squirrels, rats, mice, hamsters, guinea pigs
Order Cetacea	Whales
Order Carnivora	Dogs, bears, cats
Order Pinnipedia	Sea lions, seals, walruses
Order Tubulidentata	Aardvarks
Order Proboscidea	Elephants
Order Hyracoidea	Hyrax, coneys, dassies
Order Sirenia	Dugong, manatees
Order Perissodactyla	Horses
Order Artiodactyla	Pigs, camels, deer, cattle, sheep

if very little is known about its phylogenetic history than if there are numerous, apparently connecting links. Because of these pitfalls, experts rarely agree completely and the systems of classification are necessarily subject to change.

The Primates

By reason of his anatomy and other characteristics, man is a mammal of the order Primates. We can now proceed further to pinpoint man's position in nature by considering into what subdivision of the primates he fits. The game of subdividing the primates into special niches has given great joy to taxonomists and anatomists. The general practice today, however, is to follow the classification of paleontologist George Gaylord Simpson, who divided the order into two great groups: Prosimii and Anthropoidea (Table 2–3).

Table 2–3. The Living Primates

Order PRIMATES	*Examples*
Suborder PROSIMII	
Tupaiidae	Tree shrews
Lemuridae	Lemurs
Indridae	Indris, sifakas
Daubentoniidae	Aye-ayes
Lorisidae	Lorises, pottos, angwantibos, galagos
Tarsiidae	Tarsiers
Suborder ANTHROPOIDEA	
Callitrichidae	Marmosets, tamarins
Cebidae	New World monkeys
Cercopithecidae	Old World monkeys
Hylobatidae	Gibbons, siamangs
Pongidae	Orangutans, chimpanzees, gorillas
Hominidae	Men

The Prosimii

Tree shrews, lorises, galagos, lemurs, and tarsiers are an intriguing group of animals, veritable relics of ancient mammals. The tree shrews (family Tupaiidae), from southeast Asia, are so primitive that some zoologists still regard them as members of the order Insectivora. Anatomical, immunological, behavioral, and paleontological evidence strongly suggests that they are primates or the closest nonprimate allies. The evolution of a tree shrew ancestor into a lemurlike animal seems to have occurred some 50 million years ago. Today true lemurs (family Lemuridae) are found in Madagascar. Although their primate characteristics may not be apparent at first sight (Figure 2–5), their anatomical features are distinctly attributable to primates except for some that are found in nonprimate mammals. Their moist muzzle and projecting snout, the midline cleft in the upper lip, the large mobile ears, and the three pairs of nipples are certainly not common among the primates. Together with all the prosimians, they have been referred to by German zoologists as *Halbaffen,* or half-monkeys. Probably the lemurs became segregated from the rest of the primates in Madagascar soon after their appearance; however, how they crossed the deep and wide Mozambique Channel from the African mainland to Madagascar remains a mystery. As anthropologist J. Buettner-Janusch said about them:

They are a unique experiment in primate evolution. If we are clever enough to take advantage of this experiment, we may continue to learn more about the nature of the mammalian order to which we belong and more about evolutionary processes in general.

The most widely distributed group of Prosimii are in the family Lorisidae, which includes galagos, pottos, lorises, and the little-known angwantibos found on the African mainland and in Asia (Figures 2–6, 2–7, 2–8). Galagos, also known as bush babies, and pottos are found in nearly all parts of Africa south of the Sahara whereas lorises inhabit parts of India, Burma, and the Malayan Peninsula, Ceylon, and the islands off southeastern Asia. These animals are all tree dwellers, nocturnal, and solitary. They can be divided into two major subgroups: slow movers (creepers and climbers) with a very short tail and small ears (pottos and lorises), and fast jumpers with long bushy tails and large ears (galagos). Today, of this once large group of animals, only one genus of tarsiers exists, *Tarsius* (Figure 2–9). Fifty million years ago they existed in many different forms and sizes and occupied North America, Eurasia, and Africa; they are found primarily on the islands of the East Indian Archipelago from the southern Philippines and Celebes to Sumatra. About the size of a rat, *Tarsius* has large eyes (thus, Spectral Tarsier), hind limbs that enable it to jump more than ten times its own body length in any direction, large, constantly twitching and undulating ears, a remarkable way of rotating its head to look around behind itself, and other highly specialized anatomical features. The tarsier also has distinctly monkeylike features, e.g., the anatomy of the nose, certain reproductive organs, and the brain. If, as is occasionally suggested, fossil ancestors of tarsiers were in direct lineage to the prosimians from which the Anthropoidea diverged, such ancestors probably left the main line of descent to the higher primates and man some time in the Eocene.

The prosimians, then, are primitive primates showing certain features found today in insectivores. Three main stocks have arisen—lemurs, lorisoids, and tarsiers (Figure 2–3). Because evidence points in both directions, the tree shrews are placed among either the insectivores or the primates.

The Anthropoidea

Monkeys, apes, and man are placed in the suborder Anthropoidea. Fossil anthropoids first made their appearance in the geological epoch called Oligocene (meaning the "few of the recent" and extending for some 10 million years after the Eocene) (Table 2–3). They seem to have been successful from the first; although many of the genera that evolved have

Figure 2–5. Ring-tailed lemur *(Lemur catta)* with twin babies. This prosimian (family Lemuridae) comes from the forest areas of Madagascar and Comoro Islands.

become extinct, some forty of them are extant, some even flourishing. Most of them are active, adventurous, and inquisitive; their movements, facial expressions, and ready response to their environment show them to be alert. The very word *monkey* is a diminutive of the word *homunculus* (little man), which is now the generic name for an extinct form of the New World monkey of the Miocene.

Most monkeys live in trees, some live on the ground, and others, while retaining the ability to climb, have become mostly terrestrial. The image of a little man living in the trees conveys the wrong impression of their true method of locomotion, for both in trees and on the ground monkeys usually run on four feet (pronograde) and occasionally leap from one rock or bough to another. Their heads are set on the vertebral column so that they can look straight ahead with ease. They can sit on their haunches with their knees bent up and the soles of their feet flat on the ground, a posture that represents one of the stages through which a four-footed animal might have to pass before adopting an upright stance. A human baby does not squat in quite the same way, but one of the indications that it is going to stand up is its preliminary ability to sit unaided and to hold its head up. Monkeys can pull themselves upright and some often walk on two legs, particularly when they are holding objects in both hands. Japanese macaques *(Macaca fuscata)* and black Celebes apes *(Cynopithecus niger)*, actually close to macaques, often stand and walk bipedally. A large female gorilla captive in the Monkey Jungle near Miami, Florida, actually preferred walking on her legs and seldom moved on all fours.

Monkeys clearly display the relative enlargement of the brain that is found progressively through the primate series. Although their sense of smell is much reduced, their visual acuity is greatly increased. Specific anatomical features in the eyes give them binocular, stereoscopic vision; they sometimes show a marked appreciation of color. This advance in visual perception is associated with the partial freeing of the hands from structures for locomotion to ones for grasping, exploring, touching, and feeling interesting objects. Monkeys also acquire information by carrying objects to their mouths and biting on them, in much the same way children and dogs do. Apparently the brain of a monkey cannot completely utilize the information coming from the eyes and hands, or else the perception patterns reaching it cannot be fully associated with past events or experiences.

Certain animals are characteristically gregarious: wolves form packs, deer gather in herds, baboons travel in troops, whales swim in schools, and seals aggregate in colonies. The advantages of hunting, feeding, and breeding together or merely sharing the same sandbank are obvious. These aggregates or societies probably always have a leader who has particular responsibilities that he carries out until old age causes retirement or forcible replacement. Such communal living is further refined

Figure 2–6. The greater or thick-tailed bush baby *(Galago crassicaudatus)*, an arboreal prosimian from southcentral and eastern Africa (family Lorisidae).

Figure 2–7. The potto *(Perodicticus potto)*, a tropical forest prosimian from west Africa. A slow, climbing animal, the potto can move freely in a suspended position (family Lorisidae).

Figure 2–8. The angwantibo *(Arctocebus calabarensis)*, a prosimian from tropical forests of west Africa not often seen in captivity (family Lorisidae).

Figure 2–9. The Philippine tarsier *(Tarsius syrichta)*, a prosimian from tropical rain forests of certain islands in the Philippines of southeast Asia (family Tarsiidae). A completely carnivorous animal, it fares poorly in captivity.

in monkeys and to greater advantage since they have developed quite elaborate forms of integrated social activity. Perhaps the best examples of this are the baboons. Numerous field studies of troops of langurs, macaques, and baboons reveal specific and often elaborate social behavioral patterns.

An animal can enjoy a social existence to full advantage only if it can communicate in some way with the other members of the group. Communication among animals is now being studied extensively. Social insects, such as bees and ants, have enormously varied, subtle, and often complicated methods of conveying information. Monkeys can communicate, but not by articulate speech (see chapters 13 and 14). They express their moods through howls, grunts, or shrieks and by chirps and chatterings variously modulated, which, though seemingly crude, convey a remarkable range of signals, though perhaps not on the level achieved by Rudyard Kipling's Bandarlog.

New and Old World monkeys The Anthropoidea are divided into the following families: the Callitrichidae (marmosets) and the Cebidae (squirrel monkeys, spider monkeys), known as New World monkeys from Central and South America; the Cercopithecidae (macaques, baboons) known as Old World monkeys; the Hylobatidae (gibbons), the Pongidae (great apes), and the Hominidae (man). New World monkeys have the two nostrils separated by a relatively broad septum, which is not always clearly demarcated (Figures 2–10, 2–11). For this reason, the broad-nosed New World monkeys have been called platyrrhines, the narrow-nosed Old World monkeys, catarrhines. This rather restricted criterion of differentiation is only one of several features that distinguish the New World monkeys, e.g., three premolar teeth, a large bulla with no bony external auditory meatus, no ischial callosities, and in some, a prehensile tail.

Marmosets and tamarins (family Callitrichidae) are squirrel-sized animals, widespread in Central and South America. The smallest, the pygmy marmoset *(Cebuella pygmaea)*, is hardly six inches long; the largest, the golden lion marmoset *(Leontideus rosalia)*, seldom exceeds seventeen inches. These animals have claws instead of nails on all digits except the big toe. All Callitrichidae except one genus have two, not three, molar teeth in each half jaw.

All New World monkeys, except marmosets and tamarins, belong to the family Cebidae. These are spider and woolly monkeys (Atelinae), howler monkeys (Alouattinae), night monkeys and the dusky titi (Aotinae), capuchins and squirrel monkeys (Cebinae), and the curiously named sakis and uakaris (Pitheciinae). The New World monkeys, or platyrrhines, are generally believed to have developed from an offshoot of the main evolutionary line of descent and to have evolved independently and in relative isolation on the South American land mass. Only the howler,

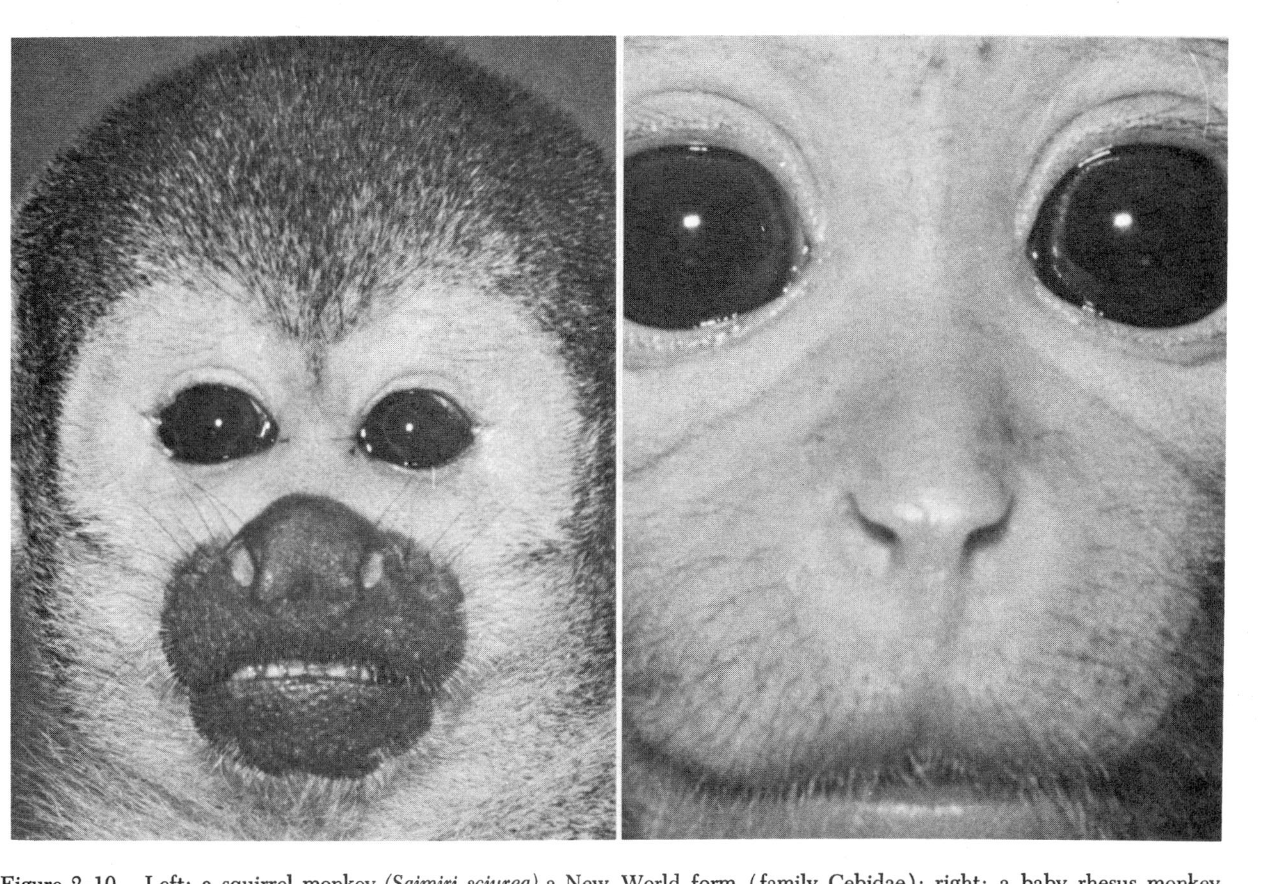

Figure 2–10. Left: a squirrel monkey *(Saimiri sciurea)* a New World form (family Cebidae); right: a baby rhesus monkey *(Macaca mulatta),* an Old World form (family Cercopithecidae). Compare the width of the space between the nostrils.

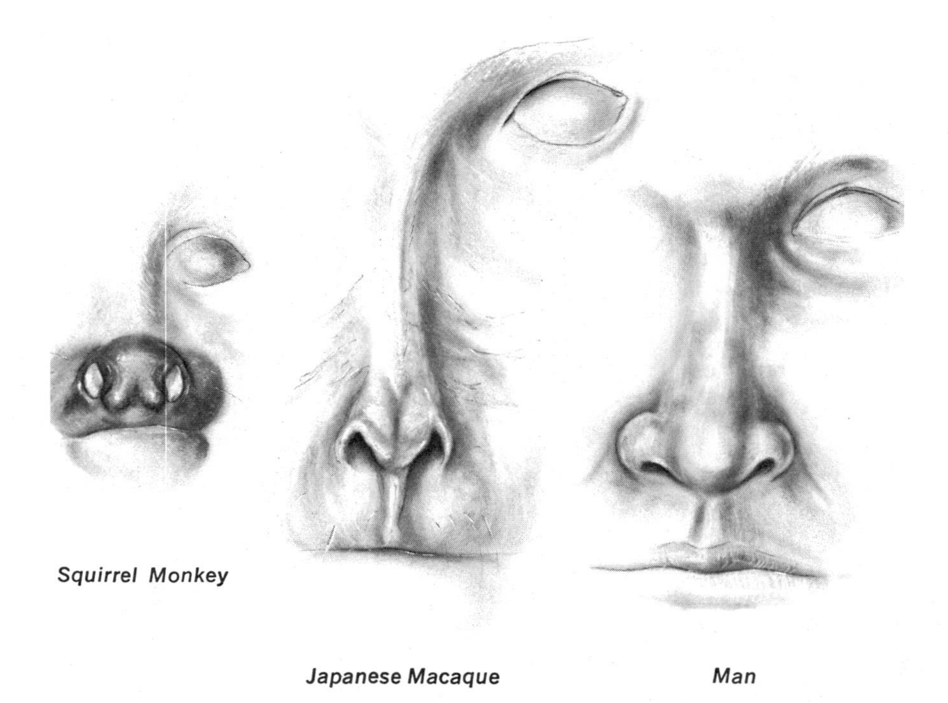

Figure 2–11. The nose of a New World monkey *(Saimiri),* an Old World monkey *(Macaca),* and man.

spider, and woolly monkeys have a true prehensile tail with a modification of the skin in its friction surface (Figure 2–12). Capuchins curl their tails around supporting surfaces, but the tail has no anatomical differentiation. The tails of squirrel monkeys have barely observable prehensile tendencies. Platyrrhines occasionally possess features of their own, such as the strikingly resonant voice of the howler monkey produced by the hollowing out of the hyoid bone in the upper part of the neck beneath the tongue. Other characteristics of these animals will be mentioned in subsequent sections. The relationships and common ancestry of Old World monkeys (Cercopithecidae), lesser apes (Hylobatidae), great apes (Pongidae), and man (Hominidae) and the time of their divergence are by no means settled. The only evidence to date of the existence of animals that might have been ancestral to the Old World monkeys has been the discovery of fossil remains in the Fayum region in Egypt of monkeys that lived during the Lower Oligocene (forty-five or so million years ago). One form, now called *Parapithecus,* whose tooth anatomy and jaw suggest that it sprang from some form of Eocene tarsioid, is a possible ancestor of Old World

Figure 2–12. A howler monkey *(Alouatta)* from South America.

monkeys; another, *Aegyptopithecus,* might have been ancestral to the great apes or even man. Until more paleontological evidence appears, such beasts must be seriously considered as candidates for the ancestry of Old World monkeys, apes, and men—the diverging evolutionary lines still extant. Some authorities believe that catarrhine monkeys soon separated from the other forms and that the evolution of ape and hominid stocks was more or less parallel.

Students are sometimes required to comment on the statement that anatomically man bears a closer resemblance to the living anthropoid ape than to any other mammalian group. Unfortunately, such academic pronouncements have sometimes encouraged the fallacy that man descended from a creature almost precisely like the ape of today. But according to Le Gros Clark this "is a gross misconception for which there has never been the slightest foundation." Failure to recognize this obvious point has led to some of the confusion prevalent in discussions on the origins of man. Le Gros Clark also emphasizes that a clear distinction should be drawn between the *biological* species, *Homo sapiens*—anatomical and physiological man—and the wider, *philosophical* concept of man.

The Old World monkeys are represented by the extremely large family Cercopithecidae, found in Asia and Africa. The family, with over seventy species, is divided into two subfamilies, the Colobinae and the Cercopithecinae. Among the Colobinae are the colobus monkey of Africa, the langur *(Presbytis)*, the proboscis monkey *(Nasalis)*, and other types of langur. They are preferential leaf-eaters and arboreal; *Presbytis*, however, is partially terrestrial. The several genera of Cercopithecinae include the mangabey *(Cercocebus)*, the guenon *(Cercopithecus)*, the Celebes black ape *(Cynopithecus)* (Figure 2–13), the patas monkey *(Erythrocebus)*, the macaque *(Macaca)*, the mandrill *(Mandrillus)*, and the baboon *(Papio)* as well as the gelada baboon *(Theropithecus)*. Old World monkeys somewhat resemble New World ones, but they are less well adapted for arboreal life. Although macaques, baboons, and patas monkeys are still eminently skilled in tree climbing when they choose or have to, they remain as troops on the ground during the day. These animals all have two premolar teeth, a tubular tympanic bone, and a hardened area of skin on their backsides called ischial callosities (*ischium* is the name for the lower part of the hip bone that has tuberosities on which the reader may well be sitting). Around these callosities some monkeys have a naked area of skin that is often highly colored during certain times of the reproductive cycle. The reproductive characteristics of Old World monkeys, which closely resemble those of apes and man, are described in a later chapter. Certain anatomical specializations in the teeth and other parts of the body are often adduced as evidence that several Old World monkeys may have deviated early from the ancestral stock.

Lesser and great apes The Hylobatidae and the Pongidae are closely related families and have been grouped together with the Hominidae by Simpson in the superfamily Hominoidea. The Hylobatidae consist of the gibbon *(Hylobates)* and siamang *(Symphalangus)*; the Pongidae include the orangutan or "man of the woods" *(Pongo pygmaeus)*, two species of chimpanzee (*Pan troglodytes* and *P. paniscus*, a pygmy form), and the gorilla *(Gorilla gorilla)*. Gibbons, the most numerous and successful, are the smallest group, seldom weighing much more than 8 kg and having a brain that often weighs less than 100 gm. They are found today in Malaya and parts of Southeast Asia. Their extraordinarily long arms enable them to swing through the trees like veritable jungle trapeze artists. Sir Arthur Keith has called this method of progression, so fascinating to watch in a zoo, brachiation (L. *brachium*, arm). Gibbons are the only apes that prefer an erect posture on the ground and that can run upright on their hind limbs for some distance. They are also the noisiest apes. The siamang possesses a huge inflatable sac beneath the chin which acts as a resonating chamber that magnifies the carrying power of its whooping cries.

The best known of the great apes, the chimpanzee, is probably a native of Angola in West Africa. An interesting history of confusion surrounds the generic names of apes. Thomas Henry Huxley, in his now famous book, *Man's Place in Nature and Other Essays*, referred to the first edition of an amusing book by one Purchas, *His Pilgrimage*, published in 1613. Purchas, in turn, had referred to Andrew Battell (Battle), an old friend who, after living in the Congo for several years, visited Angola.

Figure 2–13. A male Celebes black ape *(Cynopithecus niger)*, an Old World monkey from Celebes, an island in Indonesia.

An old soldier is reported to have told Battell of a "kinde of great apes, if they might bee termed, of the height of a man with twice as bigge feature of the limmes, the strength proportional, hairy all over, otherwise exactly like men and women in their whole bodily shape." In a marginal note he stated that these apes were called pongos. Huxley points out that the old soldier friend of Purchas was probably confusing pongo with gorilla. The name *gorilla* is alleged to be an African word for a wild or hairy woman. One is said to have been found in the Greek account of a voyage made by the Carthaginian sailor Hanno in the fourth century B.C. The name *gorilla* was finally adopted as the specific name for the ape *Troglodytes gorilla* by the missionary Dr. Savage in 1847. *Troglodytes* is now the specific name for the chimpanzee *(Pan troglodytes)* and *gorilla* for the gorilla. This confusing situation has been clarified in the last century.

Chimpanzees and gorillas come from the tropical forests of Africa, and the orangutan comes from Borneo and Sumatra. A pygmy chimpanzee *(Pan paniscus)* comes from south of the Congo River. Gorillas are the heaviest primates, a full-grown male weighing up to 600 lbs (Figure 2–14). The orang (Figure 2–15) is the next largest, weighing up to 170 lbs, and the average weight of an adult chimpanzee is about 100 lbs. (The average weight of a full-grown man is 170 lbs). Orangs, chimpanzees, and gorillas are less at home in the trees than gibbons. The latter two progress more easily on the ground than orangs, but all of them use a particular walking technique: they rest the forepart of their body weight on the knuckles of their hands, not on the palms downward. All three may have taken to the ground as an expediency because only thick-branched trees could support their great increase in weight.

The general anatomy of the great apes lacks certain specializations found in monkeys. They have no callosities on the bony ischial protuberances, and although apes, like man, have a coccyx, they have no external tail. However, this lack of a tail is not an absolute anatomical criterion of an ape, since the so-called Barbary ape *(Macaca sylvana)* and the Celebes black ape *(Cynopithecus niger)* lack tails but are monkeys. The reproductive processes of apes resemble those of man: newborn apes are well developed and carefully tended by their mothers for many months, but their period of infancy and adolescence is longer than that of monkeys. Puberty is generally reached between six and eight years of age.

The most remarkable change in the great apes is the increase in the bulk of their brain. The weight of the brain varies from about 300 gm in a female chimpanzee to 600 gm in an adult male gorilla; moreover, temperamentally chimpanzees and gorillas differ considerably, chimpanzees being more mercurial as well as more noisy and untrustworthy than gorillas. In spite of their apparent greater composure, gorillas are sometimes more slow witted. The keen sensory perception of chimpanzees is

Figure 2–14. Stefi, an adult male gorilla, from the Basel Zoo. *(Courtesy of Paul Steinemann, Zoologischer Garten, Basel, Switzerland.)*

said to be equivalent to that of man, but their brain activity seldom develops beyond that of a human child.

Like man, apes have a vermiform appendix. Although they are mainly vegetarians and eat flesh only when unable to find their preferred diet, they have large canine teeth, which they use for attack and defense.

Simpson placed man in the family Hominidae, which includes both the extinct and living races of man. The relationships between the ape and man have been studied biochemically, and serological evidence from the reactions of blood proteins suggests that gorillas and chimpanzees are closer to man than gibbons and orangutans. Such studies, however, lead to conclusions about the timing of the evolution of man and apes that are not consistent with those deduced from the fossil record. The possibility

Figure 2–15. Adolescent male orangutan.

that apes and human forms have evolved more than once is not widely accepted as an explanation. When ancient apes or men died, their soft parts were never preserved so that now only the skeletal structure can be studied. The available paleontological evidence is incomplete, but what has been discovered must be given serious consideration though only briefly here.

Fossil Remains

Human evolution cannot be traced, even in the most general way, without some consideration being given to the evolution of the entire order of Primates. Furthermore, man's relationship to living primates and his phylogenetic history cannot be fully appreciated without reference to fossil records. The earliest primate fossils are of prosimians, found in the middle Paleocene deposits of North America. During the late Paleocene and Eocene period (65 to 40 million years ago), the prosimians expanded their range and became extensively diversified in both North America and Europe. Many early members acquired enlarged, rodentlike anterior teeth, and the extinct families Carpolestidae, Plesiadapidae, and Phenacolemuridae may well have occupied the niche later taken over by the true rodents at the end of the Eocene. One member of those early primates, *Plesiadapis,* is particularly interesting because of the claws on its digits. The flattened nails of modern primates had probably not evolved at that stage. In fact, mammals from this early period are classified as primates only because of similarities between the morphology of their molar teeth and those of the recognizable primates that came later.

At the beginning of the Eocene period, the rodentlike families of prosimians were gradually outnumbered by more modern forms and probably reached their highest numbers and diversity during this time; fifty to sixty genera are reported from North America and Europe. During the late Paleocene to the middle Eocene, many of the American and European forms were quite similar, evidence perhaps that extensive faunal interchange was taking place. By late Eocene, the degree of similarity between the two areas was greatly reduced, probably because faunal interchange had been interrupted. From this period on, New and Old World primates evolved independently.

Prosimian fossils declined everywhere in the late Eocene period, either because of climatic changes or because of an evolutionary expansion of the rodents; either reason could have forced the prosimians to migrate south and could have reduced their populations. Whatever the cause, by the end of the Eocene, prosimian fossils were very rare. This paucity of fossils from the Eocene-Oligocene time zone is one of the most severe evolutionary gaps of the Primate order because at the same time in both the Old and the New Worlds, monkeys were evolving from their prosimian ancestors. None of the numerous genera and species from the previous periods can be directly traced to later fossils or living species, although the extinct genera *Nannopithex* and *Necrolemur* of Europe resemble living tarsiers.

In America prosimian fossils had virtually disappeared from the Oligocene record (ca. 28 to 40 million years ago) except for the beautifully preserved skull of *Rooneyia* from Texas. However, the characteristic features of New World monkeys had not evolved in this animal, which is clearly a North American Oligocene prosimian. In the New World, the transition from prosimian to monkey probably took place somewhere in the southern United States, Mexico, or Central America, but there is no fossil evidence of this crucial stage.

By Miocene times (12 to 28 million years ago), *Cebupithecia* and *Homunculus,* both resembling modern monkeys, had evolved in South America. However, since they are completely differentiated, they offer no evidence of a prosimian-monkey transition in the western hemisphere.

Oligocene fossils in the Old World are found only in the Fayum of Egypt. The identified divergent genera, which include *Propliopithecus, Aeolopithecus, Apidium, Oligopithecus,* and *Aegyptopithecus* are all fully evolved catarrhines and show that a considerable amount of diversification had taken place. Yet, in spite of diversity, none of these forms can be traced unequivocally to living species of monkeys and apes, nor can they be traced to ancestral fossil prosimians. During the late Eocene, early Oligocene prosimian populations became smaller and more restricted geographically, apparently in response to the evolutionary expansion and competition of the Old World monkeys. The patterns of contemporary distribution show that only such nocturnal prosimians as bush babies *(Galago sp.),* angwantibos *(Arctocebus calabarensis),* pottos *(Perodicticus potto),* and lorises *(Loris tardigradus, Nycticebus coucang)* have survived cohabitation with the diurnal Old World monkeys. All living diurnal prosimians are found on the island of Madagascar where apparently monkeys have never existed.

The evolutionary divergence of monkeys and apes can be traced in the Miocene and Pliocene deposits of Europe and Africa (Figure 2–16). *Mesopithecus* of the European Pliocene (2 to 12 million years ago) shows definite relationships to the living colobines of Africa and Asia. *Pliopithecus* remains from Miocene strata in Africa and Asia document the early stages in the evolution of the gibbon. On the basis of cranial and dental structure, *Pliopithecus* is closely related to modern hylobatids. However, the modification of arm length and the type of joints characteristic of modern hylobatids depart from that of monkeys and had only begun in *Pliopithecus.* A fossil sacrum indicates that the animal had a tail fifteen to twenty seg-

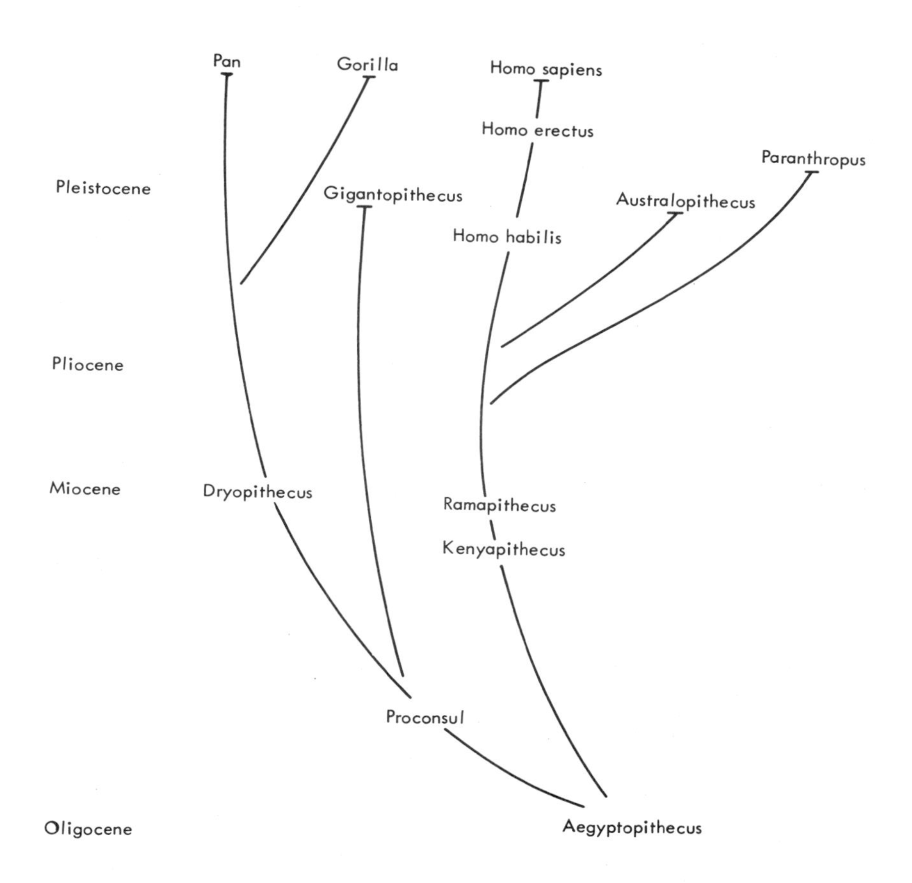

Figure 2–16. Suggested phylogenetic relationships of fossil apes, living apes, and man (after Napier, 1971).

ments long. The foot and hand of *Pliopithecus* are also monkeylike. *Dryopithecus* (tree ape) and *Proconsul* (so named because its teeth and jaws resembled those of Consul, a chimpanzee in the London Zoo) are both noteworthy. Members of the Dryopithecinae may have given rise at different times to the great apes and man. Proconsul possessed so few unequivocally human features that its ancestry to man is disputed.

The dryopithecines, a roughly contemporaneous group of Miocene and Pliocene apes, are particularly interesting because they include a possible hominid ancestor, *Ramapithecus,* and *Dryopithecus,* a candidate ancestor of the modern great apes. They also had *Gigantopithecus,* an extinct genus of large apes. The classification of *Ramapithecus* as a hominid is based on extremely fragmentary fossil remains, and its validity needs to be verified by more complete specimens. The postcranial skeleton, for example, would have to give evidence of bipedal locomotion.

Comparative anatomical and immunochemical studies of the Anthropoidea, however, suggest that man did not diverge from the ancestral stock until after the evolution of the brachiating complex in the shoulder, arm, and trunk.

The possible relationship of fossil apes, fossil hominids, living apes, and man has been discussed by J. R. Napier (Figure 2–16), who gives the following formal classification of the Hominidae:

Family Hominidae
  Genus *Ramapithecus* (including *Kenyapithecus*)
    Australopithecus
    Paranthropus
    Homo
  Species *Homo habilis*
    Homo erectus
    Homo sapiens

As Napier points out, not everyone is likely to agree with all details of this classification, which expresses his personal opinion; but nonetheless it provides an arrangement of the known material that can be considered further in the light of new discoveries.

The earliest stages of this development in the Pliocene appear in the *Pliopithecus* fossils mentioned earlier. Man shares with the apes an extremely flexible shoulder, elbow, and wrist complex, a broad, shallow chest, and a short, lower back—features possessed by none of the Old World monkeys and only beginning to evolve in *Pliopithecus*.

Not until the early Pleistocene do uncontested hominid fossil records appear. Unlike the wide Eurasian and African distribution of the dryopithecines, *Paranthropus* and *Australopithecus,* or man-apes, are concentrated side by side in Africa. Notwithstanding their proximity, the two groups seem not to have been in competition. The wear of the teeth suggests that *Paranthropus* was a vegetarian and *Australopithecus* probably a hunter. Australopithecines, who were probably less than four feet tall, weighed between 40 and 50 lbs, had a brain about the size of the living great apes and small canines, and were bipedal. Furthermore, they made crude stone tools by breaking two or three flakes from one end of fist-sized pebbles, the so-called Oldowan chopping tools, which are common in South African deposits and are associated with the fossil remains of the australopithecines at Olduvai Gorge in Tanzania, East Africa. Post-cranial remains from Olduvai and South Africa show an almost modern foot with no divergence of the great toe. This single fact must have caused considerable social reorganization, since the young of *Australopithecus,* unlike all nonhuman primates, could not cling easily to their mothers, nor could they climb trees, as living apes do. Fragments indicate that the pelvis, but not the foot, had evolved in the direction of modern man. A hand from East Africa is intermediate between that of man and of the great apes.

The East African deposits are particularly important because they have recorded a continuous history of hominid evolution for at least 2 million years. In the lower beds at Olduvai, small piles of flakes are the sites of former workshops, a small circle of stones resembles a shelter, and many of the pebble tools are made of a stone imported from an outcropping of rock some distance away.

For nearly a million and a half years, the australopithecines lived in the area of Olduvai and showed no dramatic biological changes. But, at the 500,000-year level, fossils and stone tools show a marked change. The skull cap of a larger-brained hominid, *Homo erectus,* was found, the first representative of true men. Actually, remains of *Homo erectus* have been found all over the Old World, Asia, Europe, and Africa. They come from central Java *(Pithecanthropus)*, Peking, China (Peking man, first known as *Sinanthropus*), Algeria *(Atlanthropus)*, Heidelberg, Germany (Mauer man) and other places.

It must be emphasized here that the nomenclature of the ancestors of man is confused and confusing. For example, Ernst Haeckel, a nineteenth-century German biologist, coined the name *Pithecanthropus,* meaning ape-man, even though such a creature had not been found. When the first remains of fossil man were later discovered in central Java, the name *Pithecanthropus erectus* was applied to it. However, this creature resembled the genus *Homo* so much that *Pithecanthropus* was replaced by *Homo.* Therefore, it behooves us to know what our ancestors were rather than what they should be called. Furthermore, we should examine fossil records with a view toward reconstructing behavior patterns and culture because it is successful behavior that leads to successful adaptation.

Analysis of the australopithecine material has been complicated by the finding at Olduvai Gorge of a more advanced form considered by L. S. B. Leakey, P. V. Tobias, and J. R. Napier (1964) to have reached the human grade and to merit the name *Homo habilis* (*habilis,* capable). They argue that *H. habilis* has a brain case with *Homo*-like features, at least more so than any other australopithecine pattern, and that the thumb, clavicle, and foot have salient human characteristics. *Homo habilis* is described as having a hand capable of chopping pebbles, even with a short thumb. Controversy simmers about whether the remains, mainly from a young specimen, are sufficiently distinct from *Australopithecus* to be called anything other than *Australopithecus habilis.* Nonetheless, *H. habilis* may have belonged to a population intermediate between *Australopithecus* and *Homo erectus.* It has also been suggested that the *Meganthropus* fragments from the Djetis deposits of Sangarin in Java represent this same intermediate grade of hominid.

The postcranial skeleton of *H. erectus* was completely modern, and the size of the brain had increased to a mean of about 1,000 cc. Moreover, evidence from Choukoutien in China and Ambrona in Spain shows that *H. erectus* used fire and hunted cooperatively. In the Ambrona Valley of Spain, archaeological and fossil remains suggest that populations of *H. erectus* used fire to drive elephants into a bog as a hunting strategy. The tool kit of these early men represents a significant refinement of the simple choppers of earlier man-apes. For example, flakes were removed from both sides of a core of stone to produce a longer, straighter, and sharper cutting edge in the Acheulian hand-axe style of *H. erectus*. The Acheulian tradition may have developed from the earlier Oldowan chopping tool industry. Since similar tools are found at Choukoutien in association with fossils of *H. erectus*, the Oldowan pebble industry may have been transported out of Africa by early *H. erectus*. Another chopping tool industry from the oldest archaeological site in Europe at Vértesszöllös, Hungary, also resembles the Oldowan and may prove the early spread by *H. erectus* of this trade through the Middle East. The Acheulian industry, which ranges from 500,000 to 75,000 years ago, seems to have developed from the Oldowan throughout the Old World.

Unfortunately, the record of *H. erectus* covers only the period from 500,000 to 300,000 years ago, and the tool-makers of the Acheulian industry remained unknown for nearly 200,000 years.

By 110,000 years ago, a very different and fully modern hominid, Neanderthal man, had already appeared, and the question of who was making the Acheulian tools during the later stages of that industry is further confused. Evidence from the back of a skull from Swanscombe, England, and from a face and forehead from Steinheim, Germany, favors the theory that a population intermediate between *H. erectus* and *H. sapiens* manufactured the tools during the later phases of the Acheulian. The Swanscombe skull fragments of about 200,000 to 300,000 years ago are nearly modern in size and form and are associated with Acheulian tools, and yet the face of the Steinheim man of roughly the same period is more modern than *H. erectus* and more primitive than *H. sapiens*.

By 110,000 years ago, the essentially modern fossils of Neanderthal *(H. sapiens)* had ruggedly constructed faces with large brow ridges and massive jaws. They developed a new industry, the Mousterian, which included delicately flaked Mousterian points and an extraordinary variety of tools for sawing, boring, and scraping. The Mousterian industry evolved out of the Acheulian, a transition that is documented in the cultural and fossil remains of the Tabūn Cave in Palestine. Bone and wood were used for chipping stone flakes; the extreme care with which the tools were manufactured may be interpreted as the first record of art.

The Neanderthals lived during the harsh climatic conditions of glacial advances. The complexity of their tool kit and the severity of the climate they lived in suggest that early forms of *H. sapiens* wore clothing. They often inhabited caves, where skulls of cave bears are arranged in such a way as to suggest the first evidence of religion or perhaps of the earliest hunting trophies. Fossil remains of toothless human jaws and of jaws with abscessed or missing teeth, apparently from old individuals, indicate that early *H. sapiens* cared for the aged. There is much evidence at Skhūl in Mount Carmel, Palestine, and at La Ferrassie and Le Moustier, France, that the dead were buried; and in some cases, stone tools, animal bones, and even flowers were buried with the bodies, the first indication that early man believed in life after death.

The period of the Neanderthals extends from 110,000 to 35,000 years ago, in Europe, Africa, and the Middle East.

From 35,000 to 10,000 years ago, fully modern *H. sapiens,* known as Cro-Magnon in Europe, continued living a stone-age existence. These men produced elaborate stone tools and left the first samples of tooled bone and ivory and cave paintings.

Shortly after this period, the Old Stone Age ended and was followed quickly, in geological terms, by the domestication of plants and animals and the use of metal.

The many human fossils uncovered in various areas of the Old World suggest that the numerous natural experiments, among the ancestors of the higher primates, have eventually resulted in man. Hominid fragments from many parts of the Old World point to the remarkably wide distribution of primitive man, who has been a successful animal for a long time, living in relatively large numbers in most of the habitable parts of the earth, and invading other areas when those previously occupied became in any way unsuitable for survival.

Modern man, sometimes called Neanthropic to distinguish him from ancient or Paleoanthropic man, is difficult to define in biological terms. The progressive form of Neanderthal man had some features that resembled our own, but creatures very like ourselves probably existed even before the emergence of Neanderthal man. Thus, Neanthropic and Paleoanthropic men were probably contemporaries as far back as 500,000 or 600,000 years ago even though no fossils document this contemporaneity.

Fossil evidence of modern man has been traced as far back as 200,000 years ago, but he really emerged in his present form only about 37,000 years ago. The men of that time were indistinguishable or nearly so from those of today. It must be emphasized that modern men emerged gradually and in widespread groups and were not any more alike than those of today. The fact is that since 37,000 years ago earlier forms of men were no longer seen. Since most paleontological work has been done in Europe, we have learned much about the men who lived there but not much about those who lived elsewhere. For example, Africa was populated by the ancestors of modern Bushmen, but no one knows who the ancestors of modern Negroes were or where they came from. For that matter, the same applies to Mongoloids. Furthermore, even when a skull is well preserved, it is not easy to assign it with any certainty to a specific race. Experts have made frequent errors in these difficult assessments.

It has been conjectured that modern men were separated into existing races (Caucasoid, Negroid, and Mongoloid) half a million years ago. If, as Carleton Coon believes, *H. sapiens* originated in Asia, he must have migrated east and south. Cro-Magnon man settled in Europe, and the southern invaders displaced the races of Africa. Other Asian forms probably migrated to Polynesia and Australia. Later, 20,000 to 15,000 years ago or more, Asian man reached the American continent by way of the Bering Strait, which at that time may have had a land bridge. Probably all of the distinct races of New World Indians arose from these wanderers.

The early European modern man, called Cro-Magnon after the cave in which the fossils were first discovered, had a frame about six feet tall and a cranial volume of around 1,300 cc. Like his predecessors, he was a cave dweller; extant murals on cave walls indicate that he developed stone and bone tools and a remarkably advanced art. His rule of Europe extended from about 35,000 to 10,000 years ago.

Cro-Magnon man, who must have migrated constantly, had largely disappeared from Europe when the ice finally began to retreat and the climate became milder. Man then no longer needed to be a cave dweller and became more nomadic than ever. He also greatly improved his stone tools, began to develop land for agriculture, and domesticated animals. Then, about 8,000 to 9,000 years ago, human progress, which had been advancing at a lumbering pace for a half-million years, suddenly exploded. In the incredibly brief period of 3,000 years, the number and kinds of tools that man invented almost defy cataloging. He learned to use metals, developed agriculture, domesticated animals, and invented many symbols to store his growing experiences and achievements. Thus, after making painfully slow progress for nearly a million years, man experienced an extraordinary blossoming of his mental potential that continued to burgeon into an ever-expanding transcendental culture and miraculous technological advances, the full expression of which is not yet in sight.

3

MAN'S MANY KINDS

The definition of man in niggling taxonomic or minor anatomical terms misses the genetic boat.
(Weston La Barre)

The less one knows about a group of animals the easier it is to be "positive" about its exact position in a system of classification. But a species such as man, about which we have a wealth of biologic information, presents a real problem, particularly when he exists in so many varieties of color, size and shape. One must wonder if indeed there *are* "races" of man. Formerly of considerable concern to anthropologists, racial classification has now become a political, social, cultural, and emotional rather than a biological consideration. Certainly, no one can deny that there are specific and nonspecific physical, biochemical, and even temperamental differences among groups of men; how much of the latter is due to heritage remains a moot question. A Nordic, blue-eyed blond is easily distinguished from a Negro or an Abyssinian, and all three differ from a black or brown Indian. Yet, we classify the Nordic blond, the Abyssinian black, and the Indian as one "race," the Negro as another. Thus, it must be obvious that such physical traits as skin color have little bearing upon the classification of human races. For that matter, some chimpanzees have white skin and others black; but no one has suggested that they should be separated into white and black "races."

Furthermore, we propose here that whereas it is still somewhat possible to separate into subraces the different ethnic groups of the United States and England, the gene flow between people of European ancestry and those of American, African, Asian, and now Arctic descent is such that before long all such superficial characteristics as skin color, type of hair, color of eyes, superficial body contours, bony structure, "temperament," and others will be amalgamated into one conglomerate form. This, as a matter of fact, is already taking place. There cannot be today many American Negroes free of some white genes and, to a variable extent, many whites free of Negro genes. Hence the only conclusion is that, despite their superficial differences, all men belong to the species *Homo sapiens* and exhibit the same basic inclinations, abilities, and potentialities.

Mostly because of ignorance and the outmoded belief in the superiority of one race over another, it was once possible to discuss races with considerable arbitrariness without stirring emotional conflicts. But today we know that there is no biological basis for the assumption that one race is better than another. In a statement on the study of races, Washburn (1962) writes: "The races of man are the result of human evolution . . . [and] are open parts of the species, and the species is a closed system." Thus, the important point is the evolution of the species and the factors that have contributed to this evolution, not its races, which probably resulted from isolated local factors and therefore are of minor importance in terms of evolution. Whereas the evolution of races may have come about through mutation, selection, migration, and genetic drift, the relation between human biology and culture cannot be disregarded, for, as Washburn says:

> Selection is for reproductive success, and in man reproductive success is primarily determined by the social system and by culture. Effective behavior is the question, not something else.

He points out that a population drift depends on its size, which is dictated not by purely genetic factors but by culture. The numerous causes for population migration—e.g., economy, transportation, and warfare—are all recorded in archeologic relics. Giving us, as it does, glimpses of our past, archeology contributes as much to the study of the origin of the races of man as genetics. Considerations of the human races have a tendency to bog down so that even the anthropologists, who should know better, have been so busy subdividing our species, looking for infinitesimally small details between small parts of the species, that they have lost sight of the evolution of the species. Concern should be, not with picayune differences between parts, but with who the people are, where they are, and how many there are. Political and emotional overtones must be eschewed in any serious and precise discussion of the meaning of race.

Man's tendency to migrate is such an important part of his history that scarcely any human population can now be completely separated from others. As migration brought in new genes, the differences among the races were reduced. Thus, aside from the chance entry of other factors—genetic drift or founders who differed from other members of the species—racialism is nonexistent.

The principal factor in the establishment of racial differences must have been selection. Thus, the extant differences among the races must at one time have been adaptive. Ironically enough, such adaptive accommodations survive without any adaptive role to play in the present. Take skin color, which today is still a complete puzzle, as an example. It has been suggested that when we come to fully understand skin coloration, we will understand the origin of races, although not necessarily of our species. Color must surely have been important in the past, although it should not be in a modern technical society. When we speak about races we have to keep two things in mind: first, we cannot reconstruct the past by assigning adaptive or nonadaptive properties to all human biologic characteristics; second, all men have adapted in a large number of unique characteristics: all have developed a speech and can learn any language; all perform and can learn untold numbers of skills; all cooperate, have artistic senses, have a sense of wonderment, and therefore have developed a science, philosophy, and religion. Now extremely mobile, human beings have extraordinary means of communication. Since neither of these factors favors the perpetuation of sociocultural and geographic isolation, adherence to the older concepts of fixed races becomes less and less tenable. The idea of race is so confused that up to 1965 the United States Bureau of the Census, helpless to devise its own definitions, accepted those "commonly accepted by the general public." The 1965 issue of the *Statistical Abstract of the United States* contains the following statements about race:

> It [race] does not . . . reflect clear-cut definitions of biological stock, and several categories used obviously refer to nationality. Color divides the population into two groups, white and nonwhite. The nonwhite population consists of Negroes, American Indians, Japanese, Chinese, Filipinos, and all other groups not classified as white. Persons of Mexican birth or ancestry who are not definitely Indian or of other nonwhite stock are included in the white population. Persons of mixed parentage are placed in the race or color classification of the nonwhite parent.

The 92nd edition, published in 1971, contains the terse, somewhat self-conscious, statement:

> The population is divided into three groups on the basis of race: White, Negro, and other. Persons of Mexican birth or ancestry are classified as white unless they are definitely of some other racial stock, such as Indian.
>
> The 1960 and 1970 censuses obtained the information on race principally through self-enumeration. Persons of Mexican or Puerto Rican birth or ancestry who did not identify themselves as of a race other than white were classified as white. In 1970, the father's race was used for persons of mixed parentage who were in doubt as to their classification. In 1960, persons who reported mixed parentage of white and any other race were classified according to the other race; mixtures of races other than white were classified according to the father's race.

We leave it to the reader to reflect upon the ensuing chaotic state of affairs.

Whenever in the ebb and flow of human history people of different races have been brought together, miscegenation (L. *miscere,* to mix; *genus,* race, stock) has followed. The outstanding example of miscegenation between whites and Negroes in the United States resulted not only in genetic mixture but especially in an extraordinary social and historical consequence. This was ironic since neither group was racially uniform to begin with: the whites had come mostly from different parts of Europe and the Negroes from widely separated areas of Africa. The trivial genetic isolation initially "enjoyed" by white immigrants later disappeared as the groups gradually became amalgamated. Negroes likewise underwent the same blending and mingling that finally resulted in a single "race."

The most significant stages of miscegenation in the United States occurred during the period of Negro slavehood. Early miscegenation took place primarily between white men and Negro women, and thousands of hybrid mulattoes were added to the population. Ironically, even though half their genes were from a white parent, they have always been classified socially as black. Therefore, mulattoes have tended to marry each other or Negroes with no white genes. The initial result was an infiltration of white genes into the Negro population without a reciprocal gene flow from Negroes to whites. But as time went on this one-sided flow of genes became more reciprocal and no doubt in more recent years has resulted in a nearly mutual exchange.

Fossil remains of man are too scarce to provide clues about the existence or absence of what we call races. Perhaps what we now recognize loosely as human races have existed for a very long time.

Classification of Man

The difficulties of classifying man are numerous. First, there is the problem of nomenclature, which threatens to stymie even the anthropologists. But, preceding this difficulty is another prime consideration: why try to classify man according to races in the first place? To contend that classification leads to an understanding of the history of man is to disregard the fact that races have little if any biological importance. Yet even today, the serious and impartial scientist has to resist the pressure of politicians and others to segregate mankind into arbitrary groups for nationalistic or imperialistic expediency. Despite some slight agreement that at least in a general way human beings can be separated into races, it is regrettable that the word *race* is so deeply rooted in our language that it cannot be avoided. Curiously, in a zoological sense, race often denotes a subspecies or regional variation of some animal species, but obviously such connotations would not be acceptable in a nationalistic sense. Where is the man brave enough to differentiate physically between someone of the "Dutch race" and someone of the "English race"?

Taxonomically, a race is a group of very closely related plants or animals within a species (including man), each of which possesses specific inherited physical features. When applied to human beings, this apparently reasonable and reasoned definition becomes fraught with many difficulties. What features should be selected, and should all features have equal value? Since there are obvious differences in the relative importance of the selected features, who is to decide their hierarchic order of importance?

Linnaeus classified mankind into four main continental varieties, basing his separation on the color of skin, hair, and eyes: *Americanus rufus, Europaeus albus, Asiaticus luridus,* and *Afer niger;* in a fifth group he included the forms he considered to be abnormal. Later, Blumenbach pointed out that since there are so many intermediate gradations in color, appearance, height, and other traits, all men must be related and the differences among them are only of degree. Basing his classification on the shape of head, face, and nose, the form of hair, and the color of skin and hair, he recognized five main divisions: Caucasian, Mongolian, Ethiopian, American Indian, and Malayan. Blumenbach laid too much emphasis on the extremely variable features of the facial skeleton, but admittedly his concern with the anatomy of the skull established a solid foundation for the study of craniology.

In subsequent efforts to subdivide mankind, seldom was any

thought given to the nature of the forces that caused the differences so ardently observed and measured. The whole stress was on physical features. The degree of projection of the jaws (snoutiness, or prognathism), the curl of the hair from the scalp, the cephalic index (maximum breadth of skull × 100, divided by maximum length), facial breadth and height, the slope of the nose, and the characteristics of the bony nasal apertures—all had their protagonists. W. L. H. Duckworth (1870–1956) of Cambridge University made perhaps the first serious attempt to evaluate the relative importance of the various anatomical features. His classification of man (1904) used cephalic index, prognathism, and cranial capacity as primary criteria; facial, nasal, orbital, and epidermal features were of secondary importance. Duckworth's criteria were widely adopted, and some are still used today.

Most modern anthropologists are reluctant to plot formal arrangements of types of men. As new methods are devised, older classifications based primarily on superficial physical features gradually become nonsense. The five categories that follow are presented only as examples of classification; they should be considered in the light of the preceding discussion.

1. *Europiforms*

This large, widespread population, which originally inhabited all of Europe, north and northeast Africa, and parts of Asia, has now migrated to all parts of the world. Known colloquially and with hopeless inaccuracy as "white," this group shows extremes in skin color from nearly white or pink to very dark brown (Figure 3–1). Its members have every conceivable type of hair from straight to kinky, black to nearly white, gold to coppery red. They have no prominent prognathism, and their narrow, high-bridged noses tend to be pointed. No comment is possible about the form of the head and stature because almost all gradations can be found. Loose subdivisions include Nordics from around the North Sea; Baltics, Samians, or Lapps, and Uralics from the district between the Ural mountains and the Ob basin in Russia; Alpines, or Rhetians, from central Europe; Dinarics, or Illyrians, from the Balkans; Taurics (Armenoids) from the Caucasus and Asia Minor; Mediterraneans from the countries surrounding that sea; Erythriotes, or Eastern Hamites (including Somalians and Abyssinians); Chersiotes from India and Ceylon; Aralians from southern Russia; and the Ainu, reportedly the aboriginal inhabitants of Japan. To concern ourselves with the numerous physical differences and similarities of all these peoples or, for that matter, to argue why they have been placed in this and not in another catgeory would add little to this discussion. It could only be a useless

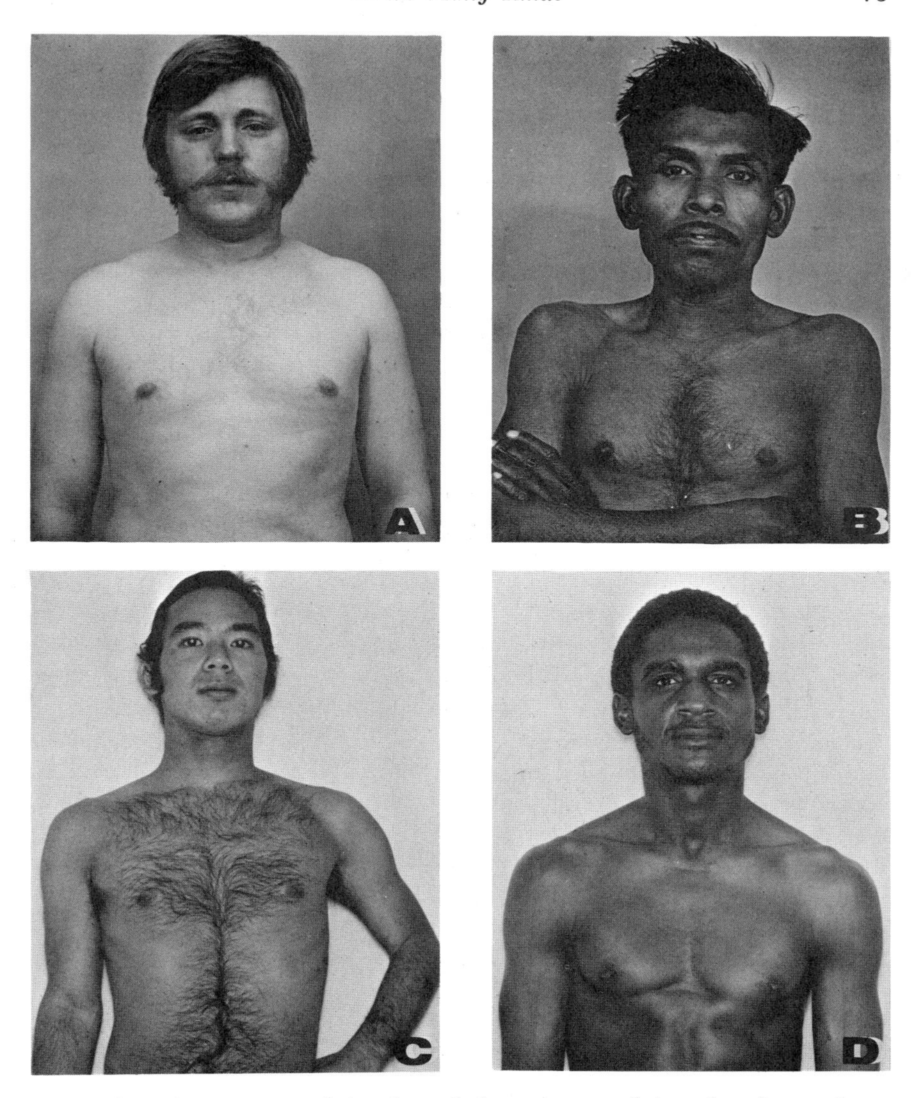

Figure 3–1. A comparison of the physical characteristics of these four figures shows as many differences between the Europiforms A: Nordic, and B: Chersiotes, from India as there are between A and C: Monogoliform and A and D: Negriform.

academic exercise, since but a few pockets remain in the world where relatively pure groups can be found. The United States, for example, with a population of more than 200 million, is composed of members from nearly every one of the subgroups listed, all of whom have had opportunities to interbreed. Although, as in England, the majority of the original settlers were from Nordic, Mediterranean, and Alpine stock, it is no longer possible to make precise groupings.

2. *Mongoliforms*

The most numerous peoples on earth today are the "aboriginal" inhabitants of Asia, the Americas, and certain parts of Europe and Oceania. Despite numerous exceptions, they have a rather broad, flat face, prominent cheekbones, straight black hair, and medium or short stature. Their skin color varies from a light olive to reddish brown. The children of Mongoliforms have a grayish or bluish sacral (or Mongolian) spot or spots on the back just above the cleft between the buttocks. Mongoliforms also have an epicanthic fold of skin from the upper eyelid to the side of the bridge of the nose, hiding the inner angle (canthus) of the eye. The Mongolians, who form a large part of this group, are widely distributed from Siberia to Astrakhan and from north China, Japan, and Korea to the East Indies. The eastern, or "true," Mongols are still highly nomadic tribes and excellent horsemen. The Huns were a Mongoliform people, who lived "riveted to their horses."

Other subdivisions of the Mongoliforms are the Nesiotes of Indochina and Indonesia; the Polynesians, and the Arctics, including the Eskimos. Much controversy still rages about the position of the Polynesians and Eskimos, some authorities maintaining that the former are better placed with the Europiforms, others that Eskimos should be considered a race in their own right. In a number of classifications, American Indians have been placed in a separate category, but since the evidence is meager they are included here, partly out of desperation, with the Mongoliforms. At least eight subdivisions of American Indians have been recognized, one of which includes the "redskins of the plains" (Sioux, Crow, and Blackfeet).

3. *Negriforms*

Originally confined to that part of Africa south of the Sahara and to parts of Asia, many members of this group were transported as slaves to the New World, and its representatives can now be found almost everywhere. They have a dark, "black" skin, kinky hair, a broad nose, a prognathous face, and thick, turned-out lips. The African Negriforms, or Negroes, are subdivided into five subraces. The tall Sudanians inhabit the grasslands south of the Sahara; the even taller Nilotes inhabit the Sudan; the Guineans, including the Yoruba, are from Nigeria; and the Congolians come from Central and Southwest Africa; the Zingians, the greater part of the Bantu-speakers, are from south of the equator.

The Negrillos, or pygmies, of the equatorial forests have been accorded separate racial status. Except for pituitary dwarfs, these are the smallest human beings known; the men are seldom over 4 feet 8 inches tall and the women are usually some 3 inches shorter. (The term *pygmy* is derived from a Greek word meaning "cubit," an ancient standard of measure extending from the elbow to the top of the middle finger, in all about 18 inches.) Their most noteworthy feature is abundant rust-colored hair on the scalp, axillae, and pubis and yellow down-hair on their backs. Pygmies have been subdivided into three subraces according to their western, central, and eastern habitats in the belt between 5° N. and 5° S. The Negritos from Malaya, the Andaman Islands, and the Philippines and the Oceanic pygmies (once widely distributed but now confined to Dutch New Guinea) resemble African pygmies, but their hair is black. The Negriform group is completed by the Papuans of New Guinea, the Melanesians from New Guinea and Fiji, and the extinct Tasmanians, who are believed to have been descendants of Asiatic Negritos. The last aboriginal Tasmanian died in 1876.

American Negroes originated mostly from the West African colonies from Senegal in the north to Angola in the south; records show that about 70,000 slaves were exported from West Africa between 1733 and 1785. Despite the different tribes, the African populations of those areas are relatively similar.

4. *Khoisaniforms*

Although commonly included with Negriforms, the Bushmen and Hottentots have been assigned to a different category. The name *Khoisaniform* is derived from the words *Khoi-khoin* (men of men) and *Sansan* (San people), by which Hottentots and Bushmen describe themselves, respectively. They have a wrinkled, yellow-brown skin and spiral hair, an elongated head and bulging forehead, prominent cheek bones on a small face, and no marked prognathism. Large deposits of fat in the female buttocks cause excessive protrusion. The Khoisaniforms are short and delicately built with relatively short limbs and an overall childlike or infantile appearance. These traits have often been cited as examples of the persistence of fetal characteristics into adulthood.

5. *Australiforms*

This term is derived from *austral,* meaning "southern." The group includes the native people of India, Ceylon, and Australia, who have yellowish to dark-brown skin and black wavy hair. Their heads are high relative to the length, with prominent ridges at the eyebrows (supraorbital ridges). The members of this group have low, retreating foreheads and an almost keel-like top to the skull. Veddians of India and Ceylon and Australian aborigines are the two main subdivisions.

The latter have particularly interesting anatomical features and a specialized culture; the description that often refers to them as the most primitive people alive today is hardly justified. Aborigines are a lean, hard people with linear build, long heads and faces, long thin extremities, and slender hands and feet (Figure 3–2). They have wide cheekbones and a broad nose with a depressed root and wide nostrils. Aborigines share with the Veddians a 30 percent incidence of a heavily haired pinna to the external ear (Figure 3–3). They have a rich, reddish-brown color (chocolate), darker in the regions north and west of Australia. Their hair is dark brown to black in adults, but some children have fair scalp hair. There is little doubt that the aborigines came from Asia; there were some 300,000 of them when the first white men landed in Australia. The impact of European civilization forced the number to fall to about 50,000, but it is now rising again. Aborigines are a genetically heterogeneous collection of populations, the heterogeneity probably arising from random genetic drift operating on small, isolated breeding populations, the occurrence and spread of new mutants, and the mixing of gene pools from different groups. The environment, especially the desert, may also have exerted a selective influence.

Disadvantages of Classifying Man by Gross Physical Features

None of the physical features summarized in this brief description can be called an accurate or satisfactory criterion. With few exceptions, physical characteristics are not inherited in a simple manner, and in many cases their manifestations are controlled by independent genes. A single physical feature may also be the result of several factors, each controlled by more than one gene. Skin color is largely the result of different quantities of melanin, and stature is determined by the length of the bones of the lower limbs, the depth of the pelvis, the thickness of vertebrae and intervertebral discs, the thickness of articular cartilages, intraarticular discs, and height of skull. It varies according to the position of the body when measured, and everyone is slightly taller in the morning. Let us not forget also that physical features vary with age, sex, nutrition, and climate. Certain modifications and deficiencies in maternal diet during pregnancy likewise exert remarkable

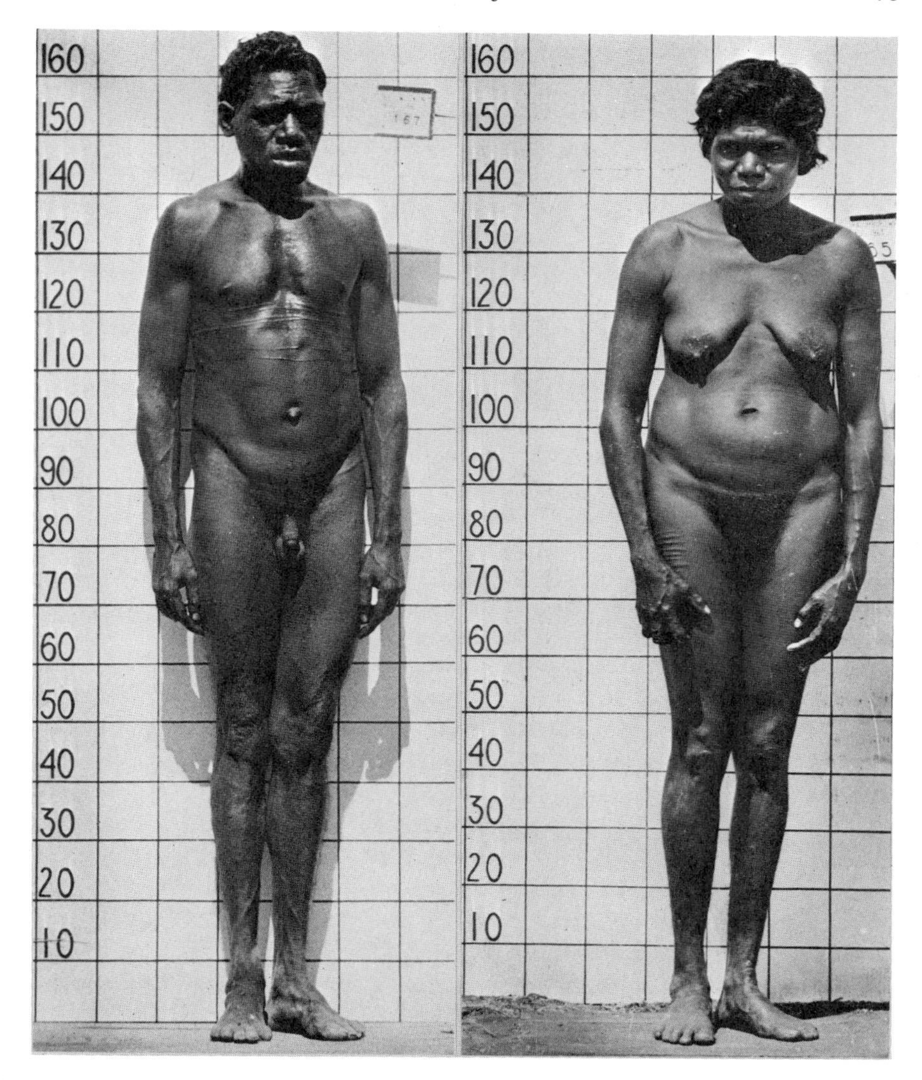

Figure 3–2. Australian aborigines, female and male. The woman, 28 years old, is 5 feet 3 inches tall and weighs 116.8 pounds; the man, who is 37, is 5 feet 4 inches tall and weighs 123.1 pounds. Both have a linear build and thin, long extremities.

effects on the form of the fetus and, no doubt, on that of the adult. Despite the most careful mathematical treatment of measurement, the institution of statistical devices, such as the coefficient of racial likeness, and the analysis of intercorrelations of groups of characters, a classification of mankind based on such physical factors still enjoys only a dubious reputation.

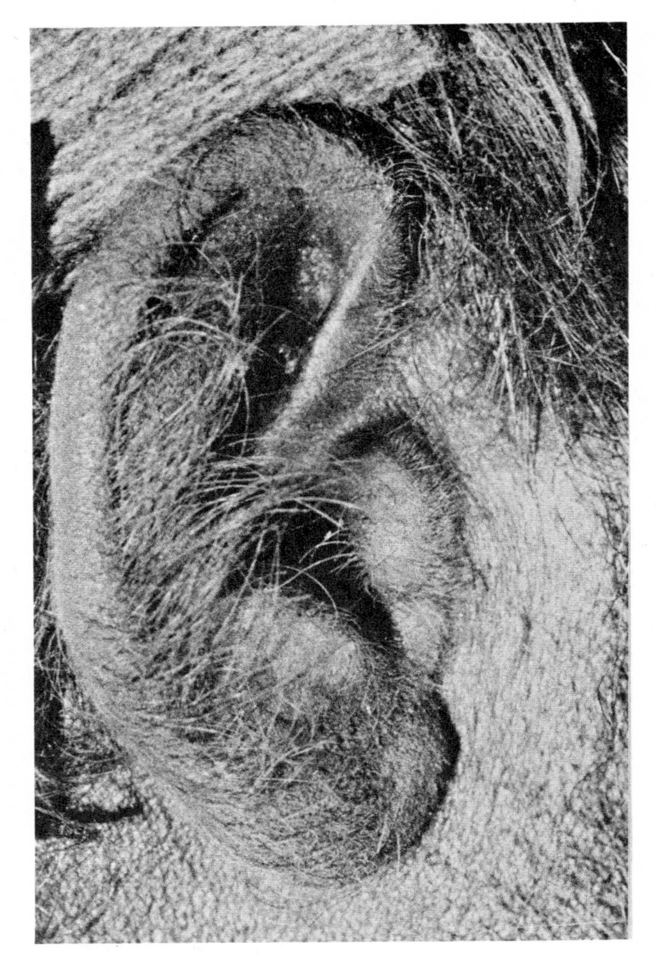

Figure 3–3. The heavily haired pinna of a male Australian aborigine. *(Figures 3–2, 3–3 courtesy of Professor A. Abbie.)*

"Nonphysical" Traits

The above attempts at classification would surely be expedited if man possessed sharply differentiated, genetically determined characters, fixed for life and measurable with unfailing accuracy. Some such characters do exist and have anthropological value, and no doubt more will be discovered. The best of these are not physical in the gross sense but are detectable by biochemical or serological methods. We briefly mention here only a few of them.

About eight percent of the men in Europe and the United States are color blind; the percentage is lower among Mongoliforms. Color blindness is more common in men than in women because it is a sex-linked recessive condition, with the gene carried on the X chromosome. When present on the one X chromosome of the male this gene causes color blindness. Unless the gene is present on each of the two X chromosomes, the condition does not affect women. Certain rarer types of color blindness, however, are not sex linked.

Most North American Indians can taste the bitter substance phenylthiocarbamide (P.T.C.) because they and other "tasters" possess the single dominant gene T. (Seventy percent of American whites are tasters, there are fewer Arabs (63 percent), and more Egyptians (76 percent). We are not yet certain that the ability to taste this and other chemically related compounds is genetically determined, nor do we know why more women can taste it. The inability to smell hydrocyanic acid is probably a sex-linked recessive characteristic also. Because of the dangers of testing, however, these traits have not been explored much.

Two known abnormalities of blood caused by single genes can be manifested in both homozygotes and heterozygotes. One is thalassemia (from a Greek word meaning "great sea"), or Mediterranean anemia, in which the pigment of red cells (hemoglobin) is of the fetal type. Prevalent in certain regions of Italy, this disease is generally fatal during childhood in homozygotes; how it is perpetuated remains a mystery. The second abnormality is the "sickling" of red corpuscles, or sickle-cell anemia, in which the corpuscles are distorted into the shape of a crescent and the hemoglobin has abnormal properties. The trait is particularly common among Negroes; in some African groups, it appears with a frequency of up to 40 percent. Sickling, and probably thalassemia, are believed to confer a compensatory resistance to malaria. Most of the other genetically determined blood abnormalities are too rare to have any anthropological value.

Blood Groups

Even the physicians of antiquity probably asked whether blood from one individual can be introduced into the circulatory system of another. Apart from some early records of blood drinking for "therapeutic" measures, it was not until 1664 that J. Denys in France and Richard Lower in England (1666) actually performed the experiment. Lamb's blood was the first to be introduced into a man's circulation, although Pepys had earlier recorded the payment of one pound sterling to a human blood donor. Since animal and human blood are incompatible, the transfusion of animal blood into a man's circulation must have had catastrophic results. Moreover, blood quickly clots after withdrawal, and until this and other difficulties had been overcome, the transfusion of blood remained impracticable.

The discovery during the early years of the twentieth century that the blood of individuals can be grouped into four main classes was the first important step in making transfusion practical. To be safe, transfusion must be carried out between individuals of the same group or with blood from the single group known as universal donors. When it is carried out among people with incompatible blood groups, the consequences are serious and even fatal. Clotting has long been overcome by the use of anticoagulants, such as heparin and sodium citrate. These findings opened the way to the discovery of a number of other blood groups, some of which are very rare. To a point, blood groupings can be used to distinguish the various groups of mankind. Anthropologists have been investigating whether any evidence of blood groups is locked within the fossil remains of ancient man. Primates and other animals also possess different blood groups.

Between 1900 and 1902, K. Landsteiner, an American immunologist, found that human blood can be divided into four classes according to the way the red cells agglutinate with particular normal human sera (blood groups of the ABO system). This reaction depends on the presence of one, or both, or neither complex substances (polysaccharide), called A and B, on the cell surfaces. The 80 or so percent of persons who have these substances, present also in saliva, gastric juice, and other secretions, are called secretors; A and B substances can be prepared in a fairly pure state from the fluids of such persons.

The behavior of red cells depends on a set of three genes, A, B, and O. Their presence and arrangement on the chromosomes ordain that there are genotypes AA or AO (Group A), BB or BO (Group B), OO (Group O, universal donor), and AB (Group AB). The percentage frequency of the groups in the United Kingdom is O: 47 percent; A: 42 percent; B: 9 percent; and AB: 3 percent. In the United States, the "white" population has about 44 percent in group O, 42 percent in group A; 10 percent in group B, and 4 percent in group AB. Forty-two percent of American Negroes belong to group O, 26 percent to group A, 20 percent to group B, and about 5 percent to group AB. There are some differences among the Orientals, but the Chinese give a fairly characteristic representation. About 31 percent belong to group O, 25 percent to group A, 34 percent to group B, and 10 percent to group AB. Unfortunately, this classification is not as clear-cut as it would seem to be. Group A can be subdivided into A_1 and A_2, and the rare groups, A_3 and A_4 and certain sera (anti-H and anti-O) that react strongly with group O cells. The cells and sera of some individuals in Bombay behave in such a way as to indicate that they possess yet another allelic gene of the ABO system.

The determination of ABO blood groups has now been carried out in numerous countries on many millions of individuals, and the distribution of the groups and the more important gene frequencies throughout the world are moderately well known. The highest frequencies of group A are found in tribes of North American Indians; high frequencies also occur in the Australian aborigines, in western Asia, the central mountain chains of Europe, and in Scandinavia. Maximal frequencies of B occur in central Asia and northern India, and in Egypt and central Africa. It is generally low in Europe, very low among the Basques, and absent in most of the American Indians and Australian aborigines. Almost all the Indians of South and Central America belong to group O. A peripheral distribution of populations with a high group O frequency is found in Europe and Africa, mostly in geographically isolated communities.

In 1940 Landsteiner and A. S. Wiener discovered the Rhesus, or Rh, factor in serum that had been sensitized with red cells from the rhesus monkey, *Macaca mulatta*. This is a complicated system, probably consisting of four adjacent loci on each of a pair of chromosomes and cannot be discussed superficially. Rh groups are involved in causing hemolytic disease of the newborn. In marriages where the mother is Rh-negative (dd) and the father Rh-positive (DD or Dd), the fetus will always be Dd in the first instance, and half of the fetuses will be Dd in the second. In most cases, no ill effect will result in the first of such pregnancies, but in the second the mother's serum has become immunized to the D antigen of the fetus, and the anti-D antibody formed in her blood crosses the placenta into the fetal circulation and causes damage to fetal red cells. Such damage may result only in slight anemia, but in many babies death can occur from the severe destruction of red cells unless appropriate treatment is given by blood transfusion. An interesting anthropological discovery is that the Basques have the highest Rh-negative (d) frequency of all known populations. From this and other determinations of blood groups, we can conclude that the pattern of Rh frequencies in Europe probably results from the mixture of Rh-positive and Rh-negative strains and that this occurred only a few thousand years ago. The fact that Mongoliforms appear to be Rh-positive was the basis for placing Australian aborigines among the Mongoliforms rather than the Negriforms.

In 1927, Landsteiner and Levine discovered yet another system of blood groups, MN, and twenty years later the subgroups Ss. These have been much used for anthropological and classification purposes; fortunately they do not complicate transfusion because there are no natural antiagglutinins in human sera. American Indians and Eskimos have high M frequencies; New Guineans and Lapps have high N values. S is almost completely absent from Australian aborigines.

The immense value of blood group determination could be illustrated in many ways, but the information is now so vast that only books of specific reference cover it adequately.

Blood groups of animals Substances very much like human A and B factors are present throughout the animal kingdom, including certain birds and many species of mammals. In a few mammals, a system as complex as that in man has been discovered.

Blood groups in primates are being widely studied. Chimpanzees, whose blood groups most closely resemble those of man, are either group A or O (11 percent), possess M and N, and appear to be Rh-negative, but some of their antigens differ from those of man. Gorillas are either group A or B and orangs and gibbons are A, B, or AB. Old World monkeys lack A and B antigens in their red cells. New World monkeys have a B-like antigen; the saliva of monkeys may, however, contain antigens that cannot be detected in their red cells. Apparently, then, ABO groups are common to animal life and were inherited by man when he took his place in the system. Part, at least of the MN and Rh systems, appears to have been inherited from ancestral primate stock.

Blood groups of skeletal remains It is not easy to determine ABO groups in bones long dead (blood group archeology); therefore only a few reliable observations are available. Blood groups determined in Egyptian and American mummies indicate that most prehistoric and early Egyptians belonged to the A and B groups, most American Indians to group O. Techniques obviously need to be extended to other ancient human remains.

Body Habitus and Somatotyping

Another method of sorting out kinds of men is to consider their constitution, or the sum total of the structural, functional, and psychological traits of individuals. Some relation may exist between the shape and parts of a man (body habitus) and his character and temperament. The writings of Hippocrates on the matter were passed to Galen and on to subsequent generations until the seventeenth century. Hippocrates maintained that there are in the body four "humors," or liquids—blood, black bile, yellow bile, and phlegm. An excess of any one of these affects the temperament of an individual, his anatomy already having certain predispositions to bring about that particular excess. The majority of men were, therefore, classified as sanguine, melancholic, choleric, or phlegmatic, though blends of the various humors were allowed. Such beliefs no doubt influenced Shakespeare when he had Julius Caesar say:

> Let me have men about me that are fat:
> Sleek-headed men and such as sleep o'nights;
> Yond' Cassius has a lean and hungry look;
> He thinks too much: such men are dangerous.

The importance of constitution to medicine was recognized by John Hunter and later by Jonathan Hutchinson (1828–1913), a surgeon of the London Hospital, who suspected that individuals with a certain build and temperament were likely to be susceptible to particular diseases. Early attempts to analyze constitution failed because only extremes were considered, whole individuals rather than composites of numerous parts were dealt with, and the many factors that cause variations were not taken into account. A German psychiatrist, Ernst Kretschmer (1921), classified all human beings into three main types: *pyknosomatic,* or *pyknic:* stocky, compact, and rotund individuals; *leptosomatic,* or *asthenic:* tall, lean men; and *athleticosomatic:* big, broad "heavyweights." He further suggested that one type was more likely to exhibit particular mental disorders and to contract certain diseases. This was a popular scheme, but it had too many defects and did not survive. Its worst imperfection was the absence of any basis for measurement.

In 1940 W. H. Sheldon used three 7-point scales to assess the amount that any individual showed of three components, *endomorphy, mesomorphy,* and *ectomorphy.* An individual high in endomorphy, scale 7:1:1, would be spherical in caricature, with roundness of head and abdomen, penguinlike limbs—an epitome of Edward Lear's adjective "runcible." Predominance of mesomorphy, scale 1:7:1, would be a muscular athlete with cubical head, muscular neck, and burly chest. The ectomorphic, scale 1:1:7, is a lean man, with the lowest mass and relatively greater skin area, thin-faced with high forehead, receding chin, and spindly limbs. Naturally, these are extremes and the majority of individuals have ratings of 3:4:4, 3:5:2, and so on. The respective rating was called the somatotype (Gr. *soma,* body). Sheldon also introduced a further component, called gynandromorphy, by which he hoped to show the degree, on another 7-point scale, to which a male body resembled a female, and vice versa. Such characters in a male as large hips in relation to shoulder breadth, much fat over the pubis and in the mammary region, and overlap of the thighs with the feet placed together would be those of gynandromorphy.

Even though Sheldon placed somatotypes on an anthropometric

basis by compiling tables of seventeen standard diameters, the original classification is subjective and open to criticism. His scale allows only limited jumps in rating, and the components are correlated. What is needed, therefore, is some kind of factor analysis that can present large numbers of measurements, taken at the same time, of many variables. We are constantly faced with such problems when we try to base a comparison of biological materials on measurement.

Sheldon has also attempted to relate physique with temperament. It is well to remember, though, that any attempt to establish relationships between body and mind should be approached with caution. Sheldon and Stevens suggest that temperament can be assessed as the result of sixty traits correlated in three groups of twenty. The first group of components, called *viscerotonia,* embraces traits of love of comfort, relaxation, sociability, gluttony, conviviality, and need of contact with people when troubled. In the second group, *somatotonia,* are vigorous, assertive, noisy, courageous individuals, indifferent to pain, and in need of expressive action when troubled. The third group, *cerebrotonia,* is composed of youthful-looking people who show restraint, inhibition, love of privacy, secretiveness, sensitivity, fear of pain, and need of solitude when troubled. In a scoring system, an extreme viscerotonic would rate 7:1:1. The results of observations on individuals over the period of a year are used to compare with the somatotyping scales. Various criticisms have been leveled at these definitions, and some have condemned the system as useless. At best, it must be regarded as incipient and experimental. Apart from the merits or shortcomings of the system, doctors know that susceptibility to certain diseases is often correlated with a particular constitution. Significantly enough the psychologists rather than the anatomists have endeavored to define constitution.

Despite the considerations presented in this chapter, the uninitiated may not be able to define the "races" of man much more clearly than before. We have attempted here only to emphasize the enormous difficulty of defining unfailingly, without prejudice and free of social implications, the subdivision of mankind into distinct groups. Regardless of social prejudice and real or apparent physical traits, each race has developed a language, society, and culture of its own. What physical and cultural differences existed could be perpetuated only as long as the races remained allopatric. In a sense, most of mankind is now sympatric, which means that people of different races breed together and that physical and cultural differences will soon disappear. This is happening all over the world. Both the concept of race and the actual existence of races are doomed to extinction, despite temporary barriers that demand unwarranted segregation.

4

MAN'S CURIOSITY ABOUT HIS ANATOMY

What geography is to history such is anatomy to medicine—it describes the theater.
(J. Fernel)

An understanding of anatomy is a prerequisite for the classification of animals. Aristotle (384–322 B.C.), who was probably the first to attempt to group animals into categories and to determine his own niche, did not arrange animals in a definite hierarchic system, but he obviously had some understanding of phylogeny. Although he probably never dissected a human body other than a fetus, he apparently realized that simple or "lower" forms of life existed as opposed to complex or "higher" ones and that man—the acme of perfection—ruled at the top.

Aristotle was no doubt the first comparative anatomist; his discussion of the adaptation of animals to their environment leads us to believe that he may even have developed a theory of evolution. The fact that he did not call a whale a fish but placed it correctly among the mammals is evidence of his perspicacity.

Early Greek Physicians

Anatomical tradition originated as a branch of rational scientific thought in early Greece. Yet, there is no evidence that the most famous ancient school of medicine, founded by Hippocrates (born on the island of Cos about 460 B.C.), had any first-hand knowledge of the structure of man. Early Greek physicians, who probably did little human dissection, were nonetheless maliciously accused of cutting open living men. Herophilus, a Greek physician of Alexandria (355–280 B.C.), said to have been the first to dissect a body publicly, distinguished among and described many parts of the nervous system; he maintained that arteries carried blood, not air as had previously been thought (the word *artery* means "air-carrier"). Several parts of the body are still known by the names he gave them, e.g., *torcular Herophili,* the confluence of cranial dural sinuses. Yet four centuries later the historian Tertullian referred to him only as a butcher of men, accusing him of having dissected 600 living persons.

After Hippocrates, Galen of Pergamum (ca. 129–199 A.D.) was a most important figure in the history of medicine. Born in Asia Minor, he attended Marcus Aurelius for several years and was one of the great physicians and biologists of all time. A perspective of Galen's contributions is not possible without some consideration of the world he lived in and of his own background. As a Stoic be believed in "final causes." Regarding each part of the body as created to achieve a specific end, he contended that each had been fashioned, by a kind of anatomical predestination, precisely for that end and hence could not be other than it was. Galen's purpose in his *De usu partium (Concerning the Use of the Parts)* was to prove in minutest detail that the entire human body with all of its parts is so perfectly constructed for the actions it performs that any alterations in its details would be deleterious. In this remarkable book, which he began in 165 A.D. and completed in 175 A.D., he attempted to define the final cause of every structure of every organ in relation to its determined function. The body of man reflects that perfection, and a knowledge of the function and uses of the organs reveals the power and wisdom of their Creator.

Thus, Galen's thesis is like that of Aristotle, who said that "Nature does nothing in vain." In fact, Galen was a great admirer of Aristotle, particularly of his *De partibus animalium,* which maintains that all animals have been designed with the best possible bodies. Yet, to justify his own work, Galen questioned some of Aristotle's conclusions, stating that since Aristotle did not know all the actions, he could not tell all of their usefulness. Galen saw "Nature" as industrious, skillful, wise, and just. Since "She" assumes a creative role, "She" is to be praised and worshiped as God. In fact, the essence of *De usu partium* is to Galen a sacred discourse, a hymn of praise to the "Creator." "Nature" decides the just and proper size, shape, location, relationship to other organs, texture, and all other qualities of every part with infinite wisdom. Yet, beyond this lies "Necessity," which dictates that certain things are impossible and "Nature" never attempts the impossible. Galen was conveniently blind to defects in the body and made ingenious excuses for "Nature" when "She" had been somewhat less than omniscient. Furthermore, Galen was not above suppressing and distorting facts when these conflicted with what he needed to see to support his theory. Still, to pass too harsh a judgment on his fallibility would be to overlook a milestone in the history of medicine and to denigrate Galen's undisputed competence and devotion to scientific truth.

Galen's line of reasoning strongly appealed to the newly emerging Christian philosophy. Later, as his writings came to be accepted as an accurate exposition of the work of the Creator, it was deemed unnecessary to observe nature since all of its problems had been arbitrarily solved. The period that followed, the Dark Ages, was characterized by a total acceptance of the Christian attitude that the body, having only a temporary value, is not worthy of study. Anatomy came to be generally considered a pagan indulgence concerned with the pitifully perishable body. Hence, by a devious kind of irony, for 1,300 years the very discoveries of the "galenical system" came to exercise a profound and stultifying influence on medicine.

Whether Galen ever dissected a whole human body is not certain, but as official surgeon at the gladiatorial arena, he presumably saw some cruelly dissected parts. He recorded many anatomical observations on man, and as a keen observer, was more often correct than not. He erred occasionally in ascribing to man structures he had found in other animals. To point out his errors in judgment in such a sketchy account as this would run the risk of magnifying them out of proportion to his very real contributions. Such blunders as thinking that the ventricles of the heart communicate by means of an opening and that the liver is the source of the venous system seem trivial when contrasted with his contributions. He made numerous correct observations on the cerebral blood vessels, and the vein that drains the inside of the cerebral hemispheres is still called the great cerebral vein of Galen. Some of his observations, e.g., his warning about the danger of hemorrhage during and after tonsillectomy, are valid even today.

Galen's most important anatomical observations helped to increase our knowledge of the skeleton and muscles (*musculus*, meaning "little mouse," from the fancied resemblance of the mouselike movements of muscle beneath the skin). He wrote *On Bones for Beginners* and once possessed a complete human skeleton to study. His advice to students not only to acquire a book-knowledge of each human bone but to examine all bones carefully with their own eyes is still valid. He named many of the muscles, e.g., the masseter, or muscle that clenches the jaw in mastication. Some of his knowledge of muscles seems to have been obtained from dissections of Barbary apes, members of a species of macaque (*M. sylvana*) inhabiting northwest Africa that are now maintained in Gibraltar as pets by the British garrison stationed there.

Since many of Galen's writings have been lost, his opinion on man's position in nature is difficult to determine. In his *Concerning the Use of the Parts,* he adopted the attitude of high priest. Having been created by an Omniscient Being on a determined plan, the anatomy of man, Galen felt, is perfect and all of his organs and their functions are in absolute harmony. The fires of anatomical inquiry and observation practically ceased after Galen's death. His word ultimately became law, and as late as the sixteenth century his works were quoted as the final reference.

The Middle Ages

After the downfall of the Roman Empire, the center of intellectual inquiry moved east to Syria, Persia, and Arabia. Many important medical texts were translated from the original Greek into Arabic and later, in the monasteries of Italy, from Arabic into Latin. In this way, Galen's influence was extended to Italy and western Europe. The works of Avicenna of Bokhara (980–1037 A.D.), a brilliant Arabian physician, were also translated into Latin. His *Canon of Medicine,* which became the main reference source of the Middle Ages, was essentially a restatement of Galen's writings with some added descriptions of bloodletting. The Arabic influence is seen in some of the current names for veins. For example, the name *saphenous* comes from the Arabic *al-safin,* meaning "hidden" or "secret." This vein is normally visible over only a small part of its lengthy course on the surface of the leg.

In the thirteenth century many universities were established throughout Europe. Their history is synonymous with the history of scholasticism, that vast system of metaphysics, which, as the historian of medicine, Charles Singer, has written so well, "however it may sharpen the wits, does nothing to develop the senses." Once again anatomy went into a temporary abeyance and only occasionally were the ancient Greek texts resurrected and perused. For a time no dissection was carried out in the medical schools of the new universities. Authorities believed that since everything had been described by Avicenna, whose book was based on the observations of the omniscient Galen, dissection would be a waste of time and would merely confirm what was already known.

At last, however, the practice of dissection was resumed in the medical school of the University of Bologna. This university, founded in 1156, was the famous seat of law, and the law school seems originally to have had control over the medical faculty. At any rate, dissection may have first been performed primarily to obtain evidence for legal purposes. Although these legal dissections, or extended postmortems, added little to anatomical knowledge, they were an important beginning that led eventually to the founding of a department of anatomy whose first head, Mondino de' Luzzi (1270–1326), earned the title "Restorer of Anatomy." Mondino is said to have carried out dissections himself; but in those days a professor usually sat high in his professorial chair, shaped like a pulpit, and read or lectured while his assistant dissected the cadaver (that of a criminal) and his students listened and watched —and slept. This method of teaching was excoriated by the great anatomist Vesalius as

> a detestable ceremony in which certain persons . . . perform a dissection of the human body, while others narrate the history of the parts; these latter from a lofty pulpit and with egregious arrogance sing like magpies of things they have no experience of, but rather commit to memory from the books of others.

Despite such bitter censure and the opposition of public opinion and religious leaders, who felt that dead bodies should not be "mutilated," dissection had returned as a permanent source of anatomical knowledge. Realizing that medicine could not advance without this knowledge, civic authorities gradually permitted limited public dissection, first in Bologna and then more widely elsewhere. However, problems were ubiquitous. Dissection had to be done quickly in the fourteenth century. There were no preservatives, and in warm climates the bodies soon decomposed. Furthermore, since the only available bodies were those of executed criminals, one could not rely upon obtaining them for a particular date.

The Renaissance: Art and Naturalism

After making practically no progress for more than a thousand years, anatomy took a sudden leap forward in the fifteenth century, that strange and wonderful period when human genius proliferated and expressed itself in the graphic arts. Naturalism, or the theory that art should conform exactly to nature, had begun in the thirteenth century and came into full bloom in the fifteenth. No longer satisfied with erecting approximate or distorted representations of human figures, artists began to study and dissect the human body, emphasizing the skeleton and muscles. The human form as it is depicted on canvas or paper, carved in stone, or cast in metal during the fifteenth and sixteenth centuries gives ample evidence that artists were well versed in anatomical knowledge.

The first illustrated textbook of anatomy, by Berengarius of Capri (1470–1530), professor at the University of Bologna, was published in 1521. It opened the way to many others, the aim of which was to substitute sketches or drawings for the distasteful and arduous task of dissection. Because of this anatomical visual aid, Berengarius was dismissed from his chair, accused of dissecting living people, using indecent language, and having profligate habits. Berengarius was also the physician of the sculptor Benvenuto Cellini, and the anatomical accuracy of both the professor and patient is above criticism.

Michelangelo (1475–1564), Raphael (1483–1520), Dürer (1471–1528), Leonardo da Vinci (1452–1519), and others left numerous sketches of dissections, often with lengthy notations. Michelangelo's drawings, like his paintings and statuary, have a powerful vitality that goes beyond a mere pictorial representation of muscles and bones. Parenthetically, in contrast to human figures, which were graceful and anatomically correct, the figures of animals during that period were often awkward and distorted, and reflected the artist's ignorance of their anatomy.

Among the artist-anatomists, Leonardo da Vinci attained the greatest heights. Even discounting his paintings, Leonardo must be considered one of the most extraordinary of men—and an outstanding biologist. Sadly, however, he had little influence upon the anatomists of his period. In fact, Leonardo was a modern, centuries ahead of his contemporaries; he stood like an island, unknown and largely ignored by others. Endowed with insatiable curiosity, limitless perception and skill, and indefatigable energy, he made innumerable observations on the makeup and function of the human body and of other animals and plants. Although he himself had dissected only a few human bodies, he must have witnessed many public dissections. He studied the carcasses of animals, but like Galen, erroneously assumed that their structures corresponded exactly to those of man. His anatomical studies are gathered into notebooks that he and Marcantonio della Torre, a professor of anatomy from Pavia, had intended to use in a book on anatomy and physiology. Della Torre, however, died and the book was never written. Had it been, anatomical studies might have advanced several centuries.

Leonardo had a consuming interest in the skeleton, which he depicted with great fidelity. His sketches of bones show them from all sides, and he was the first to describe and illustrate the maxillary antrum and frontal sinus. He identified the different vertebrae and placed them correctly in the vertebral column to obtain the proper spinal curvatures.

Occasionally he blundered in articulating some of the bones, the result, most probably, of not having a complete skeleton at his disposal. Leonardo's studies of the muscles are among his most striking achievements. In representing muscles and bones, he made use of his inexhaustible knowledge of mechanical and architectural principles, which gave functional significance to structures as he saw them. Much concerned with the circulatory system, he made remarkably clear drawings of the heart. In his delineations of the dissected heart, he indicated the intraventricular moderator band, a structure not recognized by anatomists until centuries later. (The moderator band is a bundle of heart muscle fibers containing specialized muscle cells for the conduction of neural impulses.) Leonardo also introduced a method of injecting liquid substances that solidified into anatomical spaces and thereby contributed a better understanding of their shape. Apparently he injected molten wax into the ventricles of the brain and described them accurately for the first time.

The naturalistic movement probably had sufficient impact on the approach to anatomy to have paved the way for the most outstanding anatomist of the Middle Ages, Andreas Vesalius. Born in Brussels in 1514, Vesalius studied at Louvain, Montpellier, and Paris. Obtaining bodies by devious means, he dissected them avidly, even taking parts home to his rooms. His dissections of "that true Bible as we count it, the human body and the nature of man" soon revealed to him the many errors committed by Galen. However, the publication of his famous *De humani corporis fabrica* in 1543 aroused such storms of protest, even from those who had been his teachers, that he destroyed much of his unpublished work and abandoned dissection.

Probably by order of the Inquisition, in 1563 he made a pilgrimage to Jerusalem, perhaps to atone for carrying out postmortem examinations and to learn a more orthodox behavior. He had hoped to return to academic life the following year, but after his departure from Jerusalem, he was shipwrecked on the coast of Zante, an island off Greece, where he died October 2, 1564.

The publication of the *Fabrica,* as it is often referred to, was a turning point in the history of medicine. The book was a faithful account of the anatomy of man, and the illustrations (Figure 4–1), largely executed by Jan Stephen van Calcar under the tutelage of Vesalius, were of such high quality and possessed such expressive vitality that not even the multicolored blocks of modern artists surpass them.

England and the Golden Age of Anatomy

The sixteenth and seventeenth centuries were the golden age of human anatomical discovery, and modern topographic anatomy still eponymously remembers the names of the investigators of that era. At the end of the

Figure 4–1. Anatomical drawing from *De humani corporis fabrica* (1543) by Andreas Vesalius (1514–1564). From a first edition in the Anatomy School, Cambridge University.

sixteenth century, England became the center of anatomical research. Thomas Vicary (d. 1561) was the first master of the Barber-Surgeons' Company, incorporated by Henry VIII in 1540; under his name the first textbook of anatomy in English, *The Englishman's Treasure,* or *The Anatomie of the Bodie of Man,* was published in 1577. It was not original, being based on earlier manuscripts, but it was widely used.

John Caius (1510–1573) was mainly responsible for introducing into England the thorough dissection of the human body, first in London and then at Gonville Hall, Cambridge (later renamed Gonville and Caius College). Strangely, Caius, who had spent five years in Italian universities and medical schools and had shared a house with Vesalius, remained an ardent Galenist. Although Caius contributed little to the anatomical sciences, an alumnus of Caius College, William Harvey (1578–1657), was one of the most distinguished anatomists of all times.

Harvey's influence profoundly altered the theories of the structure of man and introduced an experimental basis for functional anatomy. He too visited Italian universities, in particular Padua, where he studied with Fabricius, a pupil of Vesalius. In 1598, Hieronymus Fabricius of Acquapendente (ca. 1533–1619) built the old anatomy theater in Padua, which still survives. In 1615 and 1616 Harvey delivered anatomical lectures at the Royal College of Physicians. There, he first outlined briefly his theories of the function of the heart in relation to the circulation of blood, ideas that were no doubt influenced by the works of Cesalpinus, a professor at Pisa and Rome. Twelve years later, in 1628, Harvey finally published in Frankfurt his brief, hard-to-read, but most remarkable book *Exercitatio anatomica de motu cordis et sanguinis in animalibus (An Anatomical Dissertation on the Movement of the Heart and Blood in Animals),* a report of his observations on the circulation of blood in man and animals.

Vesalius had given an excellent, if somewhat inaccurate, description of the structure of the heart, so that by Harvey's time it was rather well known. The valves of the great arterial vessels of the heart, aorta, and pulmonary artery had been fully described and their function interpreted correctly. Everyone, from Galen to Leonardo and including Vesalius, who had studied the heart before Harvey, believed that the septum between the ventricles was perforated to allow blood to pass from one ventricle into the other. Some general hints on the circulation of blood had been given by several scholars, and in 1574 the system of valves in the veins had been fully explored by Fabricius. Apparently, however, Fabricius had no real concept of the significance of these valves.

Harvey made numerous observations on the motion of the heart in living animals. He knew that when the heart contracts it becomes smaller and hard, at the same time that the great arteries expand. He showed that the atria of the heart have the same relation to the ventricles that these have to the arteries. Thus, when the atria contract, blood passes into the ventricles, which expand; these in turn contract to force the blood into the arteries, which also expand. He logically deduced that the same blood is forced from the atria into the ventricles and thence into the arteries. The cardiac and arterial valves prevent blood from the ventricles and arteries from flowing back into the atria and ventricles, respectively. Thus, Harvey saw the flow of blood through the heart as continuous. It was evident to him that blood entered the right atrium through the vena cava, the opening of which is also guarded by a valve.

The last piece of this puzzle to be put into place was the fate of the blood from the pulmonary artery, which Harvey correctly called the arterial vein. He explained that blood from the pulmonary artery passes through the lungs and comes back to the left ventricle of the heart by way of pulmonary veins (Harvey's venal arteries). With the completion of this puzzle, Harvey further proved a constant circular movement of the blood from the heart into the lungs and back, and from the left ventricle to the rest of the body, by way of the arterial system from the aorta and back again into the right atrium through the vena cava. These observations, documented by many and confirmed by repeated experiments in living and dead animals, established the foundations of modern physiology and biological experimentation and are a milestone in the history of medicine, perhaps even in the history of man.

Harvey always insisted that any anatomical description that gives little or no consideration to physiology is almost valueless and that hypotheses must be tested by experiment. Although Harvey's work revolutionized the study of anatomy and physiology, he was himself a conservative man, much influenced by Galenism. He supposedly dissected nearly a hundred species of animals, but much of this work was lost during England's civil war. After the publication of his book, Harvey's reputation suffered and his medical practice dwindled. Although a great teacher, he gained no disciples during his lifetime.

The Microscope and the Stairway of Animals

Harvey's work opened a wide field for experiments in comparative anatomy. Yet advances were slow before the eighteenth century and the efforts of John and William Hunter. In the more than two hundred years between Harvey and the Hunter brothers, two significant events took place. First, the invention of the compound microscope, which opened a new world for anatomical exploration, was one of the earliest of the many valuable contributions made by the investigators of the physical sciences to biologists. Second, a critical synthesis of the morphological facts relating to many species of animals made a comprehensive classification possible.

The microscope led to the discovery of cells, the smallest structural units of living forms, and to the investigation of all cell types and their component parts. Thus histology, the study of tissues, was born. Except for some further spectacular discoveries, such as the existence of spermatozoa, progress in the early study of tissues was slow. Many difficulties had to be overcome, primarily the development of methods for preparing cells and tissues for microscopic examination. Since optically all fresh tissues are relatively similar, methods had to be devised for clearly differentiating each tissue element from the others. Tissues had to be preserved in as lifelike a way as possible, dehydrated, cut into very thin slices, stained, and made translucent. Real technical progress in this area did not come until the nineteenth century.

In 1838, T. Schwann and M. H. Schleiden proposed the cell theory. Although this provided a welcome stimulus to histology, many maintained, with some justification, that the microscopist examined only the chemical artifacts displayed by dead tissue. Anatomists were at first strangely reluctant to explore fully the potentialities of the microscope. Most of the nineteenth- and some twentieth-century anatomists ignored the instrument altogether.

The Swedish naturalist Carolus Linnaeus, 1707–1778, was the most famous classifier of all. He placed man, *Homo sapiens,* within the order of mammals that he called Primates. Linnaeus intended that the term *Primates* should define the first, or highest, order of the class Mammalia. He included in this order the apes, monkeys, marmosets, and even lemurs, which he likened to monkeys. (Lemurs are now classified in the separate suborder Prosimii because the evidence obtained from fossil remains indicates that they are more primitive.) Among the criteria for defining primates, Linnaeus included the hands. Nearly a century earlier the Dutch anatomist Nicholas Tulp (a figure in Rembrandt's painting *The Anatomy Lesson*) had described a chimpanzee, and in 1699 Edward Tyson had dissected another. Tyson maintained that the anatomy of his young chimpanzee resembled that of man "more closely than any of the ape kind, or any other animal in the world, that I know of." Linnaeus never saw an ape, alive or dead, but aware of Tyson's description, he concluded that apes and man should be placed together at the head of the order Primates.

Other zoologists, however, asserted that only man really possesses true hands and argued that the human hand was such a characteristic feature of his anatomy that man should be placed alone in a separate order, Bimana, "two-handed." All other primates—apes, monkeys, and their lower relatives—were grouped in an order Quadrumana, "four-handed." The apparently striking differences between the human foot and hand and the hands and feet of other primates are not now considered to be as significant as they once were. Man was also placed in a separate order Erecta under the impression that erect posture and bipedal gait are uniquely human characteristics.

The practice of arranging types of animals in hierarchies of kinship often leads to difficulty not only in nomenclature but also in deciding the exact limits of particular genera and species. Yet, in these attempts to marshal all known creatures into a "stairway of animals," or *échelle des êtres*, a special preoccupation of French anatomists about the time of the Revolution, can be seen the preliminary groundwork of the great biological theory of the nineteenth century, evolution. These tentative attempts at serial classification demonstrated that a common plan, best exemplified in a basal type, is the source of infinitely subtle modifications and variations.

All this theorizing gave rise to what was known as Transcendental Anatomy, based on the major premise of the existence of only a single animal. This theory provided the impetus for the great search for the archetype. Arguments concerning the nature of the archetype, attempts to explain the development of the entire skull from multiple vertebrae, and so on, became nothing short of a nightmare. In 1859 Charles Darwin put an end to the confusion with his *Origin of Species*. The stairway, or scale, of animals, ceased to be a series of static forms and became instead a moving row of ancestors with descendants constantly undergoing modification. Such modifications exemplified in successive generations of adult animals formed a series of stages known as a phylogenetic series.

Inevitably, embryology became involved in these new theories. Even before the early evolutionary concepts crystallized into knowledge, the nineteenth-century German embryologist Karl von Baer (1792–1876) had noticed a remarkable similarity among the early embryos of different animals. Once, when he had neglected to label two early embryos, he was unable to decide to which animals they belonged and could only say that they were the early embryos of some vertebrate or vertebrates! Similar observations, made previously by a number of anatomists, had led to various pronouncements. When in 1828 von Baer convincingly demonstrated for the first time the existence of the mammalian egg, it became apparent that mammals gradually passed through a successive series of embryonic stages beginning with the unicellular egg and continuing until adult form and shape were reached. This became known as the ontogenetic series, and to link the phylogeny of an animal with its ontogeny then seemed logical. Thus, man became man only after he had passed through transient embryonic stages resembling those of fish, reptiles, birds, and mammals. Against this background, the "Laws" of von Baer came into existence and with them the German zoologist E. Haeckel's famous theory, "Ontogeny recapitulates phylogeny." Unfortunately, this catchy phrase once learned is difficult to forget. It has enjoyed a prestige that, as Sir Gavin de Beer has aptly said, "has had a great, and while it lasted, regrettable influence on the progress of embryology." John Hunter also held views somewhat similar to those of the recapitulationists. However, he did not indulge in much theorizing and dismissed the difficulty of explaining whatever theorizing he did by saying that the whole thing was quite obvious!

Haeckel's "theory" conveys an erroneous impression of developmental history. In its defense, however, one must admit that during development embryos do show fragments of their ancestral developmental stages. But some phases are eliminated or bypassed, some shortened or prolonged, and others sometimes appear out of chronology. Some structures appear to have sprung up spontaneously, others to have arisen as adaptations of the embryo to its particular developmental conditions. The embryo "recapitulates" only *some developmental* events of only *some* of its ancestral forms. Ontogeny, then, may repeat some of the ontogenies of ancestral forms, particularly when these steps have structural and functional importance to the individual.

We must now return to the somewhat earlier period of John Hunter (Figure 4–2) because, in a way, he was the originator of

Figure 4–2. John Hunter (1728–1793). From an engraving in the Anatomy School, Cambridge University, of the portrait by Sir Joshua Reynolds.

modern anatomical methods. Born in Lanarkshire in 1728, he went to London at the age of twenty and studied anatomy at St. Bartholomew's Hospital. After some early adventures as a naval surgeon and with the army in Portugal, he returned to London where he amassed his famous collection of dissections, specimens, and pathological curios. It is said that he had a pond in his garden decorated with skulls.

Hunter was a practical anatomist and experimentalist whose ambition was to epitomize in his museum the structure and function of the organs of man and other animals. His reputation no doubt rests on the grand conception of the Hunterian museum, but he also experimented. The dictum "Don't think, try it" has often been attributed to him. He preached and practiced anatomical methods that, had they listened, would have saved his contemporaries and successors from engaging in much useless speculation on the theoretical and philosophical aspects of anatomy. The Hunterian method would have led them more directly to study causes and modifications in ontogeny and the morphogenetic forces that underlie evolution. Hunter's specimens were enclosed in glass jars (many of which were destroyed during the World War II blitz on London in 1940), but his own vital and functional approach to anatomy has been a continuous source of encouragement to research.

It is fashionable nowadays to denigrate the value of anatomical museums, but we should not forget our debt to the development of techniques designed, as Robert Boyle wrote in 1659, to make "more durable subjects for the Anatomist to deal with." Boyle should probably be credited with introducing the preservation of tissues and parts in spirits of wine (alcohol), which greatly improved the methods of injection. This technique exerted a remarkable and lasting influence on anatomical studies, providing novel, striking, sometimes beautiful, and easily verifiable results. Injection experts used water, milk, ink, oil, salts of heavy metals, and mercury, and various substances that could be "thrown in" liquid and set later, such as wax. Most famous were the preparations of Frederik Ruysch (1638–1731), whose large museums were filled with technical brilliance; his first was coveted by Peter the Great who finally bought it in 1717. Although Ruysch's museum was described as a "perfect necropolis, all of the inhabitants of which were asleep and ready to speak as soon as they were awakened" so lifelike were the injected fetuses, infants, and adults, he must also be credited with establishing the virtual ubiquity of blood vessels and of demonstrating that the larger vessels had their vasa vasorum. His work, too, was followed by the development of corrosion casts, where the tissues are removed from the solidified injection mass, and by the construction of various types of wax models. Injection techniques became of great practical value when harmless radiopaque materials were used to delineate structures in living patients who were being examined radiologically.

Evolution

In 1858 Charles Darwin exploded the medieval view of the place of man in nature. The Fifteenth International Congress of Zoology held in London in July 1958, commemorated the fact that one hundred years earlier Alfred Russel Wallace and Darwin had reported jointly to the Linnaean Society "On the Tendency of Species to Form Varieties" and "On the Perpetuation of Varieties and Species by Natural Means of Selection." The publication of *The Origin of Species by Means of Natural Selection* in 1859 was followed by storms of disapproval and emotional objection. The book was really an abridged version of a vast continuing project that Darwin wished to think about longer and explore further, but which he published under pressure from friends.

Evolution, in Simpson's succinct definition, means that "living organisms are all related to each other and have arisen from a unified and simple ancestry by a long sequence of divergence, differentiation and complication of descendent lines from that ancestry." This is the antithesis of the doctrine of special creation of each species of animal or plant, though it is not per se antithetical to the view preferred by some that the evolutionary process is a divine method of creation. According to the Darwinian theory, natural selection is an important factor in bringing about evolution; however, in the view of neo-Darwinists, such as August Weismann, natural selection became probably the only important factor.

Earlier, Lamarck had argued that acquired characteristics are inherited; neo-Lamarckists had considered that by eliminating unsuitable forms natural selection plays only a secondary role in evolution. The ancient but still popular theory of Vitalism, to which even Aristotle apparently adhered, with its idea of a vital force was still much in vogue. But, as Simpson writes, "nothing is achieved by saying that evolution is the result of an innate force or tendency." These and other theories all have their place in the history of evolutionary concepts. Since our modern theories arise from a synthesis of the neo-Darwinian school and the theories of de Vries, who held that new types of organisms arise suddenly, at random, and as the result of mutations, modern evolutionary theory is occasionally described as Synthesism. A mutation is the beginning of an inheritable variation that arises from a change in one of the genes. Mutations alone cannot govern the course of evolution; they are only rarely advantageous because, being random, they seldom fit into, or improve on, the complex workings of an organism.

At the beginning of the first chapter, the indignation with which the Darwin-Wallace announcement was received and the role that T. H.

Huxley played in the controversy were noted. Since Huxley's death in 1895, the theory of organic evolution has been accepted as a part of biological orthodoxy. Gradually, but with increasing momentum, anatomists have turned away from quibblings and discussions about names. Supplementing the scalpel, that sharp-edged fact-finder with two thousand years of service, anatomists are using increasingly more complicated instruments to determine the characteristics of man's structure. Notwithstanding the growing complexity of modern biological sciences, the days of dissection of both the dead and the anesthetized human body with a scalpel are not quite gone, and much can still be learned about man's framework. Had there been no bodies for dissection very little of this book could have been written.

Modern Anatomy

Medical students still need a substrate of first-hand information about the anatomy of man upon which to base their clinical studies. Moreover, anatomy is an important part of training to be a nurse, a physiotherapist, an anthropologist, and a human biologist of any sort. Functional investigation, whether in animals or man, must be founded on the knowledge of real structure. The surgeon's need for anatomical knowledge, as well as for such specialized knowledge as of tissue and organ transplantation, is obvious. In the development of mechanical substitutes for organs, the anatomy of specific organs down to the finest detail must be known.

Today, body-snatchers, sack-'em-up men, and resurrectionists are no longer necessary to obtain cadavers for dissection. But difficulties remain. The activities of Burke and Hare in Edinburgh during the early nineteenth century resulted in a bill, "An Act for Regulating Schools of Anatomy," which received the Royal Assent on August 1, 1832. Medical science, in Britain at least, was thereby empowered to receive a legal supply of dead bodies for dissection, the advancement of knowledge, and the study of surgery.

In Britain, no legal property remains in a dead body, which thus belongs to the state. The proper authorities can direct that the body of an individual who dies without kin be sent to a medical school. An individual can also bequeath his body for purposes of medical research after death. In both instances, after appropriate examination during a prescribed period of time, the body is buried or cremated with due rites, under the direction of an inspector.

In the United States and in many European countries, the problem of obtaining enough bodies for dissection in medical schools has always been knotty. The main source of cadavers has been the unclaimed bodies from city morgues and from institutions for old indigents or the feeble-minded. Laws in American states differ; in some states it is almost impossible to get enough bodies, whereas others require that unclaimed bodies be made available to hospitals and medical schools. However, with the increasing growth of welfare agencies, Social Security, and veteran fraternal organizations, the number of unclaimed bodies is dwindling, since these agencies provide funds for the burial of persons whose bodies would otherwise be unclaimed. At one time one could sell his body in advance for a price to certain schools, but this practice was fraught with unhappiness and chicanery. The practice of providing by will that one's body be given to a medical institution upon death has supplied a few cadavers, but it is unreliable. Moreover, in some states the closest of kin can challenge such private donations and render the will invalid.

Such obstacles indicate the necessity for each state in the future to face this problem and resolve it by proper legislation. In Michigan, legislation was enacted in 1958 whereby an individual can legally leave his body to an institution if some formalities are observed: the permission must be in writing, signed by two competent witnesses, and deposited in the probate court of the county where the individual resides. Since Americans are conformists, this worthy practice will probably grow in popularity and the medical schools will be assured an adequate number of cadavers. Morticians, however, will no doubt view this with alarm and may be prepared to fight such legislation.

The lack of bodies, however, is not the real threat to proper training in the medical schools. Rather, an increasingly disparaging attitude toward anatomical studies by those in other biomedical disciplines thwarts this research. Until recent times, anatomy was the springboard of medical training, having been taught conscientiously but painfully for two years. Gradually, as the medical educational curriculum increased in complexity, the time required to study anatomy has steadily dwindled. Until a few years ago, it was usual (and lamentable) for the medical schools to boast about the brevity of their anatomy courses.

The steadily advancing prominence of the quantitative biomedical disciplines, the deplorably inept ways in which anatomy has nearly always been taught, and the fashionableness of molecular biology have all militated against the study of the body by dissection. To avoid bother, some of the newer medical schools have contemplated dissecting stillborn infants, which are smaller and easier to store than adult bodies. Others are making greater use of models and manikins, and there is even some talk of using rhesus monkeys as substitutes. The last suggestion evokes reflections on how easily man reverts to ancient attitudes. Some schools have made at least some progress by the use of audio-visual aids, prosected specimens, and programmed teaching.

The cause of anatomy is not on trial and needs no apologies. Regardless of vicissitudes, blunders, short-sightedness, and avant-garde methods in education, to fully appreciate the human machinery at work the physician or biologist must before all else have a first-hand acquaintance with the working parts. There are no shortcuts to this knowledge. The problem before anatomists is to read function from structure since the two are inseparable. Anatomy is the tree upon which are hung the refinements of the details of molecular biology. As Sir Charles Sherrington put it, "A thing must have form to be the thing it is; but its temperament is the key to understanding its ways and works." Sherrington also pointed out that to study anatomy is to learn to read purpose in creation. Finally, no one has improved on the sixteenth-century pronouncement of Jean Fernel (our chapter epigraph): "What geography is to history such is anatomy to medicine—it describes the theater."

Anatomists have only just come to realize that one need not endow the parts of man's body with the hallmarks of uniqueness as if they were pieces of porcelain or silver. The problems to be solved are how man's anatomy came to differ from that of other animals and how the changes were controlled. No method can be ignored that could possibly help to interpret structural significance. To examine the world of matter and inanimate things, the anatomist today uses many of the techniques and instruments hitherto used only by mathematicians, chemists, and physicists. The ancient barriers between the various disciplines for studying man are crumbling. Anatomists will continue to seek for clues or links in our hopelessly fragmentary knowledge of the history of man. No doubt many more fossils will be discovered for classification and discussion, and new methods will be gradually devised to decipher what these fossils can tell us.

Notwithstanding its vital role in the history of medicine, comparative anatomy ceased to be a respected discipline as soon as the quantitative biological disciplines became fashionable. Anatomy is now being "rediscovered" and can be expected to gain in value and prestige. We have today reached a point of such specialization in biological research that we must stop from time to time to gain perspective; functional anatomy can and does provide this perspective.

Long neglected by biologists, nonhuman primates have now become fashionable animals to study; their use as experimental subjects has steadily gained acceptance and even preference. The anatomy of most of the existing forms is only sketchily known, and much work awaits the young, and old, scientist. The more we learn about the dynamics of primate structure, the better prepared we will be to understand the structure and function of man.

5

MAN'S LARGE BRAIN

The newborn baby's brain is relatively so monstrous . . . that the rest of the body seems almost an afterthought.
(Weston La Barre)

Man's cleverness, and indeed all those characteristics that have made him the most successful of mammals, are commonly attributed to his large brain. Since articulate speech, literature, art, and science as well as his complex technology all appear to depend on this brain, it is fitting that a book on the biology of man should examine how the brain appears to an anatomist.

Its size is perhaps its simplest feature. The average human brain weighs about 1,400 gm, approximately three times more than the brain of his nearest living relatives, the great apes. Only large mammals like the elephant, with a brain of about 5,000 gm, and the largest whales, with brains of 9,000 gm, surpass him. Although there seems to be a relation between body size and brain size, the ratio varies widely. Very small mammals have a high brain/body-weight ratio; in a mouse weighing 20 gm, it is about 1:30. In very large mammals, the ratio is low; for example in whales, the largest of which weighs over 100 tons, it is 1:40,000. In large sheep, goats, and dogs, whose body weights are near that of man, the ratio is about 1:300, and in gorillas somewhere near 1:200. In man, however, the ratio is about 1:40, only slightly lower than that in very small mammals (Figure 5–1). In absolute weight, then, man's brain is among the heaviest of all existing animal brains. One must conclude that regardless of the level of intellectual attainment, body size has some effect on brain size, which varies much less than body size. There is a minimum size that cannot be reduced if the brain is to perform its basic functions, and probably a maximum beyond which further enlargement would bring little or no advantage.

What is striking about man is his possession of a brain that is much larger than his body would require if his behavior were no more complicated than that of other mammals. Therefore, much of the increase in brain size that occurred during his evolution from apelike ancestors must be related to his higher intellectual level. How much is not easy to say. The fact that the weight of the brain varies widely even within the human species only serves to complicate the matter. The weight of 1,400 gm is an average; in people who live normal lives, it ranges between 1,000 and 2,000 gm. Above or below these limits, human brains are nearly always pathological. The average weight of the female brain is some 50 to 100 gm lighter than that of the male, and it is also slightly lower in the aborigines of Australia than in most living populations.

Too much significance should not be given to the crude measurements of a very complex organ. Provided its weight lies within certain rather wide limits, mere brain size does not seem to matter. Although both elephants and whales, with enormous brains even compared with man's, are among the more intelligent mammals and the elephant is credited with an efficient memory, they are not thereby necessarily more intelligent or more "human" than other mammals whose brains are much smaller than those of man.

Perhaps there are important qualitative differences between the neural tissues in different brains or differences in the way the tissue is organized, neither of which is detectable by measurements of volume or weight. Therefore, before going further, we should examine not only the general nature but also the functions of neural tissues.

General Nature of the Nervous System

In a broad sense, the tissues of the nervous system are like an electrochemical system that can receive, codify, coordinate, and store information about events outside and inside the body and can instruct the latter to respond appropriately. Unless an organism can adapt to the changes within itself and its environment without disturbing its internal stability,

Figure 5–1. The relative brain/body mass (or weight) ratio depicted from left to right in man, chimpanzee, chacma baboon, and sheep.

it cannot survive or maintain its identity. This self-maintenance, in which the nervous system plays an essential and predominant role, is known as homeostasis, meaning literally "remaining the same." Certain chemical mechanisms are important in maintaining homeostasis. In the most primitive organisms, they are the only such mechanisms; in more advanced organisms, they are represented by the endocrine, or ductless, glands. The endocrine system is described in chapter 9; it is mentioned here because it cannot be separated from the neural homeostatic mechanisms. In most activities, the two systems work together as a neuroendocrine complex.

Although the activity of the nervous system is electrochemical, it consists, like other systems, of specialized cells. Nerve cells, or neurones, are like other cells in that each has a nucleus and cytoplasm enclosed by a membrane, but they are specialized for the reception of stimuli and the conduction of impulses. The cytoplasm of nerve cells grows out into fine processes, some, known as dendrites, that usually branch freely and do not extend far from the mass of cytoplasm that surrounds the nucleus (the perikaryon, or cell body) and one, known as the axon (or nerve fiber), that branches less freely and usually extends some distance from the cell body. The most important of the many differences between dendrites and axons is a functional one: under normal circumstances the surface of the dendrites and cell body is "receptive," that is, it can receive stimuli from events in the immediate environment, whereas the electrochemical change induced in the cell is conducted away from the cell body by the axon.

Thus the basic characteristics of a nerve cell include the ability to receive stimuli in the dendrites or cell body and to conduct a wave of electrochemical change along the axon to its termination. This wave of change, or nerve impulse, is the code unit into which information derived from the stimulus is translated and is comparable to the dot or the dash of Morse code. The important difference is that the nerve impulse is of only one kind and not two. An impulse passes or does not pass, the only possible variation being frequency; a strong stimulus leads to a rapid series of impulses, a weak one to a slower and usually shorter series.

The fate of the impulses conducted by the axon is obviously important. The axon, or its branches, can end in small specialized areas of contact either with a muscle cell, capable of contracting and producing movement, or with a glandular cell, capable of secreting. The combined activity of these *effector* cells results in the behavior of the organism. At the specialized areas of contact with the axon, known as the synapses, a nerve impulse is transmitted to the effector cell, stimulating it to activity. Axons can also form synaptic contacts with the receptive surface of the dendrites or cell body of another nerve cell. Thus, a nerve impulse from the first can stimulate the second to discharge an impulse along its axon or, alternatively, can inhibit the activity of the second cell and make it unresponsive to impulses received at synaptic contacts with other cells. Since the receptive surface of the dendrites and cell body is relatively large, it provides the opportunity for many (in some cases thousands of) synaptic contacts with other nerve cells. What happens in a cell depends on the combined effect of the impulses arriving at all these synaptic contacts and, in particular, on the balance between the activity of excitatory and inhibitory impulses. Since axons often branch near their termination, they may form synaptic contacts with many other nerve cells and thus spread their effects. The other cells can, in turn, stimulate many effector cells to produce a massive reaction in a large part of or the whole organism, so that the effect is out of proportion to the original stimulus. A pin prick, for example, which at the most stimulates very few nerve cells, can lead to the massive withdrawal of the limb or whole body. The study of the structure, location, and physiology of synapses—synaptology, as it has come to be called—is an important part of the study of the nervous system and is essential for a complete understanding of its functions.

The simplest multicellular organism has a diffuse network of interconnected nerve cells. But even so primitive an organism has *afferent* (sensory) nerve cells on or near the surface, which receive stimuli from the environment, and *efferent* (motor) cells, which connect with effector organs, such as muscles. Within the network of nerve cells and fibers between them, sensory stimuli are coordinated and distributed to the efferent cells so as to produce an appropriate coordinated reaction. The multiplication of environmental messages received by more advanced animals from their remarkable complex of sense organs (organs sensitive to light, chemical senses like smell or olfaction, taste, pH, senses of vibration and of hearing, touch, heat, cold, and pressure) calls for a matching intricacy in the mechanism that integrates all the coded signals they generate. In such organisms, a diffuse nerve net is no longer adequate; nerve cells must be grouped into a central nervous system (CNS), connected to the sense (receptor) organs by afferent, or incoming (sensory) nerves and to the muscles and other effector organs by efferent, or outgoing (motor) nerves. In vertebrates the CNS takes the form of a brain within the skull at the cranial end of an elongated spinal cord inside the vertebral column.

Development of the Nervous System

One of the earliest recognizable primordial tissues in human and all vertebrate embryos is a thickened area on the dorsal part of the embryonic mass called the neural plate. The margins of the plate rise up to meet and fuse in the midline, forming first a groove and then a hollow tube, which sinks below the surface. At first the tube is open at each end (the neuropores) but later closes (Figure 5–2).

From the outset, the head end of the neural tube is much enlarged. This cranial enlargement, or brain, is intimately related to the structural development of the head. Since in all elongated, ambulatory organisms the head is the first part to be exposed to new surroundings, it is the best possible place for assessing changes and receiving information. Moreover, it is in the most advantageous position for regulating the adjustments that must be made by the rest of the body. The evolutionary advent of jaws and of the special senses of taste, smell, sight, and hearing was simultaneously associated with an increased number of nerve cells in the head region, which required a brain to coordinate them. This in turn necessitated connections between the nerve cells of the brain and those of the trunk and limbs, which would be situated in the caudal part of the neural tube, i.e., in the spinal cord. These connections are the axons of brain cells, which descend in bundles, or tracts, to form synapses with the cells in the spinal cord; similar ascending tracts from the spinal cord provide a feedback of information to the brain.

Very early in embryonic life, the expansion of the brain at the head end of the neural tube assumes the form of three swellings, or

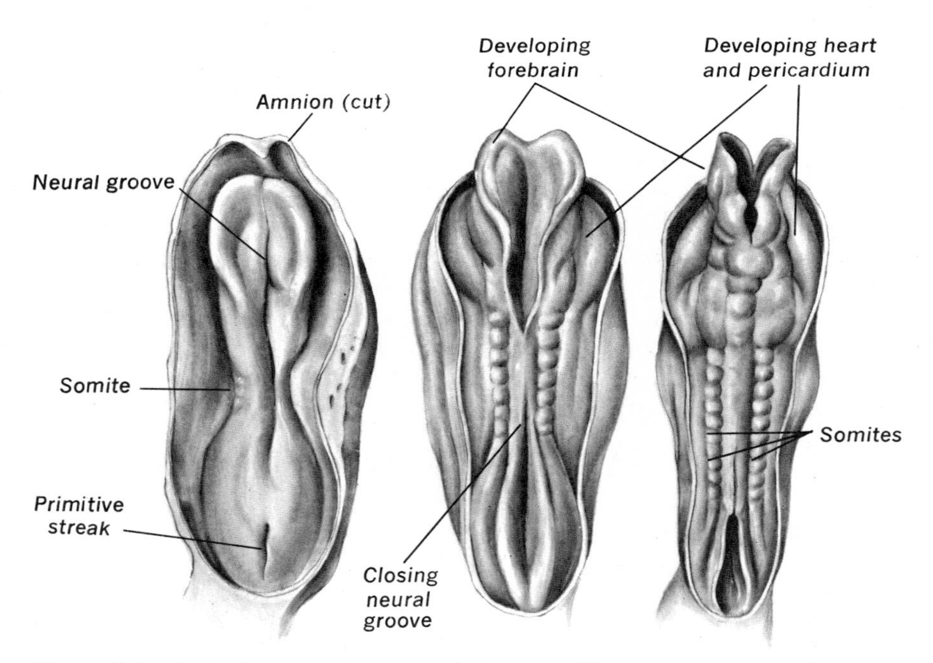

Figure 5–2. Early human embryos 21 to 22 days old (3 to 10 somites, about 1.4 to 1.7 mm), showing the development of the nervous system.

vesicles, one behind the other: the fore-, mid-, and hindbrain, or the rudiments of the major subdivisions of the brain (Figure 5–3). The forebrain is related to the olfactory apparatus, which develops from bilateral ectodermal thickenings (called placodes). These induce the outgrowth of two secondary vesicles, which are forerunners of the adult cerebral hemispheres. The midbrain, in turn, acquires a relation with vision. In all primitive vertebrates, the optic nerves discharge mainly to the midbrain where they induce the development of complex layers of nerve cells in its roof, forming the tectum (roof), or optic lobes. In mammals, and particularly in primates, including man, most of the fibers of the optic nerves are diverted to the forebrain. Only relatively simple visual reflexes, such as those that control the size of the pupil and certain eye movements, remain located in the midbrain; hence the optic lobes have diminished in importance and relative size. This transference of all the more complex aspects of vision from mid- to forebrain is an example of the process of prosencephalization by which many functions previously performed at lower levels of the nervous system were more or less completely transferred to the forebrain. This process occurred in all mammals, but went farthest in man, where it is associated with the precocious

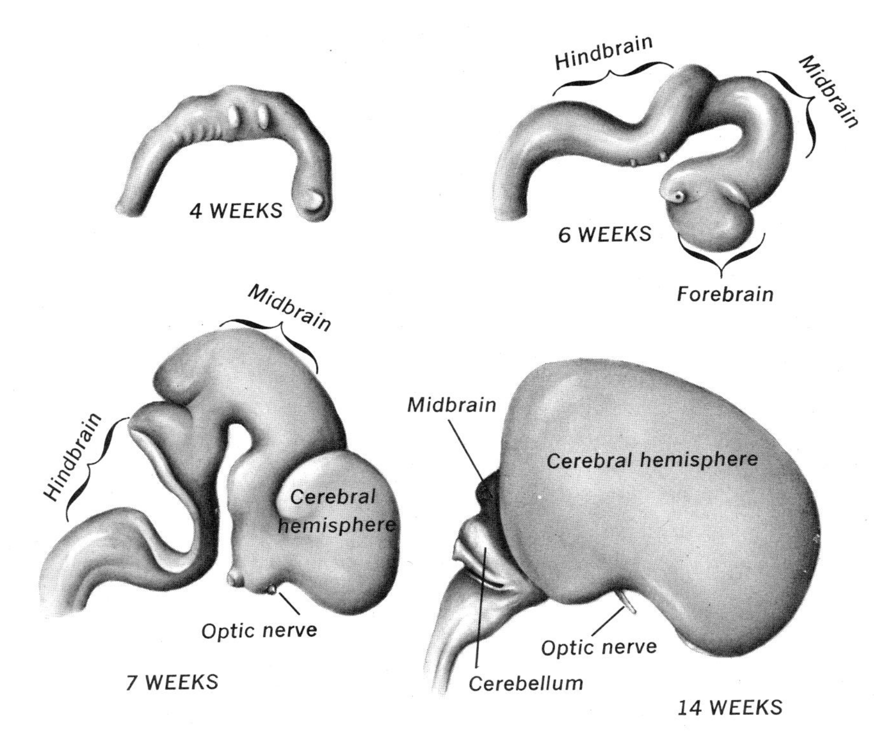

Figure 5–3. Stages in the development of the human brain.

growth and relatively enormous size of the cerebral hemispheres. Their expansion is so remarkable that even in three-month-old embryos they have already overgrown the mid- and hindbrain (Figure 5–3). Finally, the hindbrain, which is continuous with the spinal cord, acquires a relation with the special sense that provides information about the position and movements of the head and is indispensable for the maintenance of posture and equilibrium. Its peripheral receptor, the vestibular apparatus, also develops from an ectodermal placode, similar to the olfactory placodes. On reaching the hindbrain, the nerves from this placode induce the growth of a great mass of neural tissue on its dorsal side that gives rise to the future cerebellum.

The unequal growth of the three brain vesicles, even at very early stages, causes the cranial part of the neural tube to bend into a characteristic S-shaped flexure (Figure 5–3).

The main parts of the CNS in man can be recognized by fourteen weeks of gestation (Figure 5–3). From the forebrain have developed the two cerebral hemispheres connected to the olfactory receptors by olfactory nerves; a part of the forebrain, the diencephalon, has remained in the middle. The diencephalon, which means "between brain," is located between the hemispheres and the brain stem, which is made up of the midbrain and the hindbrain. The brain stem is a cranial continuation of the spinal cord, but the special development of the tectum, or optic lobes, in the midbrain and of the cerebellum in the hindbrain has no counterpart in the spinal cord itself. Anatomically, weighing the whole brain is equal to weighing a collection of highly diverse structures with coordinated but equally diverse functions; the numerical result, therefore, cannot be assigned a simple significance.

For the brain and spinal cord to function, they must be connected by afferent nerves to the sensory receptors and by efferent nerves to the muscles and other effector organs. Most of the developing nerve cells remain inside the CNS and connect with each other in many complex patterns, but the axonal processes of some push their way out of it for various distances until they make contact with muscle fibers or glands. Bundles of such axons form peripheral motor nerves, and the groups of cells within the CNS from which they grow are the motor nuclei.

The afferent nerves, which connect the sense organs with the CNS, have a different origin. When the neural tube is first formed, certain primordial cells at the margin of the neural plate are not incorporated into the tube, but remain outside it on each side as the neural crest (Figure 5–4). These cells give rise to various peripheral neural structures as well as to other tissues unconnected with the nervous system. The longitudinal columns of neural crest cells break up so that in each segment of the body two clumps of such cells, right and left, are found close to the developing nervous system. Crest cells, now recognizable as neuroblasts, or primordial

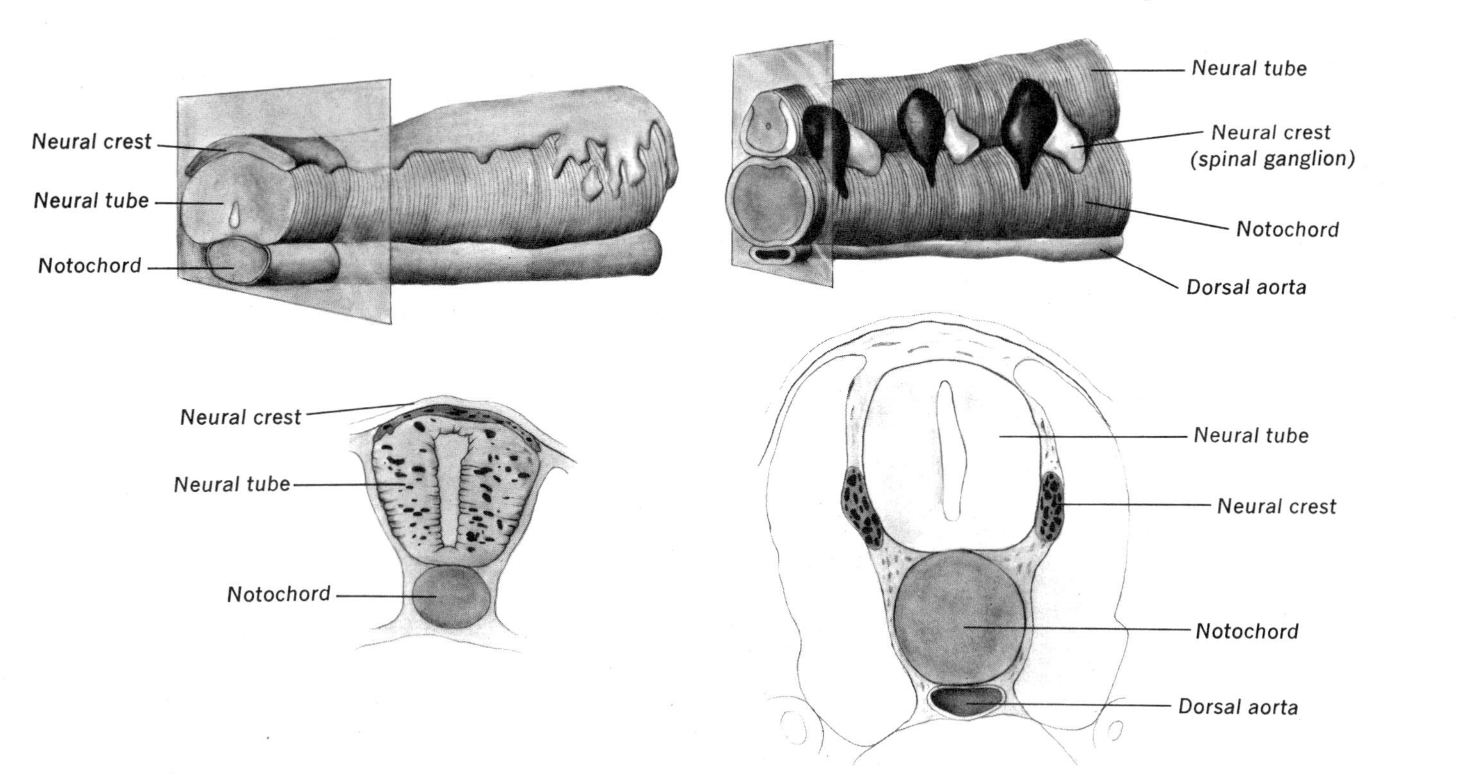

Figure 5–4. Diagrams showing two developmental stages of the neural crest. The two figures on the left, shown in three dimensions above and cut transversely below, represent the early neural tube covered by a continuous mantle of neural crest. The two figures on the right show more advanced stages, with the neural crest separated into segments.

nerve cells, each develop two processes: one growing toward and into the dorsal part of the CNS; the other, away from it, usually in association with the axons of motor nerve cells that grow out at the same time from the ventral part. The peripheral processes of the neuroblasts from the neural crest grow to the skin or to small sense organs in muscles and viscera. They carry impulses toward the central nervous system, past the cell body in the clump of neural crest tissue, now known as a dorsal root, or sensory ganglion, into the CNS (Figure 5–4). The peripheral processes of these cells conducting toward the ganglion are comparable to dendrites; the central processes conducting away from the ganglion into the CNS are comparable to axons. Other peripheral ganglia or groups of nerve cells are also formed from neural crest tissue, i.e., from primordial cells that have migrated to other parts of the body, especially those related to the viscera. These nerve cells form an important part of the visceral, or vegetative, nervous system, which regulates the activities of the viscera. Seen from a phylogenetic point of view, the visceral nervous system preserves some of the characteristics of the diffuse nerve network, which in primitive invertebrates constitutes the entire nervous system. Although important, the visceral nervous system of man has little relevance to the size of his brain and will not be described here.

These, then, are the three main parts of the nervous system: the CNS (brain and spinal cord) and the two divisions of the peripheral nervous system (PNS); one part, called somatic, supplies the skin, muscles, ligaments, and joints with both afferent and efferent fibers; the other, called visceral, supplies the heart, lungs, alimentary canal, and so forth with afferent and efferent nerve fibers. The question of what guides the growing fibers toward their appropriate connections either in the central or in the peripheral parts of the nervous system remains unanswered. Some have suggested that adjacent structures guide the fibers to their destination; others suggest other factors, such as the action of local electrical forces and gradients or specific chemical affinities between nerve fibers and the structures upon which they terminate. Genetic "instructions" are no doubt at work, but this is evading the issue; the fact remains that in spite of much experimental analysis, no clear or decisive answers have yet been obtained.

Other features of the nervous system must be briefly mentioned. The embryonic neural tube (Figures 5–2, 5–4), as its name implies, is hollow, and the canal persists throughout life as the central canal of the spinal cord, continuous with the expanded ventricles of the brain. From behind to forward, the cavities of the brain are the fourth ventricle in the hindbrain, which is connected by a narrow "aqueduct" running through the midbrain with an expansion in the forebrain, the third ventricle. The third ventricle in turn communicates with the cavities or the two lateral ventricles in the cerebral hemispheres on each side. These ventricles are lined throughout by a ciliated epithelium (ependyma), which in certain situations assumes secretory functions and is pushed into the ventricular cavities in the form of tufts that contain highly vascular connective tissue. These invaginations, the choroid plexuses, secrete cerebro-spinal fluid, a watery substance that contains little protein and resembles normal saline, and fill the whole ventricular system and the central canal of the spinal cord. The fluid can escape through apertures in the fourth ventricle into the subarachnoid space that surrounds the entire central nervous system. This space is limited externally by a thin arachnoid membrane with a tough, fibrous membrane, the dura mater, outside it. In this way the CNS is completely isolated from contact with the general tissues of the body, protected by two membranes, and cushioned by cerebro-spinal fluid in the subarachnoid space. The CNS can therefore function undisturbed by outside events except those that give rise to nerve impulses brought by the peripheral nerves through specific pathways.

Not all the elements of the nervous system are nerve cells. In the CNS are several varieties of neuroglial cells, with branching cytoplasmic processes like nerve cells; the processes, however, do not differentiate into dendrites and axons or conduct nervous impulses. Neuroglial cells form a supporting network for the nerve cells and when necessary help to insulate them. Some neuroglia (or simply glia), for example, the astrocytes, or star cells, which are closely related to blood vessels, transport material from the blood to the neurones. Other glial cells are intricately wrapped around nerve fibers to form a lipoidal sheath called myelin. The myelin sheath is discontinuous, being interrupted at short intervals by what are called nodes (first described by Louis Ranvier in 1878). The sheath not only acts as an insulating covering but plays an important role in the conduction of nervous impulses. In myelinated fibers the impulse appears to "leap" from node to node (saltatory conduction), and the overall velocity of conduction is greater than in unmyelinated fibers. The thickest myelinated axons conduct most rapidly, some with velocities up to 120 meters per second. Around peripheral nerves, the elements comparable to neuroglial cells are called Schwann cells. Unmyelinated fibers have a continuous covering of Schwann cells, which in myelinated fibers form the lipoidal sheath, similar to but not identical with the myelin in the central nervous system.

The Parts of the Brain

Figure 5–3 shows the development of the fore-, mid-, and hindbrain and the continuity of the latter with the spinal cord. The mid- and hindbrain eventually give rise to the brain stem, on which is set the forebrain. The brain stem preserves some of the characteristics of the spinal cord; for example, a series of cranial nerves is attached to it. Although these nerves innervate structures situated mainly in the head (e.g., the jaws), they have many of the characteristics of spinal nerves and are a continuation of the spinal nerve series into the head. Each cranial nerve is associated with a nucleus that contains the nerve cells that give rise to motor efferent nerves or that receive the terminations of afferent fibers. These correspond to part of the gray matter of the spinal cord where motor and sensory nuclei are arranged in continuous columns, known as ventral and dorsal horns because of their appearance in a transverse section of the cord. In the brain stem these columns are mostly broken up into individual islands, or nuclei, each associated with a particular nerve; for example, the fibers of the twelfth, or hypoglossal, nerve (supplying the muscles of the tongue), which emerge from the most caudal part of the hindbrain (medulla oblongata), arise from the hypoglossal nucleus. In the same way, fibers from the cochlea, the organ of hearing, end in the cochlear nuclei in the medulla oblongata. In addition to the nuclei of cranial nerves, there is in the brain stem a somewhat diffuse network of nerve cells and their processes, the so-called reticular formation. This primitive arrangement, present in all vertebrates, including man, has connections with all parts of the brain and spinal cord and is concerned with inhibiting or facilitating the activity of any part of or the entire nervous system. For example, it mediates the so-called arousal reaction, which recalls the organism from a somnolent condition to active awareness.

Anatomically and functionally, the striking and characteristic features of the brain are the cerebellum, the optic tectum, or roof, of the midbrain, and the forebrain. These centers integrate information received from the special senses (smell, vision, vestibular information or sense of position and movement) with more general information (e.g., tactile) received from all parts of the body; on the motor side, they are responsible for the detailed regulation of the behavior of the whole animal (Figure 5–5). Because these parts are not so directly concerned with the activity of the individual parts or segments of the body but with what can be called a pattern of behavior advantageous for the life of the animal, they are sometimes called suprasegmental structures. They are interconnected and receive afferent connections from the spinal cord; they are also provided with efferent fibers which, through the reticular formation or by more direct connections with the motor nuclei, regulate general behavior.

In primitive vertebrates the suprasegmental structures are dominated by the special sense with which they are concerned (hence the appropriateness of the term *rhinencephalon*, or "smell brain," for the cerebral hemispheres). Only in the warm-blooded groups, birds and mammals, have they become integrative centers for all kinds of sensibility and have themselves acquired dominance over almost all aspects of behavior. This rise to dominance of the suprasegmental structures in the brain is the result of a process called encephalization.

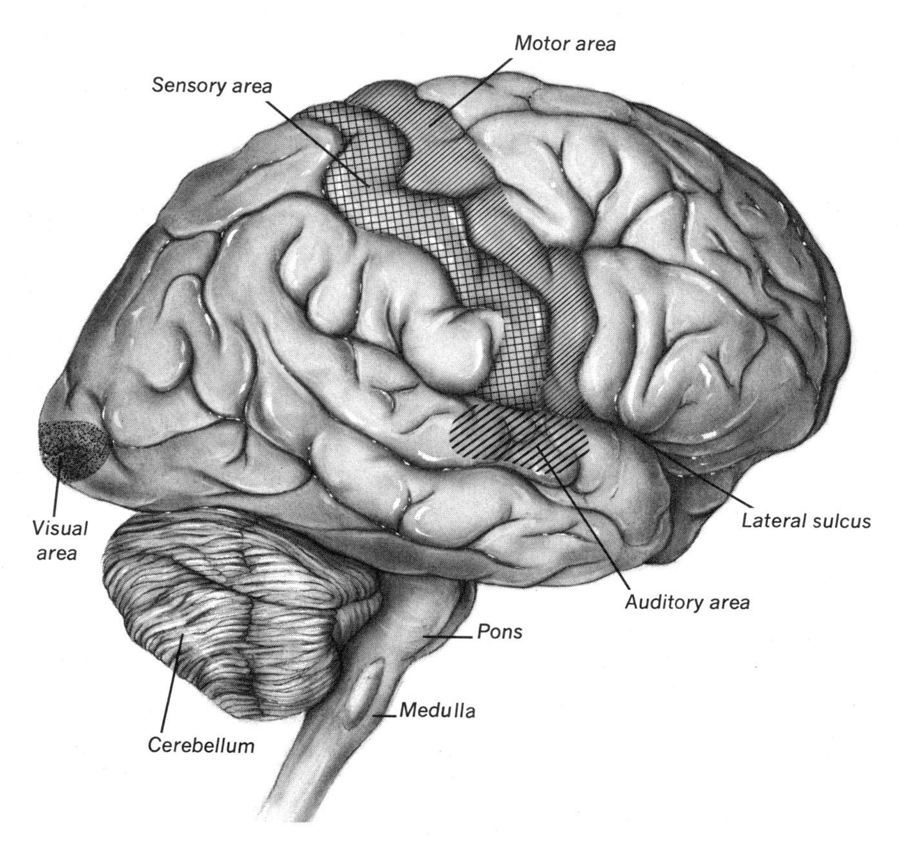

Figure 5–5. The surface of a human brain. Compare with Figure 5–7.

In birds, encephalization has led to a kind of brain that integrates the more stereotyped behavior patterns such as the sequences of activities in courtship rituals, nest-building, raising of young, and so on. Apparently the neural basis for this behavior is built into a bird's brain, which is genetically programmed in such a detailed way that learning from post-natal experience is comparatively unimportant. In mammals, on the other hand, encephalization has produced a neural mechanism with far greater plasticity. The play patterns of young mammals are essentially a process of postnatal learning, necessary for a nervous system that has not been programmed genetically to perform complex stereotyped behavior. There is little doubt that anatomically and physiologically the development in mammals of a particular region of the cerebral hemispheres is chiefly responsible for the plasticity and ability to learn from experience. That region is the cortex, which, because it is new and characteristic of the mammalian brain, is called the neocortex. This part of the cerebral hemisphere is only slightly developed in birds. The relatively enormous development of the cerebral cortex in more advanced mammals, particularly in man, is mainly responsible for the large size of their brains. Although encephalization has occurred to some degree in all vetebrates and has reached a high level in birds, the enlargement of the forebrain resulting from a disproportionate growth of the cerebral cortex is characteristic of mammals.

In describing the parts of the brain, we will begin with the part that is most important in this book—the forebrain, particularly the cerebral hemispheres, or cerebrum (Figures 5–5, 5–6). Cerebral hemispheres are basically bilateral hollow vesicles, which began as olfactory centers or a rhinencephalon. In mammals the fibers that carry information about hearing, smell, taste, touch, and position have access to the cerebrum and cause a disproportionate expansion of the dorsal region of the vesicle called the pallium (mantle). During its development, neuroblasts migrate

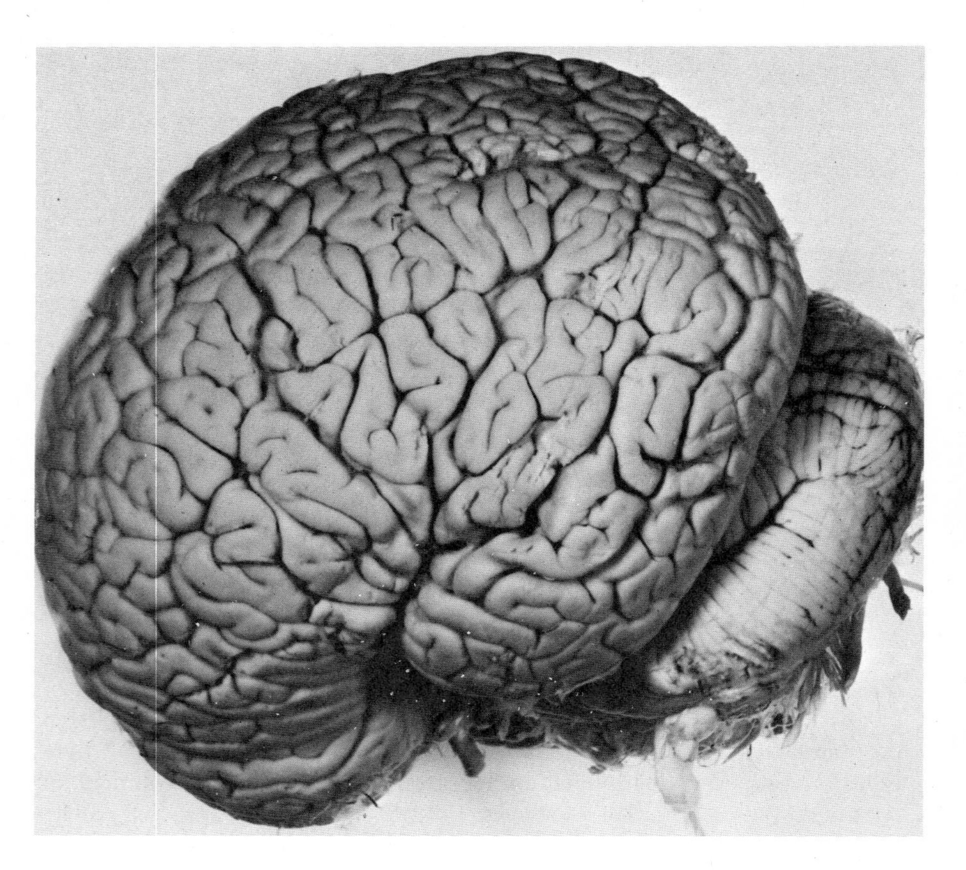

Figure 5–6. The brain of a dolphin, with its complex patterns of gyri and sulci.

from the lining of the cavity of the vesicle, the lateral ventricle, and form a layer of gray matter close to the surface, usually differentiated into six sublayers. The type of lamination and types of cells formed differ in the various regions of the cortex. For example, in man the area of the cortex that receives visual information is so different from the others that it can be distinguished by the naked eye. In a section cut perpendicular to the surface, a conspicuous white line can be seen in the gray matter, known as the stria of Gennari, after an eighteenth-century Italian anatomist from Parma, who first observed it. Structurally, the cortex (or better, the neocortex) is highly heterogeneous, a reflection of its functional complexity. The ventral, or basal, region of the cerebrum develops differently. The cortex formed here is simpler, with fewer laminae, and remains dominated by olfactory connections; it is, therefore, known as the olfactory lobe and its cortex as the paleocortex.

Large masses of nerve cells beneath the surface remain close to the ventricular lining to form the basal nuclei of the cerebrum, or basal ganglia, as they are more often called. In birds, the basal ganglia form the greater part of the cerebrum and are extremely complex structurally and functionally. Apparently the birds have exploited their possibilities to the full and neglected the dorsal pallial regions. By contrast, the mammals have exploited the possibilities of the dorsal pallial regions to the full, as the highest level of the nervous system, leaving the basal ganglia in a relatively undifferentiated and subservient condition. This fundamental difference between the avian and the mammalian cerebrum is related to the behavioral differences already mentioned.

In primitive mammals, such as the Insectivora, neocortical expansion has not proceeded very far, and the basal olfactory areas are relatively large. In such *macrosmatic* animals, the cerebrum is still largely a rhinencephalon. The sense of smell is important in most mammals, even though the neocortex has become much more extensive than the olfactory cortex. In two orders of mammals, however, the fully aquatic Cetacea and the Primates, the olfactory apparatus and its neural counterparts on the base of the brain have retrogressed so much that the animals are *microsmatic*. This condition is particularly marked in the great apes and man. Some Cetacea have regressed even further, and some dolphins have lost the sense of smell altogether and become *anosmatic*.

The cerebral hemispheres do not comprise the whole forebrain. There is a part in the middle, the diencephalon, which surrounds the third ventricle and is continuous caudally with the brain stem. No peripheral nerves are connected directly with the diencephalon, but a mass of nuclei called the thalamus develops in its lateral walls and another mass called the hypothalamus in its floor. An outgrowth from the floor of the diencephalon gives rise to the posterior lobe of the pituitary. The hypothalamus is connected with the pituitary by nerve fibers and blood vessels, and the two form a neuroendocrine apparatus that regulates many basic functions of the body (see chapter 9). The hypothalamus is the center for the integration and regulation of probably all visceral activities and for the expression of such emotions as anger and fear. The thalamic complex of nuclei above the hypothalamus is particularly important in the evolution of the mammalian brain. The thalamus receives ascending tracts from the spinal cord and brain stem and is a center for sensory integration; it also relays sensory information to the cerebral hemispheres, in mammals mainly to the cortex. The fibers radiate from the thalamus to the visual, auditory, and general sensory areas known collectively as sensory projection areas. The latter receive information from the receptors in the skin and in the joints and ligaments, which collect information about bodily posture and movement. Because it passes sensory information on to the cortex, the thalamus has been called, "the gateway to the cortex," and the mammalian forebrain has been aptly described as a thalamocortical mechanism.

The forebrain, then, is concerned not only with receiving, integrating, and discriminating sensory information but with storing this information in the form of memory. A complex of structures in the cerebrum, some cortical and some basal, known collectively as the limbic lobe, is thought to be connected with memory. The limbic lobe is strongly influenced by olfactory stimuli, and it is well known that certain odors evoke memories. Olfaction is the only sense with direct access to the cerebrum, without passing through the thalamus.

From the thalamus we return to the cerebral hemispheres and will consider their structure and functions, particularly those of the neocortex, for it is clear that the development of the neocortex is the most important factor in the evolution of the human brain. Thus far, we have dealt with the forebrain mainly in its relation to sensory experience. If it is to make use of either immediate or remembered experience, efferent pathways must be present through which it can influence motor neural mechanisms at lower levels and through them behavior in general. The most primitive of these pathways arise in the basal ganglia of the hemisphere, pass to the reticular formation of the brain stem and, after relay, along reticulospinal tracts to motor cells in the spinal cord. In submammalian vertebrates this is a most important efferent pathway from the forebrain. It is still present in mammals, even in man, where lesions of the basal ganglia can lead to serious disorders of motor activity. Such lesions, however, do not interfere with sensation, memory, or intelligence, and do not cause complete paralysis; movements can be performed, but clumsily and inefficiently. Thus, the basal ganglia, although important in mammals, are accessory to other "higher" centers where the movements necessary for a planned course of behavior are decided and coordinated. These higher centers are in the neocortex, where, in addition to the sensory projection areas that have been mentioned, there is a motor projection area.

Early in the nineteenth century it became fashionable to hold that particular functions or faculties are localized in specific parts of the cerebral cortex. The idea was associated with phrenology, a pseudoscience developed primarily by a Viennese physician, Franz Joseph Gall. Gall believed that the different areas of the cerebrum have corresponding bumps on the surface of the head, and that the prominence of these bumps indicates the degree of development of particular faculties. His ideas were popular during the nineteenth century when phrenological charts showing the location of the bumps and their presumed meanings were common. Even though phrenology was eventually discredited, Gall left a legacy: the idea of localization of function in the cortex.

In 1863, John Hughlings Jackson, a clinician at the London Hospital, observed that a kind of epileptiform seizure was associated with lesions localized in an area of the cerebral cortex and suggested that this area controls limb movements. Later, the French neurologist Broca noted that a discrete lesion in a small cortical area in the left hemisphere of right-handed people leads to an inability to speak (motor aphasia). In 1870, R. Fritsch and E. Hitzig first applied electrical stimuli to the cortex of an anesthetized animal and showed that when a certain region was stimulated, movements were elicited in the limbs on the opposite side of the body. This first demonstration of a specific motor cortical area was confirmed by Sir David Ferrier in England in 1876. At about the same time, Vladimir Betz, a Russian histologist, showed that the electrically excitable area has a specific structure, with particularly large pyramidal-shaped nerve cells still known as the giant cells of Betz.

The motor area, located near the anterior pole of the hemisphere, was extensively studied by the physiologist Sir Charles Sherrington and his pupils and more recently by the neurosurgeon W. G. Penfield. It gives rise to a tract of nerve fibers that can be traced through the brain stem and throughout almost the whole length of the spinal cord. Through these fibers, which together form the pyramidal tracts, the cortex acquires direct control of the motor nuclei of the brain stem and spinal cord. A notable feature of these pyramidal tracts is that in the hindbrain the tract from the left hemisphere crosses to the right side of the spinal cord, and vice versa. Such a crossing, called a decussation, is characteristic of most ascending and descending tracts in all vertebrates (a *pyramidal* decussation is found only in mammals, since only these animals have a neocortex). Why one side of the nervous system should be connected mainly with the opposite side of the body is not known.

The motor cortical areas and the tracts that arise from them are sometimes called the pyramidal motor system. This system is concerned with the performance of skilled, learned, or voluntary activities that require the coordination of many small movements, such as those of the hands and fingers in writing or in playing a musical instrument. Although this system is present in all mammals, the degree of its development varies. It is extremely small and functionally of little importance in hooved mammals, or ungulates, whose limbs perform limited movements and are used almost solely for locomotion; such repetitive movements can be coordinated in the lower levels of the nervous system. In primates, most of whom use all four limbs in varied ways, the pyramidal motor system is much more important and is correspondingly well developed, particularly in man in whom the motor areas and pyramidal tracts are absolutely and relatively larger than in any other mammal. This *new* pyramidal motor system differs widely from the motor system mentioned earlier, which utilizes the basal ganglia of the hemisphere, the reticular formation, and the reticulospinal tracts. The latter is commonly referred to as an *extrapyramidal* motor system, which is probably capable of regulating the simpler patterns of behavior, like facilitating or inhibiting the movements of progression. It lacks the plasticity of the cortex and is not itself capable of coordinating new and complex patterns of behavior to suit new circumstances. Nevertheless, it provides an essential background of postural and other adjustments without which the pyramidal system would be unable to function efficiently.

Sensory and motor projection areas in the neocortex are separated by ill-defined regions to which it is difficult to assign a specific function. These "association areas" are connected with the projection areas. Here all modalities of sensory information are integrated so that elaborate concepts can be built up. On the basis of such concepts, patterns of behavior appropriate to a total situation can be coordinated and adjusted in conformity with experience. To achieve this, the association areas must have further connections with the motor area and the memory store, wherever that is situated. All parts of the cortex are interconnected by many "association fibers." Moreover, since there are two hemispheres, really two brains, the two are connected by "commissural fibers" so that the right hand *can* know what the left hand is doing! Mammals have a special neocortical commissure called the corpus callosum, small in insectivores, but particularly large in man, which crosses dorsal to the diencephalon between the medial surfaces of the two hemispheres (Figures 5–5, 5–6). Submammalian vertebrates have no corpus callosum, and monotremes and marsupials at best have only the rudiments. In all the more successful placental mammals, it is a defining characteristic of the brain; even so, its presence does not necessarily mean that the higher mammalian brain cannot function as two separate halves.

To designate the cortical areas in the cerebral hemisphere that have specific functions, a new terminology that goes beyond the simple recognition of basal and pallial parts is required. The expansion of the neocortex has led to a forward growth under the frontal bone, now called the frontal lobe, where the motor area is situated. The remaining parts of the cortex are also named, not from their functions but after the bones under which they lie: the parietal lobe (with the general sensory area) behind the frontal is followed by the occipital (with the visual area) and the temporal (with the auditory area). In higher mammals, expansion has caused the occipital lobe to overlap the midbrain and cerebellum, and the temporal lobe (originally the posterior pole of the hemisphere) has been deflected downward and forward, probably because the expansion of the cerebellum prevents further caudal growth. This deflection causes the hemisphere to be bent into a U-shape, with a crease in the concavity of the U, called the lateral, or Sylvian, fissure, which separates the temporal from the parietal and frontal lobes. Further neocortical expansion within the confined space of the skull leads to a crumpling of the cortex in higher mammals and the formation of an elaborate pattern of grooves, or sulci, between which are convex elevations, or gyri (Figure 5–5). Such brains are called *gyrencephalic.* Primitive mammals, like the insectivores, have a smooth cortex, and the brains are said to be *lissencephalic.* The pattern formed by the sulci and gyri becomes increasingly elaborate in more advanced mammals and is particularly complex in man, the great apes, elephants, and whales (Figure 5–6). Some sulci and gyri have definite relations to known projection areas and sometimes form their boundaries. For example, the lunate sulcus, particularly characteristic of the ape brain, forms the boundary of the visual cortical area. In a general way, the patterns formed by sulci and gyri are characteristic of the order or genus of an animal, but they vary even among individuals of the same species.

The parts of the brain below the forebrain are not especially characteristic of primates, and, throughout, the vertebrates have many features in common. For our purpose, the most important are those referred to earlier: the tectum of the midbrain and the cerebellum of the hindbrain.

In lower vertebrates the tectum is the main visual center. In birds it has large and elaborate optic lobes that receive other sensory information by way of spinotectal and other connections. Avian optic lobes, which are laminated somewhat like the mammalian neocortex, may perform similar functions. In mammals the optic lobes have diminished both in size and function. Most optic connections have been transferred to the lateral geniculate body in the thalamus, whence they are relayed to the visual cortical area in the occipital lobe. Hearing has fared similarly. In submammalian vertebrates the part of the tectum behind the optic lobes may be the main auditory center. In mammals this region consists of two small elevations, the inferior colliculi or the inferior corpora quadrigemina. Like the elevations of the optic lobes in front of them (called the superior corpora quadrigemina in mammals), they appear to serve only relatively simple reflexes since the main auditory connections have been transferred to a thalamic nucleus (the medial geniculate body) for relay to the auditory area in the temporal lobe of the cerebrum. The mammalian tectum, including that of primates, shows no progressive features like those in the forebrain and may even have retrogressed.

The situation is different in the cerebellum, which in primitive animals developed as an extension of the nuclei into which the nerves from the vestibular apparatus discharged. A flocculo-nodular lobe, consisting of a median elevation (the nodule) and two lateral extensions (the flocculi) was formed. This is practically all there is to the cerebellum of the primitive cyclostomes (lampreys and hag fishes). Its function is to regulate the position and movements of the trunk on the basis of information received from the vestibular apparatus. A feedback of information from the spinal cord (along the spinocerebellar tracts) and from other sensory nuclei in the brain stem of higher vertebrates is associated with the growth of a body, or corpus, of the cerebellum, cranial to the flocculo-nodular lobe. The corpus regulates the performance of complex motor activities, including the limb movements, which must be coordinated with those of the trunk. The corpus cerebelli is particularly large in birds. In the more advanced mammals additional lateral expansions on each side of the corpus form the largest part of the cerebellum. The size of of these expansions, sometimes called the hemispheres of the cerebellum, is generally in proportion to the extent of the neocortex of the cerebrum; hence they can be regarded as a neocerebellum. Our knowledge of the functions of the parts of the cerebellum is incomplete, but the lateral enlargement of the corpus may be involved in the performance of the complex skilled activities associated with the extensive motor area in the cerebrum; this is a striking feature of the brain of primates, especially that of man.

The connections from the neocortex to the cerebellum are abundant in man and in the great apes. The bulk of these tracts is so large that they can be seen in the brain stem with the naked eye. Cerebro-cerebellar fibers pass from the neocortex through the basal ganglia and diencephalon to the ventral surface of the midbrain where they form most of two ropelike strands known as the cerebral peduncles; these disappear into a prominent swelling on the ventral surface of the hindbrain called the pons (L., bridge) because it seems to cross from one side to the other (Figures 5–5, 5–7). In the pons, cerebro-cerebellar fibers end in synaptic contact with groups of cells (the pontine nuclei) whose axons run around the side of the brain stem to end in the neocerebellum; thus nerve impulses are relayed from the cerebral cortex to the cerebellum. The cerebral peduncles also carry the pyramidal motor tracts, which in man occupy about one-third of their bulk; these tracts pass through the

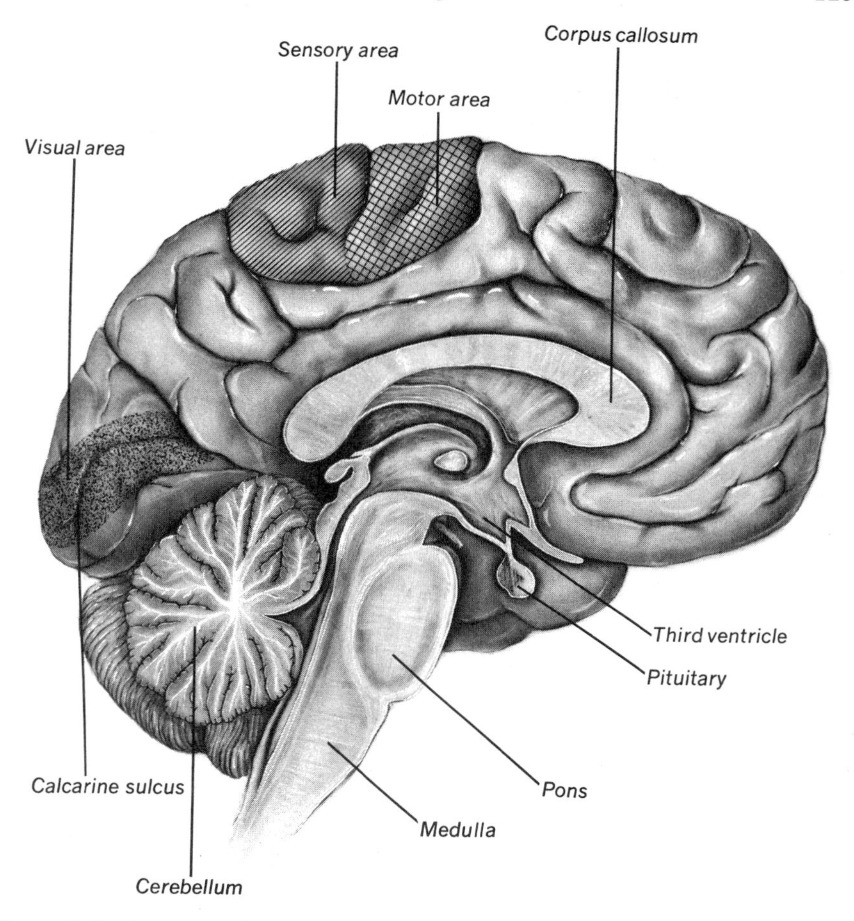

Figure 5–7. Diagram of a human brain cut exactly through the center. Compare with Figure 5–8.

pons, go to the surface of the medulla oblongata, and disappear from that surface where they decussate to pass into the spinal cord.

An important, probably the main, function of the cerebellum is to regulate the contractions of muscles so that movements are performed with a precise control of their range and force and in the proper sequence to constitute an activity that is smooth and quick. Like the basal ganglia of the forebrain and the extrapyramidal system in general, it is subservient to the neocortex, being, as Hughlings Jackson might have said, at a lower level. In a somewhat fanciful analogy, it can be compared to a skilled executive whose work is essential if the policy and orders of a board of directors are to be carried out efficiently. Since the policies and activities of men are so complex and varied, it is not surprising that the cerebral peduncles, pons, and pyramids form larger and more conspicuous features of the brain stem than in most mammals.

Blood Supply of the Brain

The gray matter is especially well supplied with blood vessels conveying the oxygenated blood needed by neural tissue (Figure 5–8). A quantity of blood nearly equaling the weight of the brain should circulate through it every minute. In a man at rest about one-third of the blood leaving the left ventricle of the heart passes to the brain. Blood reaches the brain through two internal carotid (Gk. *karos,* stupor; so called by Galen who noticed that their compression causes stupor) and two vertebral arteries. Aristotle believed that compression of the carotid arteries resulted in deep sleep. However, it has been shown that the ligature of both carotids in some animals does not produce sleep or stupor, because only their vertebral arteries carry blood to the brain. An instantaneous suppression of blood flow to the human brain results in unconsciousness in about seven seconds, and, if the interruption continues beyond a critical time,

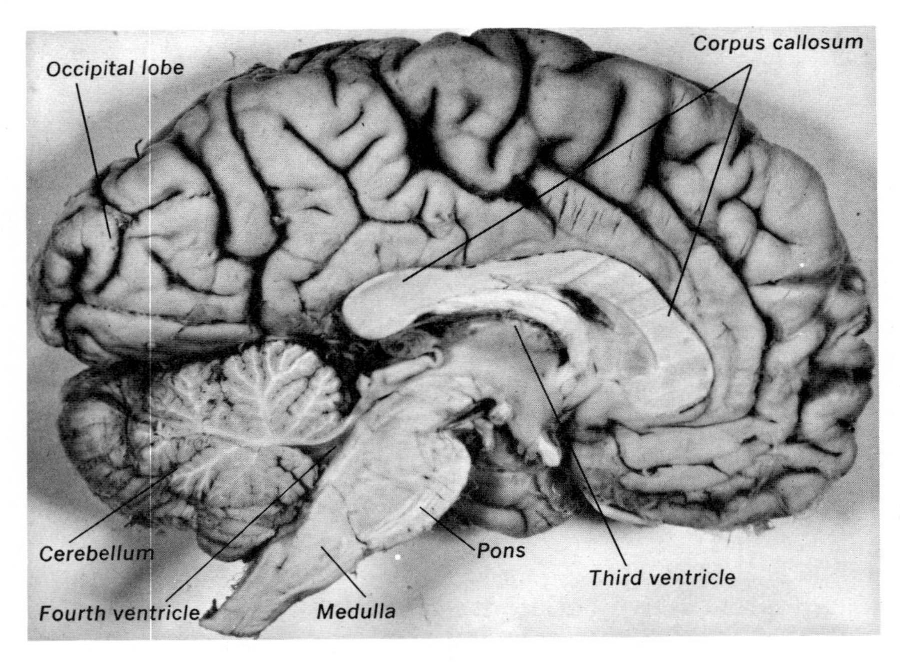

Figure 5–8. Sagittal section of an adult human brain.

the brain is irreparably damaged. This damage can be avoided if a condition of hypothermia (low temperature) is first induced by cooling the whole body, as is done in some surgical procedures.

As they enter the base of the skull, the two vertebral arteries unite to form the basilar artery (Figure 5–9). This divides again, and its branches, together with those of the internal carotid arteries, form a circle at the base of the brain (circulus arteriosus), first described by Thomas Willis (1664) and illustrated by Sir Christopher Wren. In this way the circulation in the brain is safeguarded since blockage of the main channels on one side can be rapidly compensated by flow from the other. From the circle of Willis emerge large cerebral arteries (anterior, middle, and posterior) that encompass the brain and send out numerous branches to all parts of the cortex. In some mammals, such as cattle, branches of the vertebral arteries form a network of arteries at the base of the brain known as rete mirabile, which probably stabilizes blood flow. Small but vital central and basal arteries from the circle of Willis penetrate the base of the brain. From each middle cerebral artery, central arteries supply the main fiber tracts (internal capsule) passing to and from the cortex. In man these vessels are peculiarly prone to lesions that cause severe disability and hence are often called the arteries of cerebral hemorrhage. A cubic millimeter of human gray matter contains a quantity of capillaries that if put end to end would extend over a meter.

Blood from the brain empties into the venous sinuses within the dura mater, an outer fibrous capsule around the entire central nervous system. In man these sinuses eventually discharge into the two internal jugular veins. In seals, however, the jugulars carry blood only from the face and neck. In these animals blood leaves the brain chiefly by sinuses that empty into a large extradural intravertebral vein, dorsal to the spinal cord; from here it drains into a venous plexus around the kidneys. This remarkable modification in seals is probably associated with a muscle cuff around the posterior vena cava. Seals survive the effects of the drastic heart slowing (bradycardia) that occurs during diving because of a general peripheral vasoconstriction that diverts oxygenated blood to the brain rather than to the rest of the animal.

The Brain of Primates

Extant primates include a wide variety of living animals, ranging from primitive tree shrews (originally classified with the Insectivora) through monkeys and apes to man. As might be expected, a wide divergence in the degree of brain development is seen in each of the different species of Primates.

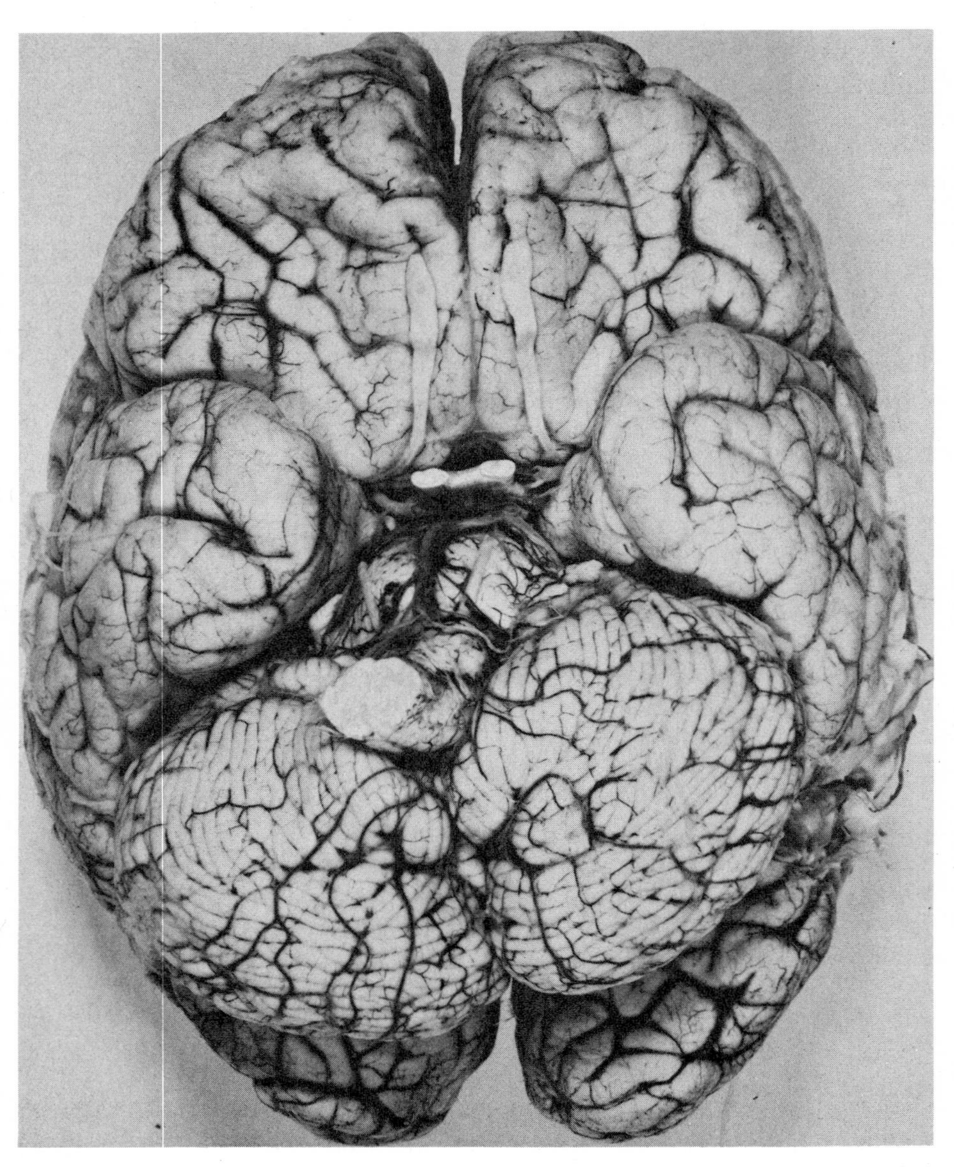

Figure 5–9. Superficial blood vessels on the underside of a human brain; only the major coarser vessels are shown.

Probably the most general statement that can be made about the evolution of primates is their tendency to develop a brain out of proportion to the size of the body (Figure 5–1). The small tree shrew *(Tupaia javanica)*, for example, has a brain/body-weight ratio of 1:35. In the mole, an insectivore of similar size, the ratio is about 1:50. However, we have seen that a simple consideration of gross size does not tell much. The brain consists of many parts that vary in size and function so that it is more significant to examine their relative changes in the parts than the overall weight or volume of the brain.

In all mammalian orders, there is an evolutionary tendency for the brain to increase in size, especially in the extent and bulk of the cerebral neocortex. Cortical growth is reflected in the diencephalon by the growth and differentiation of the thalamus, through which all modalities of sensory information, except olfaction, reach the cortex, and in the cerebellum by the development of lateral lobes of hemispheres, necessary to carry out the complex motor activities made possible by cortical enlargement and differentiation. Cortical growth has also led to several new or greatly enlarged fiber systems: a dorsal commissure (the corpus callosum) that links the two cerebral hemispheres; enlarged cerebral peduncles in the midbrain; a pons linking the cerebrum and cerebellum; and pyramidal motor tracts in the hindbrain that extend throughout the spinal cord. Corresponding additions to the ascending sensory tracts carry information beyond the brain stem to the thalamus and cerebral cortex. Changes in the fundamental mechanisms of the basal ganglia, brain stem, and spinal cord (present in all vertebrates) are for the most part secondary to the development of the cerebral cortex, and perhaps their bulk has altered mainly in relation to variations in the size of the body. These considerations lead to a discussion of primate brains and an emphasis on the cerebral hemispheres and the cortex in particular.

We have seen that even the primitive tree shrew has a relatively large brain. Compared with those of most nonprimate mammals, the olfactory bulbs and the basal olfactory cortex in the tree shrew are relatively and perhaps absolutely reduced. The neocortex is extensive, but still mostly smooth (lissencephalic), and the occipital lobe containing the visual cortex expands over the entire midbrain and part of the cerebellum. The visual cortex is associated with the animal's only neocortical sulcus, a short calcarine sulcus bordering the visual area on the medial side of the occipital lobe. The cortex below the occipital lobe has grown downward and forward to form a definite, if rudimentary, temporal lobe, which contains the temporal association cortex as well as the auditory cortical area. The temporal association cortex receives fibers from the auditory and visual areas and probably also from most, if not all, the parts of the neocortex. It is anatomically close to the basal olfactory area and to an important part of the limbic lobe (see p. 120); almost certainly it has functional connections with both. The temporal association area is therefore in a position to build up elaborate concepts based on many kinds of sensory information and perhaps to store them in memory. There is some evidence for this in man. W. Penfield, who has shown that stimulation of the temporal cortex in conscious patients can lead to the recall of a memory, speaks of it as an interpretative mechanism for complex experiences. The neocortex of *Tupaia* shows a higher degree of differentiation than that of Insectivora in that functionally different areas are more distinct and lamination is more clearly marked.

Primate trends are therefore well established in *Tupaia,* particularly in the enlargement of the occipital lobe, associated with the increasing importance of vision and the reduction of basal olfactory centers. The development of a temporal association area is also characteristic, but the absence of a significant increase in the parietal and frontal association areas must be regarded as a primitive feature. The frontal lobes are still notably small. The large optic lobes in the midbrain (superior corpora quadrigemina) probably show that transference of visual function to the forebrain is less complete than in higher primates.

From the human point of view, tarsioids and lemurs are somewhat aberrant primate types, having branched away at an early date from the stem leading to the anthropoids and man. Both groups show a further reduction of the olfactory centers, and *Tarsius* (the only living genus of the tarsioids) has a really enormous visual apparatus, shown in the brain by an extremely large occipital lobe and an extensive visual cortical area, which makes up almost half of the total neocortical area. In other respects the tarsioid brain is remarkably primitive; it has no visible sulci and gyri on the lateral surface and the temporal lobes are poorly developed. The lemurs, which are more successfully represented than tarsiers by the number of living species, seem to be neurologically more progressive. Their advance is shown most clearly in the neocortical expansion that has led to the formation of a fairly complex pattern of sulci and gyri in the larger species. Their sulci tend to run longitudinally from the frontal through the parietal to the occipital and temporal lobes, as is common in nonprimate mammals, such as the cat, and even in Cetacea. Lemurs have at least one characteristically primate sulcus, the retrocalcarine, an extension of the calcarine sulcus, caused by an axial folding of the visual cortex on the medial side of the occipital lobe. Lemurs, with a well-developed temporal lobe, have a true lateral, or Sylvian, sulcus. In spite of these characteristically primate features, the brains of lemurs have general sulcal patterns similar to those of nonprimate mammals.

It is in the Anthropoidea—both New and Old World monkeys, the great apes, and man—that the primate evolutionary trends come to full fruition. Even the most primitive small South American marmosets have brains some three times the weight of those in lemurs of comparable size, and the hemispheres have a form similar to that of man. The large frontal lobes completely overhang the diminutive olfactory bulbs. The occipital lobes cover most of the cerebellum, and the temporal lobes have grown downward and forward so that a typical lateral (Sylvian) sulcus is formed between them and the overlying frontal and parietal lobes. In spite of the lateral sulcus and a calcarine and retrocalcarine sulcus on the medial side of the occipital lobe, the hemispheres are smooth and, in this respect, more primitive than those of some lemurs.

The most striking advance of the Anthropoidea, compared with the more primitive primates, is the development of the frontal and parietal lobes of the hemispheres, resulting mainly from the growth of their association areas. In all larger monkeys, whether New or Old World, this leads to the appearance of what can be called the anthropoid sulcal pattern, characterized by sulci that run transversely to the long axis of the cortex. In the brains of monkeys and apes, two such sulci, the central and the lunate, are particularly conspicuous. The central sulcus, which runs downward and slightly forward, divides the frontal from the parietal lobes and separates the motor area of the precentral gyrus (in the frontal lobe) from the general sensory area of the postcentral gyrus (in the parietal lobe). This is also one of the most characteristic sulci of the human brain, where it was first described. The lunate is a slightly curved sulcus on the lateral aspect of the occipital lobe where it forms the anterior boundary of the visual projection area. It is so characteristic of monkeys and apes that it is sometimes known as the ape sulcus. It can be identified only occasionally in the human brain, where the greater expansion of the parietal association area has displaced most of the visual cortical area to the medial aspect of the occipital lobe.

A marked feature of the hemispheres in all apes, particularly the gorilla, whose brain is almost 600 cc in volume, is the increasing complexity of the sulcal pattern, resulting from the appearance of secondary sulci. In man, whose brain in much larger, the disproportionate expansion of the neocortex, especially of the association areas in the frontal parietal and temporal lobes, and the formation of secondary sulci and gyri have gone so far that it is often difficult in a superficial examination to identify the more fundamental sulci like the central sulcus.

The Human Brain

It is now clear that the large size of the human brain is due first to the great expansion of the neocortex, and second to the increase in size of those systems that supply it with sensory information and of those that enable it to influence neural mechanisms at lower levels. The ascending tracts of the spinal cord, particularly those on its dorsal aspect, are extremely large, and much of the information they convey, after relay and decussation, is taken directly to the thalamus of the forebrain for further relay to the general sensory area of the cerebral cortex. The thalamus is also an important integrative center in its own right, and although the details of its functions are not very well understood, there is little doubt that sensory information is selected and "processed" in the thalamus before being passed on to the cortex. Some aspects of conscious awareness may depend on its activity. As one might expect, the thalamus is considerably enlarged in man, and its posterior part, the pulvinar (from the Latin meaning "couch made from cushions"), characteristically projects over the tectum of the midbrain.

The corpus callosum, which joins the cerebral hemispheres, the neocortical commissure, is extremely large (Figures 5–5, 5–7). The transference of information from one hemisphere to the other and the coordination of their activities is an obviously important function, but it is surprising how little behavioral deficiency has been observed in those rare individuals in whom the commissure is absent. Cutting the corpus callosum in an experimental animal leaves essentially two independent brains capable of learning different and even incompatible reactions to similar sensory experiences. In addition to commissural fibers, the white matter of the hemispheres is greatly increased by the abundance of association and projection fibers. Observations by Sperry (1958) seem to show that, at least in experimental animals, the projection fibers are more important than the association fibers for cortical function. A lattice of vertical cuts through the cortex, which must divide most of the horizontally disposed association fibers but leaves intact most of the projection fibers that run at right angles to the surface into or from underlying white matter, interferes remarkably little with function. In the human brain the result might be very different. The efferent projection tracts form massive peduncles on the ventral side of the midbrain and enter a very large pons whence many are relayed to the neocerebellum, which is correspondingly enlarged. Large pyramidal motor tracts are visible in the medulla oblongata, but these, as well as the other features mentioned, differ only quantitatively from similar features in other animals.

It is remarkable that there are no parts or structures in the human brain that cannot be found in the brains of the great apes. Anatomically, all differences are in relative or absolute size. Physiologically, the situation is not very different although cortical damage in monkeys is in general much less serious than in man. Injury to the motor cortical area, for example, provided it is unilateral and does not involve the whole area, causes only transient paralysis in monkeys; in man the loss of function is much more severe and can be permanent. This severity results partly from the fact that skilled and learned movements, particularly of the hands, play a much smaller part in the life-style of a monkey than of a man. But this is a quantitative difference and represents not the acquisition of a new function but further progress in prosencephalization, with the more complete dominance of the cortex over lower level mechanisms.

Perhaps the most characteristic feature of the human brain is the enlargement of the association areas. Neurologists have given much attention to the frontal area, which is responsible for the disproportionate size of man's frontal lobes. This enlarged frontal association area seems to be linked with man's characteristic exercise of caution, restraint, and inhibition in his actions and perhaps with concentration on the prolonged pursuit of his desired ambitions. But here, too, the difference between the apes and the other anthropoids seems to be essentially one of degree. Even structurally, human neural tissue is fundamentally similar to that of other mammals; in the cortex the cell bodies of nerve cells are more widely separated by the tangled network of dendritic processes and the terminal arborizations of axons (known as neuropil), which make possible more numerous interconnections between neurones or increased "connectivity." The small, so-called granule cells with short axons that do not extend much farther than their dendrites are abundant in many parts of the human cortex, but they are not peculiar to man.

The essential similarity between the brains of anthropoids and of man (except in the matter of the relative size of the whole or of its parts) has led anatomists and physiologists to speculate whether it is the association of a large brain with certain nonneural characteristics that is responsible for the development of specifically human behavior. Two such associations are undoubtedly significant. First, there is the adoption of full bipedalism, with the freeing of forelimbs and hands, which have acquired no marked specialization to limit their range of activity. The possibility of using them as sense organs to explore the immediate environment by touch greatly increases the amount of sensory information available. There is also the possibility of using the hands in complex ways for the manipulation of external objects and the fabrication of tools. All this leads to is the extension of man's activities and powers, but for their exploitation freeing the limbs alone was not enough; it was also necessary to attain association with an efficient neural mechanism to analyze the additional sensory information they supply and to control and coordinate their movements. It seems plausible that an increase in the size of the neural apparatus in the brain and in its connectivity should be adequate, without any fundamental change in its nature.

The other association, which may be even more important, is that between the development of a large brain and a very long period of postnatal immaturity during infancy, childhood, and adolescence. The length of this period is characteristic of man and separates him sharply from all other primates. At birth the human brain weighs only 350 gm so that much of its growth takes place while the infant and child are being exposed to an enormous variety of environmental stimuli and while they are learning to adapt to their complex environment. By the age of ten years, the brain has reached about 95 percent of its adult weight. Growth is not the result of the multiplication of nerve cells, all of which appear to have been formed before birth and to have lost the power of division. To a large extent it is due to the increase in neuroglia and the development of myelin sheaths; there is also the growth and branching of dendrites around the cell bodies of neurones and the formation of collateral and terminal branches of axons. Growth is, therefore, accompanied by a remarkable increase in the potential connectivity of neural tissue, notably in the cerebral cortex. Although we know virtually nothing of the detailed processes involved, it is at least plausible that growth and ultimate function are substantially affected by environmental stimulation and learning, and that the combination is highly important in developing those specific forms of connectivity that make human behavior possible. Such a combination could also be important in determining the extraordinary differences in behavior among different individuals, which is such a marked feature of man and which seems to have no close correlation with the size of the brain or its parts. However, such speculation anticipates the difficult and controversial questions associated with assessing the relative importance of hereditary endowment and environmental influence on the final result in a mature adult human being. What we know of the postnatal growth of the human nervous system suggests that environmental influence is much more important to the development of the human individual than to the individuals of other animal species.

6

Locomotion

Human walking is a unique activity during which the body, step by step, teeters on the verge of catastrophe.
(John R. Napier)

Locomotion, that is, the movements of an animal from one place to another, cannot be separated from all the other body movements that accompany it. When in walking an animal moves from place to place, it has to carry its own mass, shifting its center of gravity with each movement and thereby demanding enormous adjustments of the entire body.

In locomotion, each animal embodies a unique compromise among the numerous factors that enable it to move in space, through water, and on earth; these include the center of its own mass in relation to its anatomical features—bones, muscles, joints, viscera, central nervous system—its size and shape, as well as the physical factors of gravity, resistance, and friction. There is a decided purpose in the design of the limbs of every animal. Contrast, for example, the almost pathetically thin anterior limbs of most hooved animals with the grossly muscular upper parts of the propulsive posterior ones. The former are really well-integrated struts that support the anterior part of the body without much physical demand being placed upon them. Such animals move about by pushing forward with the large propulsive buttock and thigh muscles. In a bipedal animal, like man, the problems involved in shifting the center of gravity during locomotion are greatly magnified because the hind limbs, primarily designed to support the body, must alternate in bearing the entire weight of the body. Fortunately, man can move his legs in all directions: forward and backward, sideways and upward. In walking, the supporting leg is always in contact with the ground, whereas in running or jumping the body remains suspended for a short time.

Standing erect and walking bipedally are two remarkable feats of mechanical coordination. Infants take 10 to 15 months to master them and actually do not learn to walk in an individually characteristic way until between 12 and 20 years of age. One's particular pattern of walking, his gait, is often so integrated with his personality that it can reflect some facets of that personality. However, even after it is established, gait can change according to a person's habits. One who in early manhood had been an active athlete and was then reduced to a desk job is not likely to retain more than a semblance of his former gait. Age not only changes the gait but also decreases the elasticity of the ligaments and so limits the movements of the joints that disequilibrium ensues. Old people adjust to these changes by taking shorter steps and enlarging the base of support by spreading the legs farther apart.

Human beings, then, are unique among all other mammals in that their upright posture and method of locomotion have modified nearly every detail of their anatomy and have made an enormous impact on the development of human society.

A number of other mammals can also assume an upright posture and stand bipedally for various lengths of time. Kangaroos and wallabies bound on their hind legs, using the tail for balance and perhaps as a springboard in leaping and a shock absorber in landing. Bears, most rodents, rabbits, weasels, and others can sit up to expand their field of vision or to hold food with their front legs while eating. Dogs and rats can be taught to "walk" on their hind legs, particularly if both front limbs are defective or experimentally removed.

Nonhuman primates have adopted a wide variety of locomotor patterns. Tarsiers and galagos have a saltatory type of locomotion, progressing either on the ground or in trees in prodigious leaps—kicking, like kangaroos, with their long, powerful hind limbs. The shorter front limbs are equipped with long-fingered, skillful hands. This type of saltatory locomotion has freed the hands of these animals for the manipulation of food. The main locomotor adaptation has taken place on their hind limbs and feet, which have become elongated. Although superficially and functionally similar, the feet of tarsiers differ remarkably from those of galagos. Many strictly arboreal primates have relatively long arms, short legs, and a long tail for balance. Some monkeys move about by using their long arms to swing by pendulum action on the limbs of trees (brachiation); their shorter hind limbs are used only for holding firm when at rest, for the initial propulsion, and for landing. Brachiation has been defined by John R. Napier as "a form of locomotion in which the dominant component is arm-swinging by which the body, suspended from above, is propelled through space by means of a rapid alternating movement of the arms." Whereas this is correct, it does not take into account the pendulum action referred to above. A brachiating animal capitalizes on this function and can move about effectively with a minimum amount of effort. The long prehensile tail of some South American monkeys serves as a balancing mechanism during brachiation and as a fifth limb and hand during rest. On the ground these monkeys have an awkward quadrupedal gait and often assume a bipedal stance.

Among them, spider monkeys *(Ateles)*, for example, have been called semibrachiators because of their occasional arm-swinging. *Semibrachiation* is not an altogether satisfactory term because it does not tell us whether the performance is only half as good or is carried out for only half the time, or both. Anthropologists enjoy arguing about these details to the point of absurdity, but few of them agree on what is a true brachiator. True brachiators—gibbons and siamangs—have long front limbs, shorter hind ones, and no tail. When they walk bipedally they hold their arms partially outstretched for balance. Their front limbs are so disproportionately long that quadrupedal walking becomes awkward.

The large lemur, the sifaka *(Propithecus)*, jumps and leaps through the trees semierect and often "walks" on the ground on its hind legs like a gibbon. Among the terrestrial monkeys, Celebes black apes and Japanese macaques, both quadrupedal, walk easily on their hind legs when both hands are occupied or when attempting to raise their level of vision (Figure 6–1). All the great apes can and do walk bipedally, including the orangutan whose limbs are specialized primarily for moving about on trees. Chimpanzees and gorillas are known as knuckle-walkers because they make contact with the ground with the middle phalanges of the middle three fingers rather than with their flat palms. Orangs, on the other hand, walk on their semiclenched fists. Much of the locomotor activity of chimpanzees and gorillas requires an erect posture, chimpanzees often carrying armloads of food while walking bipedally. Young gorillas at play run about effectively on their legs. An occasional gorilla may even prefer an erect bipedal stance to a quadrupedal one.

The feet of tarsiers, lemurs, and lorises are small in relation to the supports to which they have to cling. Their adaptation is an enlarged great toe (hallux) and elongated fourth and fifth toes, which give the foot an enormous span for grasping and clinging (Figure 6–2). In pottos and lorises the second finger is greatly reduced (Figure 6–3).

Figure 6–1. A Japanese macaque *(Macaca fuscata)*, an Old World monkey, shows his ability to "walk" bipedally.

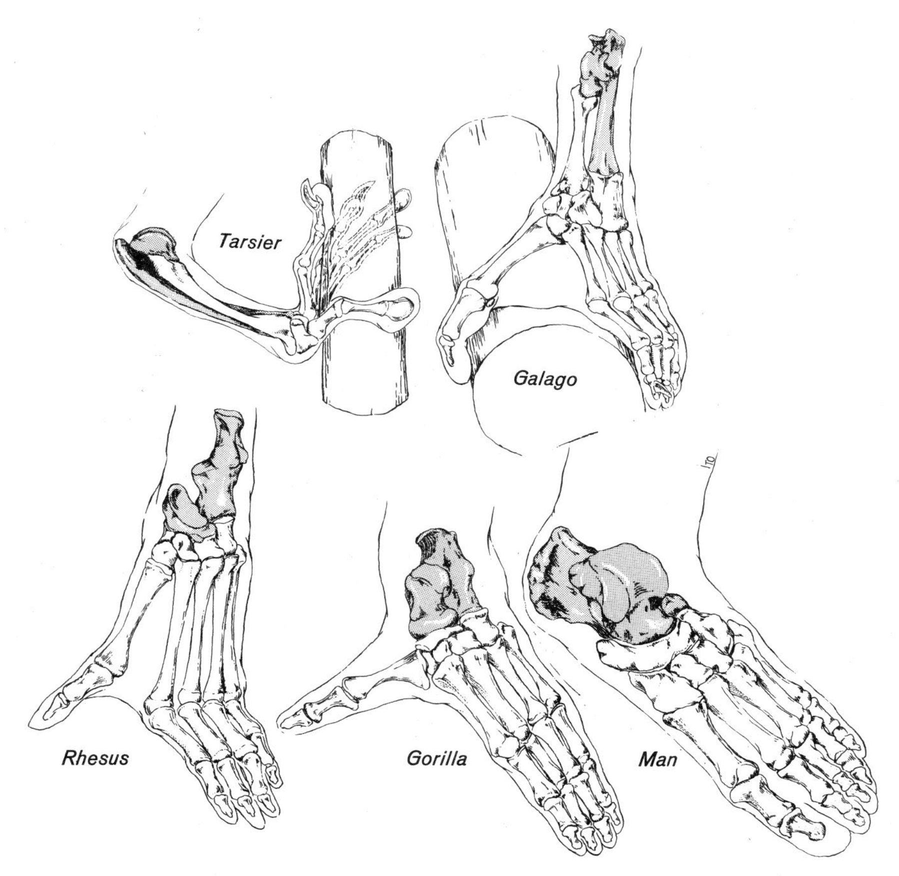

Figure 6–2. Comparison of the skeletal elements of the foot of several primates showing various adaptations. The talus and the calcaneus have been stippled.

When their foot is outstretched, their hallux and fifth toe are set at about 90° to one another. By contrast, some fully arboreal South American monkeys and the orang make no use of their great toes, but rather support their bodies on hooked lateral toes. Hence, the hallux of these animals is vestigial. Marmosets and some lemurs have a reduced hallux and sharp, clawlike nails for digging into tree bark when climbing. As most arboreal monkeys oppose their big toe toward their other toes, these other toes bend toward the big toe simultaneously. Thus, when the hallux is placed on one side of a branch, the other toes grip the opposite side somewhat like a hand.

Man's attainment of bipedalism became associated with a profound effect on the skeleton and muscles. Many bones have characteristic features that indicate at once that the animal could have walked up-

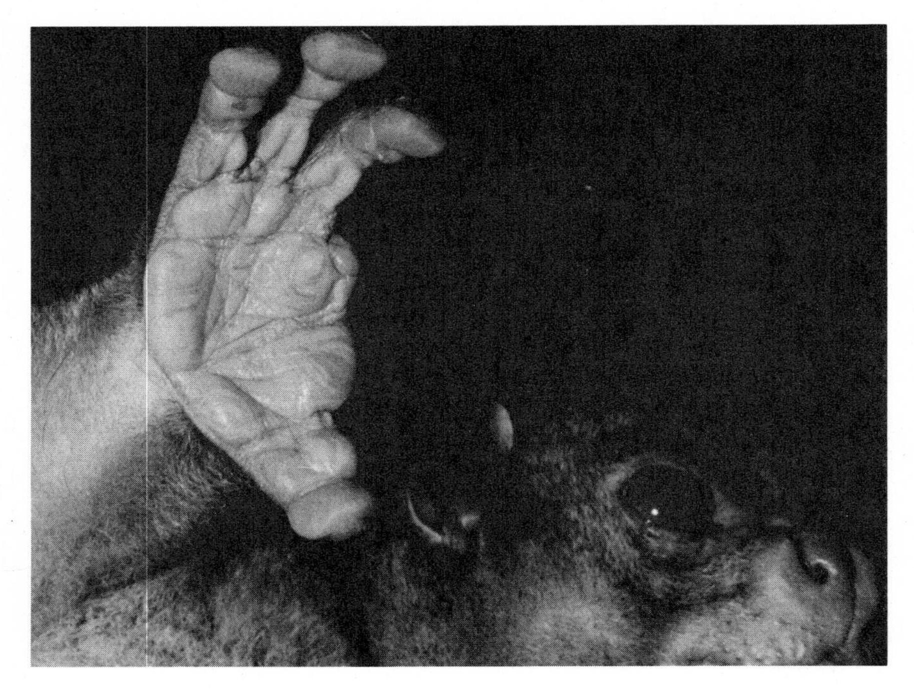

Figure 6–3. Hand of a potto (*Perodicticus potto,* family Lorisidae) showing exceptional reduction of the index finger.

right on its hind legs. One of the chief things desired in bipedalism is to achieve a balance of the head so that the eyes look in the direction one is going and are maintained on a horizontally oriented plane. None of the changes that have taken place in the skeleton, however, are quite so striking as those in the hip and foot.

Walking

Walking erect, with arms swinging, is a peculiarly human gait. Although unusual and apparently complex, walking is a remarkably efficient type of locomotion, since it makes excellent use of gravity and conserves energy. However, human beings have had to sacrifice speed for endurance. The top speed recorded for a human runner was during the 1964 Olympics, when Bob Hayes attained a maximum speed of 26.9 miles per hour. By contrast, a greyhound can run 40 miles per hour and a cheetah, 70, albeit for short distances. Yet, because of the effortlessness of walking, man's endurance is probably greater than that of any other mammal. The relatively sedentary nature of modern life causes one to forget the vast territory that ancestral man covered while hunting for food and searching for better places to live.

In recent years John R. Napier has vividly described the dynamics of man's walk:

> Man's bipedal mode of walking seems potentially catastrophic because only the rhythmic forward movement of first one leg and then the other keeps him from falling flat on his face. Consider the sequence of events whenever a man sets out in pursuit of his center of gravity. A stride begins when the muscles of the calf relax and the walker's body sways forward (gravity supplying the energy needed to overcome the body's inertia). The sway places the center of the body weight in front of the supporting pedestal normally formed by the two feet. As a result, one or the other of the walker's legs must swing forward so that when his foot makes contact with the ground, the area of the supporting pedestal has been widened and the center of body weight once again rests safely within it.

Thus, walking starts with the weight of the body being transferred to one leg. As the body's center of gravity shifts forward, the other leg must advance a step. The mechanisms involved in achieving this apparently simple act leave characteristic imprints on man's skeleton. When the weight is transferred to one leg, the pelvis tilts sideways (Figure 6–4). An appropriate swinging of the arms that balances the stride and helps in the forward thrust accompanies the pelvic movements. The full criterion of human bipedal gait is its long stride, with a long phase when one foot is on the ground and a heel-toe propelling action is apparent. In running, the cycling of the limbs is accelerated and the increased momentum reduces the effect of gravity. Running is actually fast walking with longer, quicker strides and longer periods or phases of floating in air when the body is not supported at all (Figure 6–5).

The anatomy of walking can be briefly analyzed. Suppose that one is advancing the left leg: the gluteus medius and minimus muscles on the right outer side of the loin contract and tilt the left side of the pelvis so that weight is taken off the left leg. These muscles pass from the ilium of the hipbone to the greater trochanter of the femur. The gluteal muscles, including the gluteus maximus, form the muscular mass of the buttock, a characteristic human protuberance indicative of bipedalism. That part of the hipbone, the ilium, is broad in man to give wide attachment for these muscles. Since the greater trochanter of the femur is some distance from the point of rotation of this bone with the hip joint, it gives a distinct advantage to the muscles that tilt the hip joint. The human femur has a distinctly elongated neck between head and shaft, and the neck is set at an angle of 120° to the shaft.

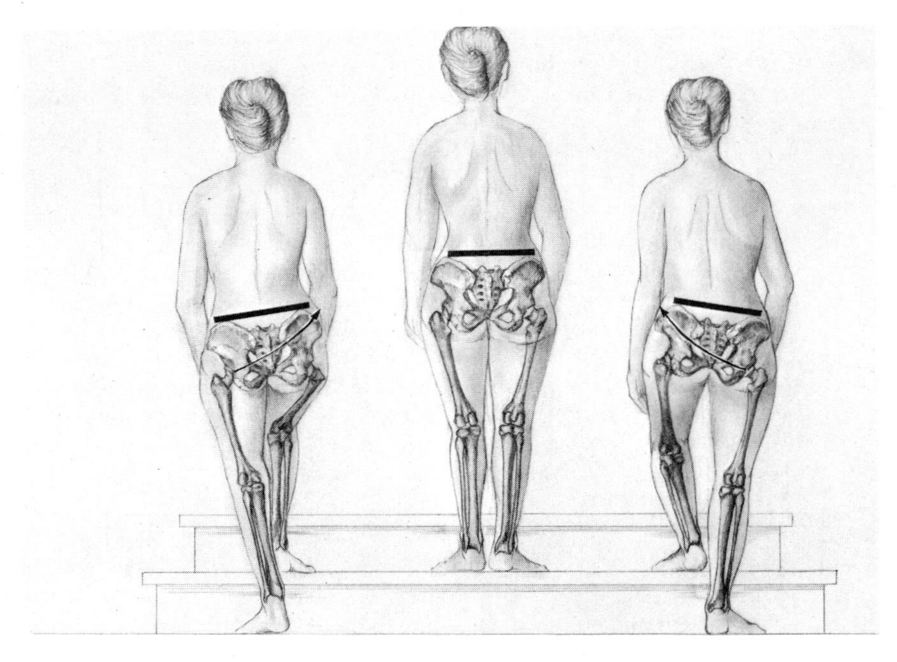

Figure 6–4. The tilting of the pelvis and shifting of body weight during walking or climbing stairs. The rotation of the pelvis with each step cannot be shown at this angle.

If the neck of the femur should for any reason be absent or off its angulation with the shaft (coxa vara), a serious disability would result that would reduce walking from a series of easy oscillatory movements to an awkward waddle.

A propulsive "take-off" thrust of the foot on the ground pushes the center of gravity forward. Weight is transferred from the heel on to the ball of the foot, and calf muscle action raises the heel from the ground. The swelling of the gastrocnemius and soleus muscles gives the human calf its characteristic shape. Consequently, the calf is much enlarged in ballet dancers as well as in those people who spend much time squatting or walking great distances. Australian aborigines, who are among the greatest traversers of territory, surprisingly have slender calves but compensate with an increased length of the inferior extremities.

The gastrocnemius and soleus are attached at the back to the bottom of the femur and top of the tibia and fibula and to the large heel bone (calcaneus) by way of a stout tendon (tendo Achillis) that can be easily felt at the back of the ankle. The tendo Achillis received its name from the myth that when Achilles' mother, Thetis, dipped him into the river Styx to make his body invulnerable, she held him by this tendon, which was thus kept from contact with the magic waters. Consequently, Achilles' heel became the fatal spot where Paris shot him with an arrow. The gastrocnemius muscle assists the soleus in plantar-flexion of the foot and also flexes the knee joint. The calcaneus juts out behind the ankle joint, and thus the calf muscles can exert a strong thrust when the heel is elevated.

As the take-off develops, the foot must be rigid. Long tendons passing to the tarsal bones from other muscles of the leg and small muscles located entirely within the foot ensure this rigidity. The knee joint must also be firm against the developing thrust. To ensure its rigidity, certain thigh muscles, in particular the four-headed quadriceps femoris muscle in the front of the thigh, contract. This well-developed muscle forms a stout tendon in front of the knee in which develops the

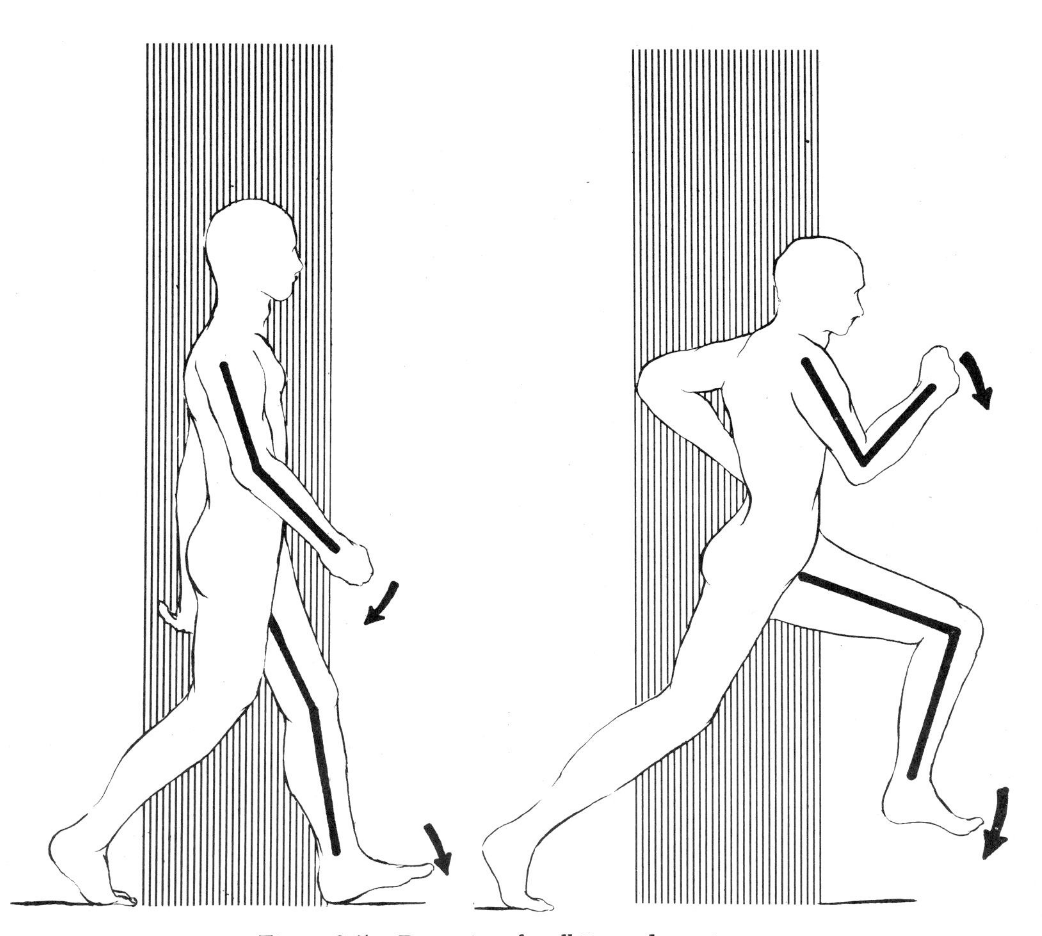

Figure 6–5. Dynamics of walking and running.

patella, the largest of sesamoids—those bones that form in a tendon. The tendon of the quadriceps continues over the patella and attaches to a marked tuberosity that can be felt on the upper part of the front of the tibia. Tapping this part of the quadriceps tendon in the resting limb stimulates receptor nerve endings that bring about a reflex "knee jerk." The quadriceps and muscles at the back of the thigh (hamstrings) help to fix the knee joint. The thrust from the take-off propels the body forward and also raises it (Figures 6–6, 6–7).

After the thrust, the advancing leg is free to swing forward, partly because of the action of gravity and partly because of the impetus developed by the take-off thrust. The tilting of the pelvis, the flexion of the knee and hip joints, and the elevation (dorsiflexion) of the foot lift the leg clear of the ground. A contraction of muscles in the front of the tibia, of which the muscle tibialis anterior is one of the most important, elevates the foot. The tendon of the tibialis anterior can be easily felt under the skin where it is covered by the tongue of a shoe. The tibialis anterior has more work to do if one's shoes are heavy or caked with mud, and under aggravated conditions its tendon sheath can become inflamed at the front of the ankle, giving rise to painful sensations in walking.

As soon as the leg has swung past the line of the center of gravity, the muscles that brought about elevation of the foot relax and the sole returns to the ground. The pelvis rotates with the swinging leg, increasing its forward movement and giving a characteristic side-to-side wobble to the body when seen from the back. This rotary wobble is particularly pronounced in women because of certain differences in pelvic proportions and can be deliberately exaggerated. The weight of the body is now transferred on to the leg in front. Both its foot and knee must be held rigid as the forward momentum of the body comes on to them, and once more the quadriceps contracts to keep the knee extended. The importance of the quadriceps is particularly appreciated in walking downhill for some distance, when the muscle may begin to ache.

The body now moves forward easily, swinging over the resting foot. Weight is usually taken first on the heel and then on the ball of the foot. The leg left behind now exerts a propelling force similar to that which initiated the first step, and the series of movements are more exaggerated, the body leans forward more, and the heel does not touch the ground. In walking uphill or in climbing stairs, the gluteus maximus plays an important part in raising the body on the leg. A foot is placed on a stair, and the muscle straightens (extends) the hip joint. At the same time the knee is straightened and the body is raised. As already noted, a large gluteus maximus and rounded buttocks are peculiar to man.

The Human Foot

Even though its five toes and a relatively large number of tarsal bones indicate the retention of a generalized primitive pattern (Figure 6–8), the human foot, like the hoof of a horse, has become specially adapted for a specific type of locomotion. It is so characteristic and the print of the bare foot so easily recognizable that a glance at Friday's footprint on the sand told Robinson Crusoe that his island was being shared by another man. The big toe, or hallux, is larger than the other four and nearly always projects beyond them. When occasionally the second toe is longer, the foot is known as a Grecian foot. The other toes are all smaller, the little one being so recessive that it may even lack a nail.

A footprint shows that weight is borne on the heel, the ball of the foot, a ridge along the outer side of the sole (Figure 6–9), and, to a lesser degree, on the thickened undersurfaces of the tips of the toes. It is as if each foot were a flexible tripod, the heel being one base, the outer margin of the sole the second, and the ball of the foot, aided by the big toe, the third. When a man balances himself on one foot, he can feel these three areas press against the ground or sole of his shoe. The fact that these regions develop hard calluses after long continuous walking further indicates the area where the bulk of the weight is borne.

However, the skeleton of the human foot is characterized not by struts of bone in the form of a tripod but rather by articulated "blocks" of bone (tarsals), arranged in a broad arch, extended and expanded by five longer bones (metatarsals) and the short phalanges of the toes. The bones fit together in such a way as to form a marked longitudinal arch along the inner side of the foot (Figure 6–10). A less-defined longitudinal arch along the outer edge, and a third, transverse one across the heads of the five rodlike metatarsals complete the three characteristic arches of the foot. The ancient Arabs regarded these arches as attributes of feminine beauty, a beautiful girl being one whose "feet are so arched that little streams can run beneath."

Several ligaments support the arches of the foot: the short interosseous ligaments, the strong short and long plantar ligaments, and the spring or check ligament below the head of the first tarsal bone, or talus. Actually the ligaments do not give the foot its springiness; that is provided by the action of short muscles within the foot and of calf muscles whose tendons pass round the ankle to insert into tarsals, metatarsals, and phalanges. All of these muscles, together with a thick, fibrous plantar band, or aponeurosis, help maintain the arches.

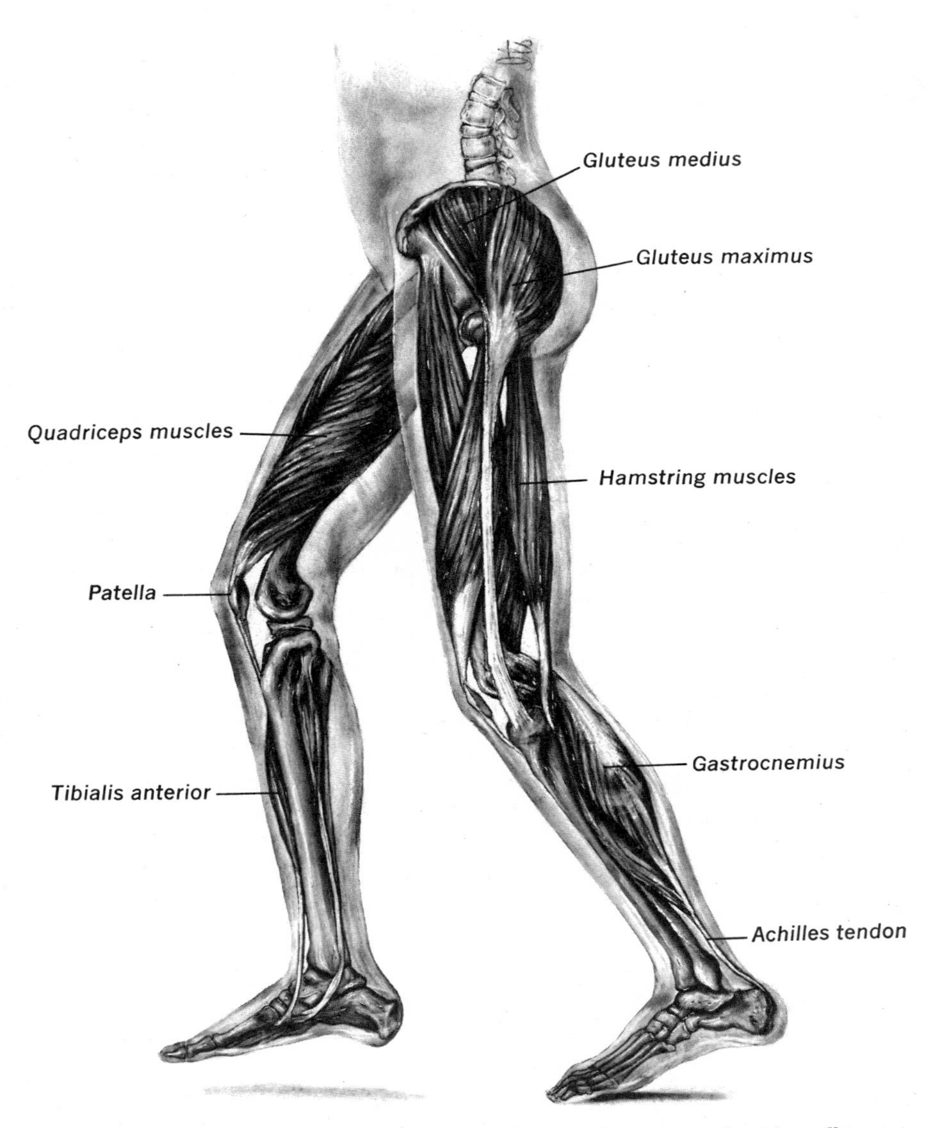

Figure 6–6. The muscles of the leg mainly concerned with walking.

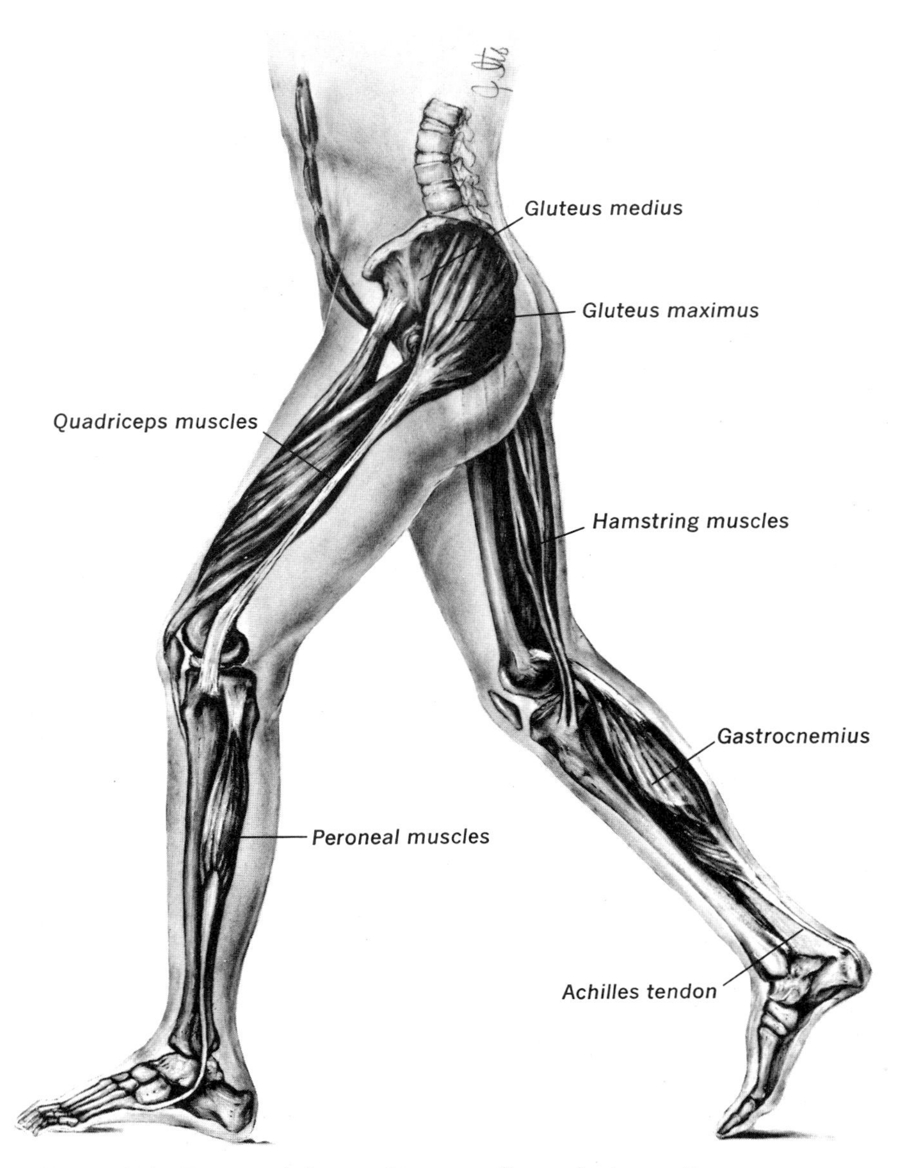

Figure 6–7. Position of the muscles principally involved in walking.

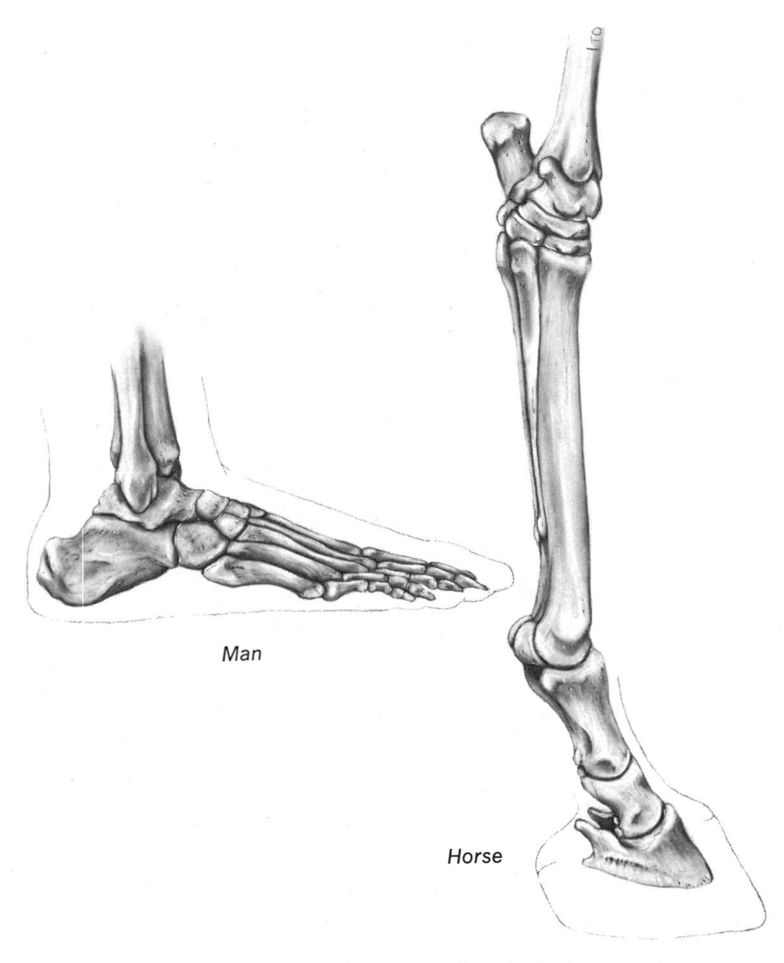

Figure 6–8. Comparison of a human foot with that of a horse.

The arches perform several useful functions. In standing and walking they ensure the even distribution of weight over the foot, according to a principle similar to that of arches in a building. They allow the foot to be used as a flexible but strong lever and result in springiness in the stride. Arches minimize shock or jar to the body during running and jumping and provide sufficient space beneath the foot to accommodate small muscles, tendons, and blood vessels without compression.

The human foot is composed of bones so shaped and articulated as to allow only certain special movements. The bones that compose the ankle joint are the tibia and fibula of the leg and the talus from the

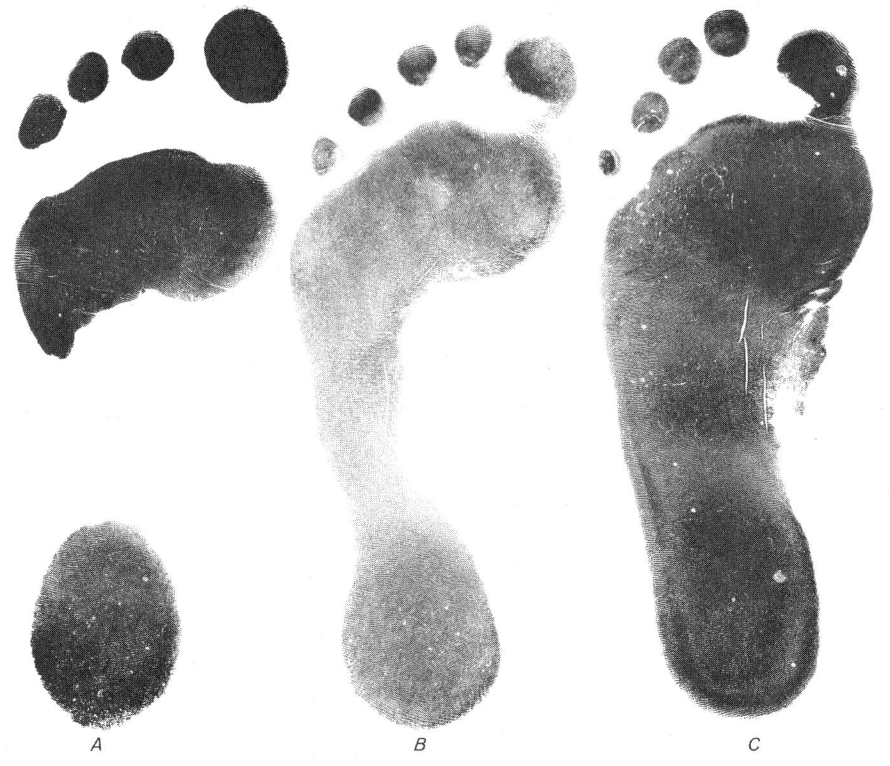

Figure 6–9. Human footprints. A: very high arch; B: moderate arch; C: flat foot.

foot. The lower ends of the tibia and fibula are expanded in a socket in which rocks the articular surface of the talus, the three bones forming a hingelike joint. The socket is deepened at the sides by flanges of bone projecting downward (malleoli) from the fibula on the outer side and from the tibia on the inner side of the ankle. The malleoli of the tibia and fibula encase the talus in a grip that prevents most movements of the talus other than rocking to and fro. Thus, the construction of the ankle joint allows mostly flexion (plantar-flexion) and extension (dorsiflexion) of the foot with very little side movement. The words *plantar-flexion* and *dorsiflexion* are used because it is not easy to determine what extension is at the adult human ankle joint.

The to-and-fro movement at the ankle joint is obviously of primary importance in walking. When a step is taken, the ankle joint of the foot that remains on the ground is plantarflexed. The ankle joint of the advancing foot is dorsiflexed and then plantarflexed when it returns to the ground. As the body swings over it, the joint becomes dorsiflexed

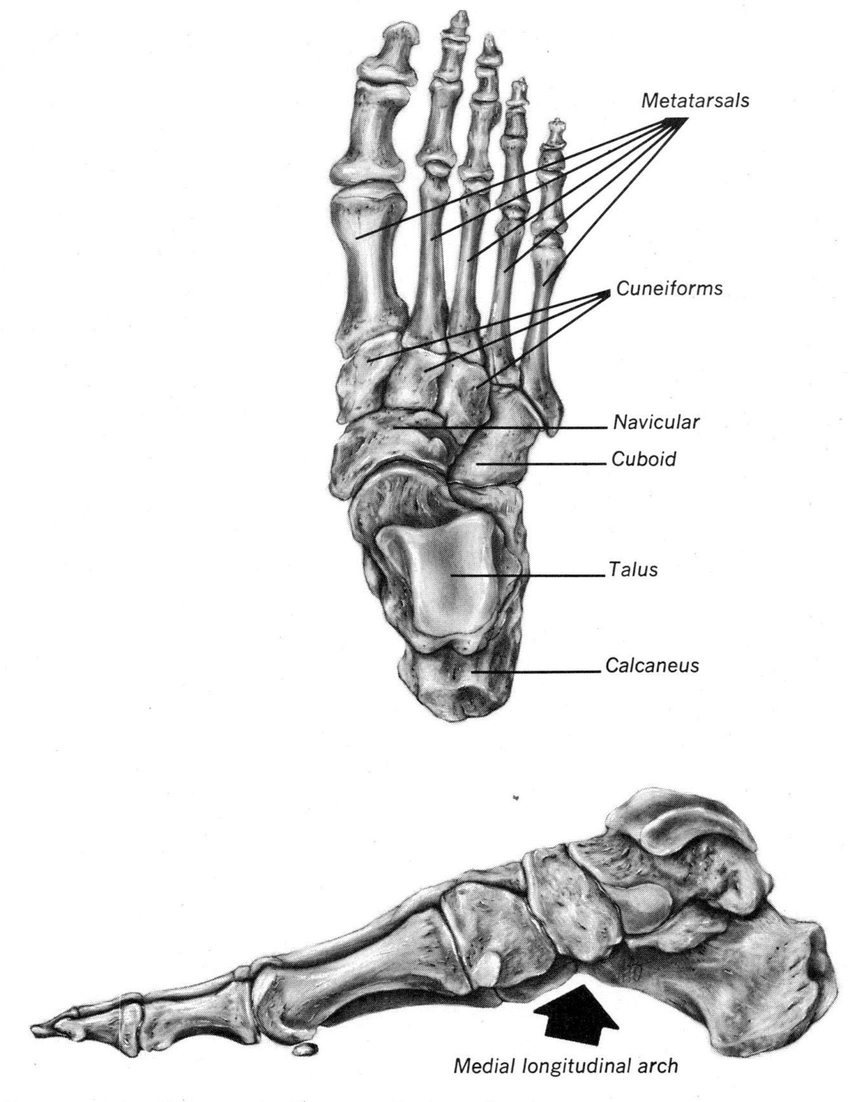

Figure 6–10. Skeleton of a human right foot; dorsal and side views. The tarsals and the metatarsals form the longitudinal arch.

again. The articular surface of the talus is wedge-shaped so that it is broader in front. Thus, the grip that the tibia and fibula exert is increased as the ankle joint becomes dorsiflexed. This is useful in preventing dislocation of the joint in jumping or in a sudden stop. There are, however, certain disadvantages in this griplike function of the tibia and fibula at the joint. Forcible sideways twisting of the foot may result in fracture of one or both malleoli or in dislocation of the ankle joint, as sometimes happens in skiing or slipping off the edge of a high step. This fracture-dislocation is known as Pott's fracture after the eighteenth-century English surgeon Percivall Pott, who first described it. When Pott fell from his horse in 1756, he broke his tibia. Some doubt, however, exists about whether he sustained the type of fracture now called after him.

The ability to perform lateral movement depends on the existence of foot joints that allow the sole to turn inward with its inner edge elevated. This *inversion* is most marked when accompanied by plantar-flexion of the foot at the ankle joint. The opposite movement, *eversion*—turning the sole outward with its outer edge elevated—has more limited range. These movements, which occur at the joints between calcaneus and talus, and those between these two bones and the navicular and cuboid are brought about by groups of muscles called invertors and evertors. Those that invert and plantarflex, such as the tibialis posterior, are characteristically large in man, partly because in the erect posture man's center of gravity lies in front of the ankle joint. A contraction of the plantarflexors counters man's tendency to tip forward. The ability to invert and evert the foot enables man to walk over sloping or uneven ground and to perform subtle flicks with his feet. Of the two movements, that of eversion is distinctly human and is produced by a group of peroneal muscles arising from the outer side of the fibula. One of them, peroneus longus, has a tendon that stretches around the outer side of the ankle and passes across the undersurface of the foot to reach the base of the metatarsal bone of the big toe. This muscle not only everts the foot but also exaggerates the transverse arch. Some of these muscles are illustrated in Figures 6–6 and 6–7.

Of the many changes affecting hominid ancestors when they became terrestrial, none seem so pronounced as those in the foot. This appendage had to be converted from a mobile, grasping structure to a relatively fixed, supporting one. Adaptive differences in the feet of existing primates are significant from the point of view of the evolutionary changes in man.

The feet of ground-dwelling human ancestors must have been large, grasping structures like those of the chimpanzee or gorilla, with opposable halluces and long toes. For such a grasping foot to be converted to a supporting one, the big toe and its metatarsal had to be brought into the line of the long axis of the foot to avoid sticking out and being constantly stubbed when walking. The toes had to become shorter and the tarsal bones stout, wedge-shaped, and arched. Mobility was consequently sacrificed for strength and resilience. The heel bone, or calcaneus, small and insignificant in the grasping foot of arboreal primates, became long and stout to serve as the attachment of the powerful Achilles tendon. Thus, contraction of the calf muscles lifted the heel off the ground, shifting the weight to the ball of the foot in walking.

Morton has shown that the foot of the gorilla is intermediate between a prehensile and a supporting one. This animal, mostly arboreal in its infancy, later becomes almost exclusively terrestrial, probably because the enormous size of adults cannot be supported by branches. Hence, this animal has a large great toe, diminished lateral toes, and a better developed calcaneus than any other ape. The toes of gorillas remain opposable. As E. A. Hooton has noted, "Each individual gorilla becomes painfully and imperfectly adapted for a terrestrial gait late in life."

Interesting comparisons have been made between the heel bones in club feet and flat feet of human adults and newborn and those of adult orangs, baboons, and gorillas. The orang, being arboreal, has feet adapted very differently from those of the terrestrial baboon and gorilla. Judging from its gross structure, the calcaneus of the orang resembles that of the human newborn, that is, the calcanei of the newborn infants are similar to those of climbing animals. Baboons and gorillas are ground walkers, but being quadrupedal and flat-footed, they transfer the weight of their body to different points on the calcaneus from those of man, and the shape of this bone, therefore, is different from that of man. Interestingly, when the arches of the human foot are broken, the stress of weight on the calcaneus changes, and this bone gradually reshapes itself to assume a form somewhat resembling that in baboons and gorillas.

Although man's foot is unique, it is not functionally infallible. When he evolved into bipedalism, man had to pay the premium of certain potential weaknesses and disadvantages inherent in it. For a number of reasons the arches of his foot can become flattened. The strain of increasing weight may stretch the tendons and ligaments of the arches until they collapse. When this happens, the resulting distortion of the tarsal bones reduces the functional efficiency of the feet, and the skeleton is adversely affected. The tendency of flat feet to splay outward leads to knock-knees and a shuffling, waddling gait like that of Charles Chaplin's Little Man. The big toe may deviate outward (hallux valgus), the second toe may become crowded out (hammer-toe), and bunions may develop.

Man's Knee Joint

Man's large knee joint is particularly well adapted for bearing weight and for a to-and-fro movement. It is formed by the expanded ends of the femur and tibia and a large sesamoid bone, the patella ("little pan"; the former name for it was *kneepan*), which develops inside the tendon that slips over the joint. The four large muscles in front of the thigh, the quadriceps femoris (Figure 6–6), form this tendon, and the combined action of the muscles and tendon is to extend the leg. The articulating surfaces of the femur and tibia do not fit very snugly; two thin, crescent-shaped (semilunar) cartilages lie between these surfaces facilitating movements, distributing pressure, and acting as shock absorbers. When a step is taken, the rotary movement involves a final straightening of the joint. With the foot on the ground, the bones are "screwed home," the joint is locked, and the knee becomes a secure weight-bearing strut. The unlocking of the joint is effected by the popliteus muscle from the back of the thigh (in man a particularly large muscle), which rotates the femur when the foot is firmly on the ground or the tibia when the foot is off the ground.

The semilunar cartilages move with the tibia but, being resilient, change their shape during rotatory movements. In sudden violent movements, parts of the cartilages may be caught between the moving joint surfaces of the femur and tibia so tightly that they can be torn or wrenched from their attachments. Man has an excellent and characteristic knee joint, but the grueling tasks to which he subjects it often require performances that tax this structure too severely.

Man's Pelvis

Like almost all mammals (except cetaceans, whose pelvic bones are rudimentary), man has a pelvis composed of three fused bones. The largest of these, the ilium, is the most superior and serves as a foundation to support the fused vertebrae at the base of the vertebral column (sacrum). Inferior to the ilium is the ischium, the expanded free ends comprising the bones upon which we sit. On the ventral side are the two pubic bones that fuse on the midline. These three bones join on each side, where they form a hemispherical socket into which fits the femur head (Figure 6–11). The pelvis, or girdle, that supports the inferior limbs and the vertebral column underwent profound modifications in man's evolution toward an upright posture.

The pelvic bones are essential for locomotion that involves the hindlimbs, and the attached muscles play a leading role in walking and in maintaining the body's upright position. The pelvis transmits the weight of the body from the spinal column to the limbs; its floor supports viscera inferiorly while the abdominal muscles that stretch between the pelvis and the thoracic cage support them anteriorly. The external genitalia are attached to the pelvis, which is so constructed that it supports the urethra for the escape of urine and the anal canal for

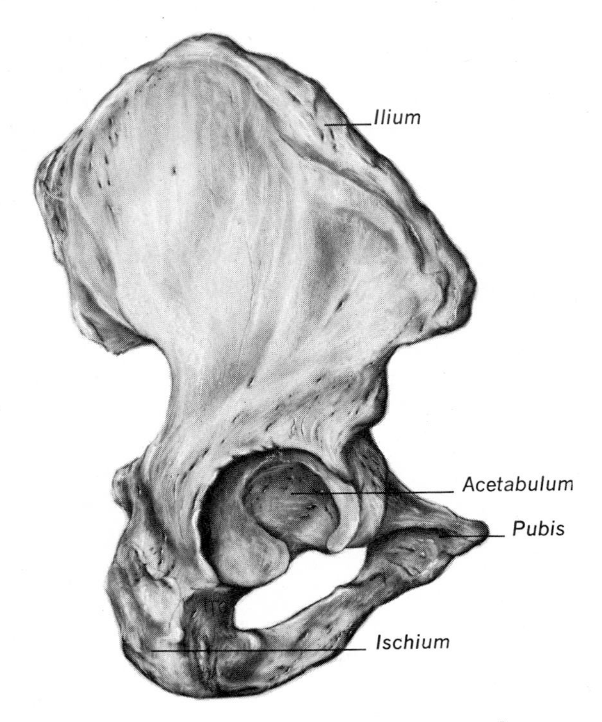

Figure 6–11. The bones of the human pelvis.

the evacuation of feces. In women it must also allow the passage of at least a 7½ lb baby at childbirth without disarranging the maternal anatomy or damaging the baby. Most anatomists agree that structure and function have been mated with considerable ingenuity in the pelvic region and that it would be difficult to suggest improvements. Even so, certain deficiencies in pelvic structure can and do cause distress to modern man.

The upper part of the human pelvis is peculiarly expanded into flaring iliac bones, which help to support viscera and offer a wide area of muscle attachment (Figure 6–12). In front, the pelvis has pubic bones set high to give better advantage to those muscles (adductors) that draw the legs together. This elevation of the pubic area increases the agility of man's legs and enables him to ride astride a horse with greater ease than could an ape. The lower part, or "true" pelvis, is large and capacious and has a characteristic outline. In the erect posture, the lowest part of the abdominal cavity rests against the pelvis (the French word for pelvis, *bassin,* aptly describes its form). However, its basinlike form and function are incompatible with a need for large

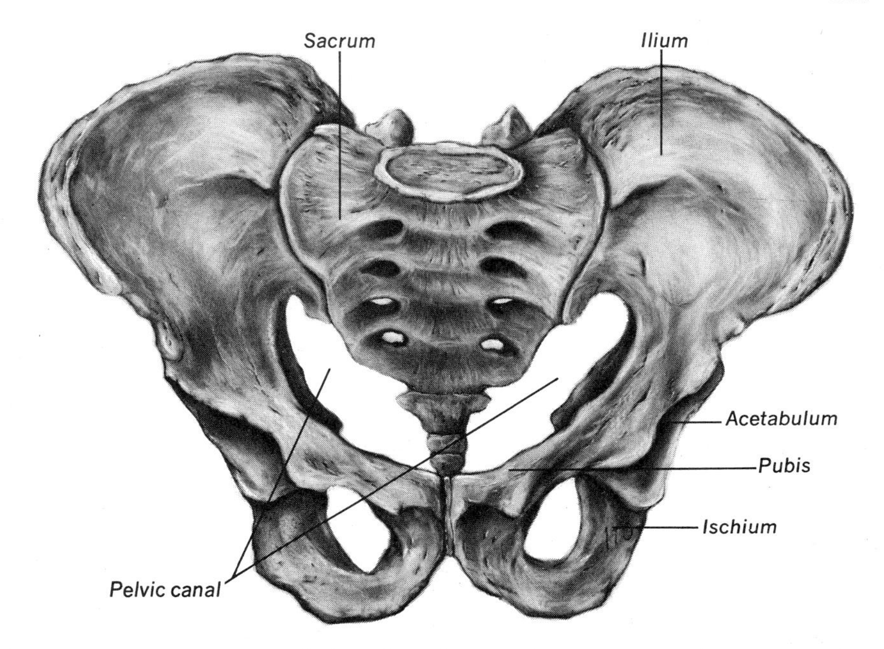

Figure 6–12. Front view of the pelvis showing the sacrum completing the pelvic canal.

openings through its floor so a muscular and fascial pelvic diaphragm perforated by the urinary, alimentary, and reproductive ducts stretches across the bony floor of the pelvis. The levator ani muscle, the most important component of the pelvic diaphragm, is slung like a hammock from each side of the pelvis.

Part of this muscle, called the ilio-coccygeus, is involved in moving the tail in monkeys, but in man it has become adapted to supporting the viscera in the upright posture. Although this sling has numerous functions, none is so important as that of retaining and supporting the viscera.

In an erect position the viscera exert a constant pressure on the pelvic diaphragm even while motionless, and the pressure increases with the slightest movement of the arms. When one bends to pick up a heavy object, pressure increases fivefold. If the movement is violent or sudden, pressure rises tenfold. Inflating a balloon or explosive coughing also raises the pressure. During these activities, the muscles around the abdomen contract and increase the intraabdominal pressure, squeezing the viscera against the pelvis. Every straining movement increases pressure on the abdominal viscera, which, being exceedingly flaccid and malleable, act as water-hammers seeking out weak points in the containing walls. Organs sometimes slide out, or prolapse, from the pelvis, or, in the event of small weak points, protrude, or herniate, through them.

Man's Backbone

Sir Arthur Keith wrote,

> The human spine is a mechanism of the utmost complexity; between the sacrum and the skull are incorporated twenty-four vertebrae, each provided with three short levers—the spinous and (two) transverse processes; each lever is furnished with, not a single muscle, but a group of them (416 in all).

Man's spine is obviously adapted in many ways to his upright posture, but many weaknesses exist along with some remarkable adaptations.

The vertebral column is a long axial pillar composed of interlocked vertebrae separated and held together by intervening fibrocartilaginous discs and ligaments. It extends from the tip of the tail bone (coccyx), which can be felt through the skin in the cleft between the buttocks, to the base of the skull. In a newborn infant the vertebral column is 8 inches long and composed of 33 separate vertebrae. It is about 28 inches long in an average adult man and 24 inches in a woman. In both, the number of separate bones is only 26. The last 4 bones are fused into a coccyx and the 5 before them become expanded and solidly fused into the sacrum. The length of the spine is remarkably constant and is said to be the least variable component of the human skeleton. During daytime activities, the length of the spine contracts by 3 to 4 percent. It is generally a little longer after lying down for any length of time primarily because of the swelling of the fibrocartilage discs between the vertebrae. The healthy adult spine can bear a third of a ton without being crushed.

The individual vertebrae, although similar in morphology, differ considerably in their shape and mobility in the various regions of the spine, a fact reflecting the different functions of each of these regions. Typically, a vertebra has a weight-bearing body and an arch. The space within the arch is the vertebral foramen containing the all-important spinal cord and the nerve roots and their covering membranes (meninges). The arch articulates above and below with the arches of adjacent vertebrae. It has transverse processes at the side and a spine at the back. The spines are not so large or long as those of mammals whose vertebral column forms a girder horizontal to the ground.

Except for those of the coccyx, sacrum, and atlas (which supports the skull), the bodies of the vertebrae are separated by intervertebral discs, which in man are thick enough to account for one-quarter of the column length. Wedge-shaped, intervertebral discs are responsible for much of the curvature of the spine. Each disc has a thick fibrocartilaginous outer part (anulus fibrosus) with densely packed fibers arranged in concentric rings stretching obliquely between two vertebral bodies with opposite obliquities in each layer. In the young and the middle-aged, the center of the disc consists of a soft, almost gelatinous substance called the nucleus pulposus, believed to derive in part from the notochord (p. 285). The composition of the nucleus pulposus varies throughout life until in old age it becomes fibrocartilaginous and eventually fibrous. The gelatinous substance found in the nucleus pulposus has the property of binding water, making the youthful discs much more efficient shock absorbers or water cushions than the entirely fibrous discs of the aged. The discs tend to equalize the distribution of pressure over the surfaces of vertebrae. They permit the rocking of one vertebra on another depending upon the arrangement of the articular processes on the arches. In old age, with changes in the nuclei pulposi, much of the flexibility as well as the resiliency of the column is lost.

In any position other than perfectly prone or supine, the nucleus pulposus is under various degrees of pressure, which is greater in the erect position; occasionally it ruptures its fibrous surroundings and protrudes, or herniates, in various directions. This is the anatomical basis of a "slipped disc." Should protrusion occur posteriorly, it may press on a spinal nerve causing different degrees and types of discomfort. Slipped discs are rare in other mammals, if indeed they occur at all. Although it is possible that factors other than posture cause slipped discs, the condition represents another potential disadvantage of bipedalism.

Viewed laterally, the adult human spine exhibits characteristic curvatures that give it the appearance of two S's stretched out one above the other—two concave in one direction, two in the other. On the other hand, the concave curvatures of the spine of a fetus and a newborn child are directed forward (Figure 6–13). The upper and larger of these primary curvatures is made by cervical, thoracic, and lumbar vertebrae; the short lower one is made by sacral and coccygeal vertebrae. The secondary curvatures, which are convex forward and develop as a child adopts an upright posture, are called compensatory curvatures. To obtain compensatory curvatures, the intervertebral discs must become wedge-shaped. The secondary cervical curvature appears two to three months after birth, when the child is able to lift its head unaided, and is well established by nine months, when he can sit. This

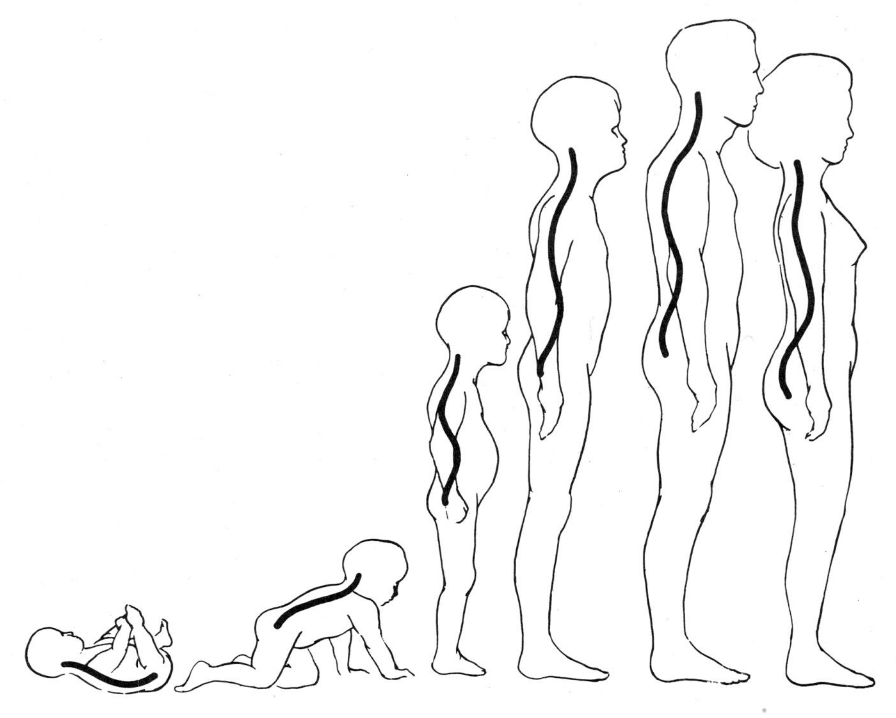

Figure 6–13. The development of spinal curvatures.

curvature represents the first in a series of skeletal adaptations to an upright posture and erect walking. It permits the head to be brought vertical to the ground and the eyes to look forward without lifting the head. The forward shift of the foramen magnum and occipital condyles in man enables his skull to balance more securely on his spine. To maintain the head erect, muscular effort, almost entirely unconscious, is required. Anyone who goes to sleep sitting upright has experienced the sudden jerking back to consciousness as his head falls forward on his chest.

The other secondary curve in the lumbar region has its point of maximum curvature almost opposite the umbilicus and extends to the articulation of the last lumbar vertebra with the sacrum. The curve is best seen in children and in women in whom fashion occasionally decrees that it be artificially exaggerated. The lumbar curve is most characteristic of human beings, especially of Khoisaniforms.

Man's lumbar region of the spine forms a flexible lever on which is poised the whole weight of the upper part of the body. Without the lumbar curve, the graceful carriage of the body would be lost, the position of the center of gravity would be altered, and one would tend to topple over backward. The lumbar spine is just right for man: if it were longer, the body would sway; if shorter, insufficient room would exist in front of it for the abdominal viscera. If the lumbar spine were slighter, it would be unable to bear the body's weight, and if stouter, would prohibit sideways bending. Walking on two feet demands that the loins be long to permit greater latitude to the muscles that move the legs. Primitive mammals had about three lumbar vertebrae, man has five. The rapid growth and elongation of the lumbar vertebrae during the first two years of a child's life seem to indicate a lengthening of the human lumbar region to accompany a plantigrade posture. In about eight percent of human skeletons, the first sacral vertebra is free and forms a sixth lumbar, whereas the fusion of the fifth lumbar vertebra with the sacrum is much less common.

Below the lumbar curve is the sacro-coccygeal curve, which is concave forward and represents the original lower primary curve. In some races, particularly among women, the sacrum is often at right angles to the lumbar spine. The backward tilting of the sacrum provides a wide area for the attachment of the large muscle mass that keeps the trunk upright. The tendency of the downward pressure of the upper part of the body to depress the front of the sacrum and at the same time to rotate its coccygeal end upward is prevented by the strong ligaments of the sacroiliac joint and others that pass from sacrum to hipbone. Because of the extraordinary curvature, the lumbar spine tends to be driven forward and downward off the sacrum into the pelvis, but the way in which the articular processes on the last lumbar vertebra are gripped within those of the sacrum averts this. The rare type of dislocation that sometimes occurs as a result of violent force or maldevelopment of vertebrae is called spondylolisthesis (from *spondylos,* a vertebra, and *olisthanein,* to slip). The rarity of this condition clearly indicates how well the mechanism of the spine is constructed.

The normal side-to-side curvatures (scoliosis) deviate to the left in right-handed people and vice versa. Shortness of one leg exaggerates scoliosis. Abnormal curvatures can result from the collapse of the bodies of vertebrae (kyphosis, or hunchback) or of the arches (lordosis).

Upright Posture and Man's Abdomen

Parts of the alimentary tract and many of its associated viscera are almost entirely enclosed within a peritoneal covering and retain contact with the abdominal walls only by double-layered mesenteries. Other parts, such as most of the duodenum and both kidneys, are attached to the back wall with only a covering of peritoneum in front of them. Most of the viscera are anchored by connective tissue, which, however, does not immobilize them completely but allows them to move up and down an inch or more during respiratory and postural movements. Those parts of the alimentary canal slung in the body cavity by peritoneal mesenteries are only partially held in place by them. Anyone unfortunate enough to have seen a large wound of the abdominal wall with viscera extruded knows that in the intact body mesenteries can be stretched.

The muscles and fibrous sheets (aponeuroses) that compose the abdominal wall are the chief viscera supporters. Musculature of the flank is arranged in three layers that blend into fibrous aponeuroses at the front. The three muscles are the external and internal oblique and transversus. Their fibers run in different directions, either obliquely or transversely, to form a living abdominal corset. The fibrous sheets from these muscles come together and blend into one, which then splits in front to form a sheath alongside the midline that encloses each of the two, long, vertically oriented, straplike muscles (the recti). The two rectus sheaths come together on the midline in a toughened fibrous band called the linea alba. The abdominal wall is therefore composed partly of muscle (mainly above the umbilicus) and partly of fibrous aponeuroses (mainly below the umbilicus and in midline). The function of both is to retain the viscera.

The lumbar curve ensures that the abdominal wall is so set in the erect posture that it lies somewhat beneath the lower abdomen. The outline of the front of the abdomen is not (-shaped but ⌊ -shaped. Thus the lowest part, attached to the rim of the pelvis, has to support the viscera directly. In middle age the lower abdominal wall often slackens and sags, and its retaining and supporting functions are further tested by increase in weight. The efficiency of the lower abdominal wall is also weakened by openings, e.g., the inguinal canal to the scrotum, which allow the passage of the spermatic cord into the scrotum. If the fibrous supports of the inguinal canal are weak, abdominal viscera can be forced down it into the scrotum to produce one of the types of inguinal hernia. Thus, an upright posture aggravates the weaknesses in man's structure.

Freedom of the Upper Limb

Bipedalism is associated with remarkable modifications in the legs and trunk. The front limbs, having been freed to carry out skilled movements, have retained considerable length. Each human arm and shoulder weighs ten pounds or more. To be free in its movements, the arm must not be rigidly attached to the body by bony struts. Rather, the suspensory mechanism of the shoulder region must be muscular. The most conspicuous muscle on the flattened, broad-shouldered back of man is the large trapezius muscle. Viewed from behind, a man's neck splays out at its root as if a monk's cowl of muscle lay just beneath his skin (the old name for the trapezius was *musculus cucullus,* a cowl). This appearance is derived from the upper fibers of the trapezius muscles that stretch from the back of the skull and neck to the shoulder girdle (Figure 6–14). Other muscles deeper in the neck assist them in their

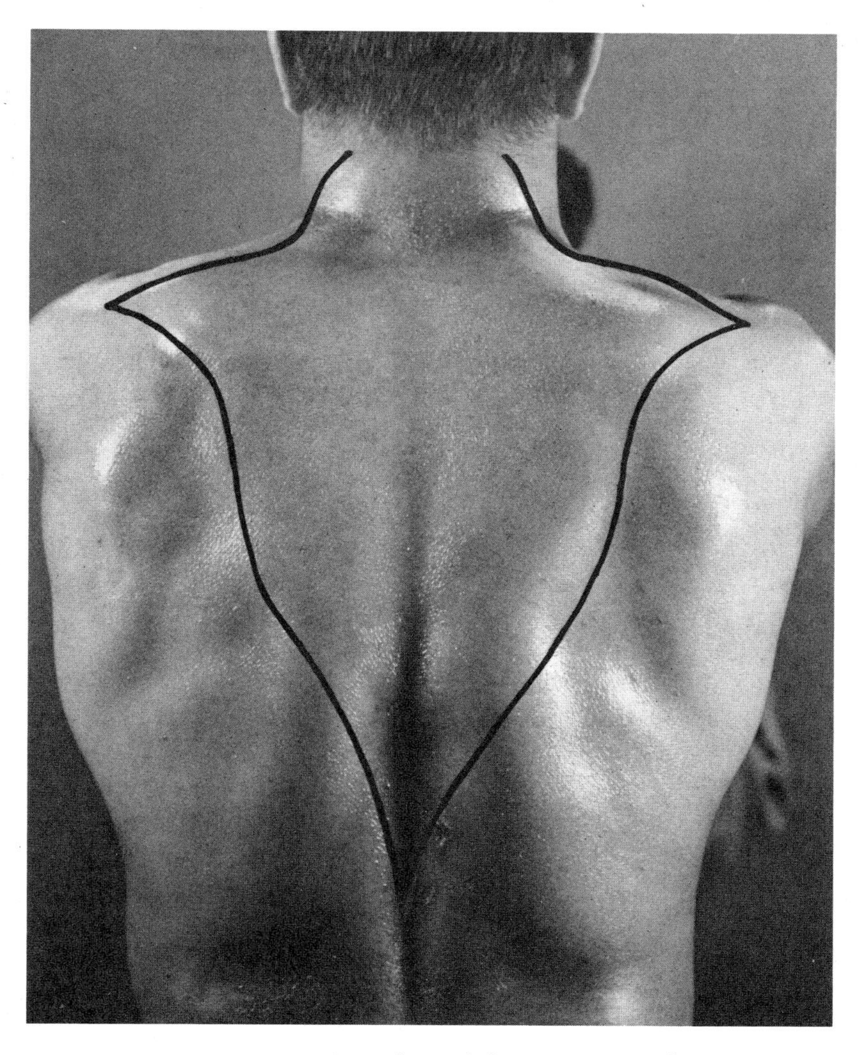

Figure 6–14. The outlines of the trapezius muscle.

postural action. All are constantly at work maintaining or fixing the position of the shoulders and the head. Although some have regarded these muscles as inadequate for their purpose, in general they operate efficiently.

Even though man's upper limb shows few anatomical specializations, the remarkable freedom and mobility of this shoulder joint and whole limb allow the hand to be brought into almost any position. The number of positions in which the hand can be placed is unique. Only one of the muscles that enable such freedom of positioning will be mentioned here. The serratus anterior arises from the side of the thoracic cage and is attached to the hind edge of the shoulder blade. Contraction of the muscle can cause a forward movement of the scapula around the thorax as in pointing, reaching, pushing, and punching. It can also rotate the scapula, as in reaching upward or at the back of the neck. Some men can touch the fingers of one hand when that arm is doubled up in the small of the back with the fingers of the other reaching down from behind the head.

If the arms and hands were used in locomotion, they would be too loosely jointed. If they had been better adapted for locomotion, as in the knuckle-walking chimpanzees and gorillas, they would have had to lose their mobility and agility.

Movements of the Hand

Above and elsewhere in this book we have noted that the human hand, notwithstanding its remarkable dexterity, has remained relatively generalized in structure. No other reference to its architecture will be made here but rather to its function.

Napier has classified the specific movements of the hand into two prehensile actions: a precision grip and a power grip (Figure 6–15). The nature of the object handled, however, and not the activity itself determines the kind of grip. Anthropoid apes, and occasionally man, employ a type of grip that Napier calls a "last resort," or hook grip. Apes use it habitually in brachiating (Figure 6–16) and man for hanging onto the edges of cliffs, or straps in a subway, or for carrying suitcases (Figure 6–17).

An inspection of the hands of other primates suggests interesting evolutionary trends. That of tree shrews is more like a foot, whose thumb, though more divergent, is hardly distinct from the rest of the fingers, all of which have keeled nails or claws and expanded touch pads. Perhaps these hands exemplify ancestral stages in the evolution of man's hands. The thumb of living lemurs is mobile but not opposable like that of man, which can oppose against the ventral surface of all

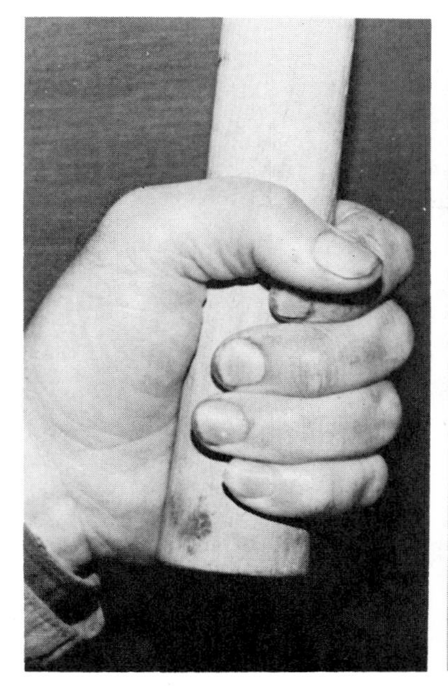
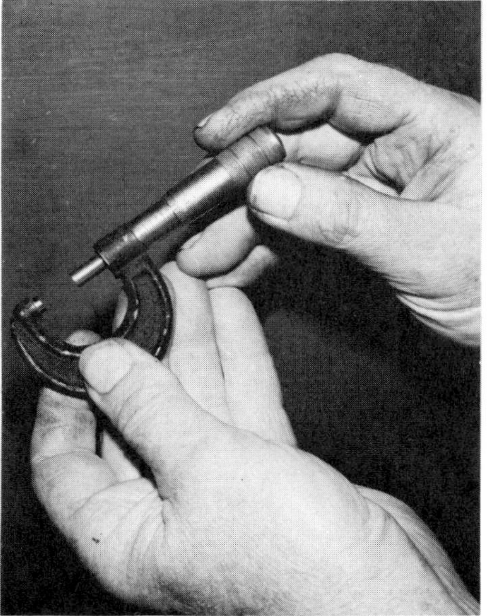

Figure 6–15. Power grip (left) and precision grip (right).

fingers. According to Napier, lemurs have a foot-hand endowed with power grip but no precision grip. In the hand of baboons, which have poorly developed precision grip, functional emphasis is shifted further to the thumb (thenar) side. The opposable thumb of these animals is short and slender.

Perhaps power grip, which all primates have, was the first to appear in human evolution. Human infants start out with only power grip and gradually develop a precision grip as they learn to play ball, write, and manipulate tools and objects. From these observations and evidence obtained from artifacts, Napier also suggests that *Homo habilis,* a fossil form discovered by anthropologists Dr. and Mrs. L. S. B. Leakey at Olduvai Gorge, Tanzania in 1960, had a well-developed power grip. Some authorities believe that this creature may be the earliest known member of the genus *Homo.* It had a short thumb with a stout, broad, terminal phalanx, unlike that of any known ape or monkey, and its hand was capable of a precision grip. Stone tools, called Oldowan chopper tools, were found together with the fossil remains of *H. habilis,* an indication that he was a toolmaker. Napier was able to make tools similar to those of *H. habilis,* who used only his power grip, and concluded that the hand of early man may not have yet evolved a precision grip. Perhaps by the time the Aurignacian culture developed in Haute-Garonne, France during the upper Paleolithic epoch,

Figure 6–16. A male white-handed gibbon *(Hylobates lar)*. Note the hook grip.

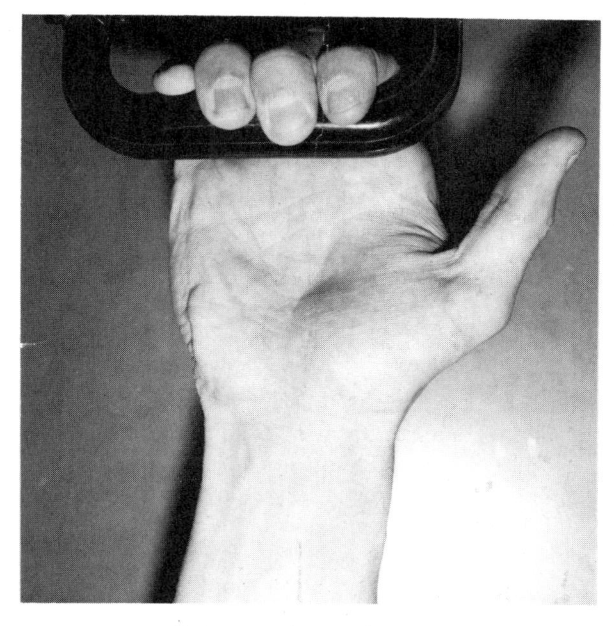

Figure 6–17. Hook grip.

the human hand had evolved into a model that was essentially modern in design and function.

Speculations

How some quadrupeds gradually evolved into erect, bipedal animals and what adaptations were involved over what length of time are still among the major perplexing questions asked by anthropologists. Some maintain that our ancestors came down from the trees more than a million years ago, went through a more or less quadrupedal phase of locomotion, and then, through modifications in pelvic structure, gradually adopted bipedal running and the emancipation of the hands for other than locomotive purposes.

Others consider that man's ancestors could not have lived in the conditions of the open grasslands until they had become fully bipedal, could run, and could carry objects such as food and weapons. It would seem to have been a critical period in human evolution. The phenomenal increase in brain size probably occurred soon after or concurrently with the change to bipedal posture.

We still have no detailed knowledge of the successive stages through which man passed before he walked upright with ease. Each of these stages may have had its own tree of ancestors, but, as R. L. Stevenson says, "at the top of all sits Probably Arboreal." The Miocene and Pliocene apes were very likely arboreal, but perhaps they progressed effectively on all fours along the ground. *Proconsul major* of the Early Miocene was about the size of a gorilla. If arboreal, it could have moved safely only in the lower branches; for this and other reasons several authorities believe that it must have been mainly terrestrial. The evolution to an upright posture can be associated with the disappearance of thick forests and their replacement with small woods separated by tracts of open scrub or savanna. The examination of fossil insects, fruits, and seeds in the Miocene strata of Kenya by L. S. B. Leakey suggests that such a terrain existed there. The morphology of fossil hipbones and some remnants of the lower limb skeleton of the Australopithecinae suggest that they adapted to bipedal posture and locomotion, but not so completely as man. Paradoxically, the effort to retain an arboreal way of life may have resulted in the appearance of upright posture, which would enable a more rapid and effective passage across ground between two woods. Although others disagree, we believe that Africa, sometime during the Miocene age, may have been one of the places where man's ancestors stood and walked upright.

7

MAN'S SKIN

The skin aids man in "perceiving" his environment and performs or initiates many of the essential adaptive adjustments resulting from this perception.
(A. M. Kligman)

The most important function of skin is to protect the organism from its environment and at the same time to maintain its contact with that environment. Although important, all other roles are secondary. The seemingly innumerable differences in the structure and function of the skin of various mammals become intelligible and certain basic patterns begin to emerge only if we keep this functional hierarchy in mind. The structural features of skin that is covered by a heavy coat of hair are very different from those of skin with scant or no pelage. For example, heavily haired skin does not have much tactile sensibility, and even if it did, the hair would shield the epidermis from contact with external stimuli. Animals with highly sensitive skin must not only receive and translate impulses but also respond to them. An acute cutaneous sensibility places the burden for action on the brain, which in most large animals is probably not large enough to accommodate it. Therefore, a heavy pelage allows only limited cutaneous sensitivity, and the more naked an animal, the more sensitive its skin is likely to be. But there are exceptions, as we shall see.

The surface cutaneous area of large, bulky animals such as whales and elephants is so great that if it were endowed with nerve receptors comparable in number to those in man, the animals would have to have brains many times their present size to respond to the countless stimuli that bombard the skin and are transmitted to the brain. But such superlative cutaneous sensibility could serve little purpose in whales and elephants, rhinos and hippopotamuses, whose size and corporal rigidity militate against their ability to alleviate skin irritations and render them relatively helpless to rid themselves of them. Feral rhinoceroses, for example, have perennially scarred skins with open ulcerating wounds; neither the wounds nor the tick birds that keep them raw by constant picking seem to bother the rhinos much, since, like elephants, they are literally encased in a nearly impenetrable cuirass and have, we must assume, high thresholds of cutaneous sensibility.

Such considerations are pertinent if we are to understand the relevance of structure to function in man or in any other animal (see chapter 4). Basic similarities as well as profound differences characterize the structure and function of skin in the major classes of vertebrates. In every order, suborder, or even closely related species of mammals, skin is often so uniquely adapted that it can only be discussed in broad generalizations. Some of its numerous functions will be discussed here.

Skin primarily protects the body from injury, preventing body fluids from escaping and external fluids from penetrating; in different degrees it also puts animals in touch with their environment. In birds, bats, and aquatic mammals, it enhances or forms outright locomotory devices; its glands secrete substances that attract or repel. Pelage and other cutaneous structures can camouflage or attract attention, and many serve the organism in its social or sexual relationships.

The types and numbers of cutaneous structures in different vertebrates are nearly limitless. Fish have scales and fins, spines, membranes, glands, and other structures; amphibians are equipped with many different glands about which very little is known. When snakes and lizards shed their outer horny layer intact, its protective role is assumed by another, already keratinized, reserve layer below. The skin of the turtle and of the crocodile manufactures bony plates covered by various keratinous structures, many of which consist of tough β keratin. Feathers, cutaneous structures with an architecture of incomparable complexity and beauty, maintain the homeostasis of birds in hot or cold environments, on the earth, in burrows, in water, or in the air. Feathers are also ornamental, sensory mechanisms, devices of communication, and the principal means of locomotion; without them, birds could not fly, swim, or survive the environment. Birds are also equipped with another cutaneous structure, a keratinous beak; their feet are covered with horny plates, and their toes terminate in claws of various sizes and shapes. Feathers, beaks, scales, and claws—all are uniquely fashioned to enable birds to adapt to their environment and way of life.

Among the vertebrates, only mammals have hair, which varies in density, shape, thickness and length, color, patterns of growth, and texture. In addition, mammals have spines, spurs, scales, horns, hooves, claws, and nails. The number and variety of mammalian skin glands have recently come under increasing study and investigation. The significance to the organism of the substances secreted by these glands—watery or viscid, fatty or proteinous, colorless or pigmented, odorless or fetid—their composition, and the mechanisms that control their secretion—all are for the most part still unknown. However, the rising interest in the role of pheromones (see chapter 14) in communication will doubtless result in a better understanding of their role in the well-being and survival of the species.

What immediately distinguishes man from all other large land mammals is his apparent lack of body hair. Regardless of individual or racial differences, man's body has little visible hair, and it has become fashionable to refer to him as a naked ape, although he is neither naked nor an ape. The hair over most of his skin is vestigial and practically invisible, but on men's faces and the axillae, mons, and scalps of both sexes, it grows profusely (Figures 1–7, 1–8). We suggest that these are epigamic (Gk. *epigamos,* marriageable) areas, since the hair they bear has no other purpose than that of visual sexual stimulation. Deprived of an adequate coverage of hair, human skin has developed adaptive structural changes that give it strength and resilience.

The characteristic features of man's skin change from the time of birth to old age. In infants and children, it is velvety, dry, soft, and largely free of wrinkles and blemishes. At adolescence, the hair grows thicker and becomes more pigmented, particularly in the areas mentioned above. General skin pigmentation increases, and the dreaded acne lesions often develop; hair growth, sweating, and sebaceous secretion gain momentum in adolescence and reach their fullest development in adult individuals. With aging, anatomical and physiological alterations, as well as exposures to sunlight and wind, all leave indelible marks, particularly on skin not protected by clothing. The dry, wrinkled, flaccid skin of old people, which has suffered many abuses, is the relic of a once-fine organ system.

Anatomically and physiologically, human skin, more than that of any other mammal, exhibits many gross and subtle topographic differences; note, for example, how the skin on the palms differs from that on the backs of the hands and fingers. The skin surfaces of the chest, pubic areas, scalp, axillae, abdomen, soles, and ends of the fingers of any one individual are as different structurally and functionally from each other as if they belonged to different animals.

Thicker than the skin of many other mammals, man's skin is also thicker on the dorsal than on the ventral parts. Skin elasticity is an individual property, largely dictated by regional, racial, and genetic factors; taut in some people, it is easily stretched in others. It is elastic enough in all to accommodate increases and decreases in the size of the body it encases and to allow the many subtle movements that characterize human activity. If one cuts out a piece of skin, the wound gapes and the piece shrinks. If one pinches intact skin, it snaps back when released. It is greatly distended over the abdomen during pregnancy or in excessively fat people. When these conditions terminate, the skin returns to normal but may bear the scars of permanent damage (Figure 7–1). Skin expertly repairs the perennial minor injuries it suffers and, given an opportunity, repairs major injuries almost equally well. Since its surface is constantly shedding, it has to constantly renew itself. Its numerous nerves record many modalities of sensibility—touch, pain, and temperature—and skin is easily the largest sensory organ of the body. Finally, since it is marked differently in every individual and endowed with subtle and gross epigamic adaptations, skin is the most important organ of personal identification and sexual attraction.

The architecture of skin is so fashioned that it attains maximal strength, pliability, and protection from the environment (Figure 7–2).

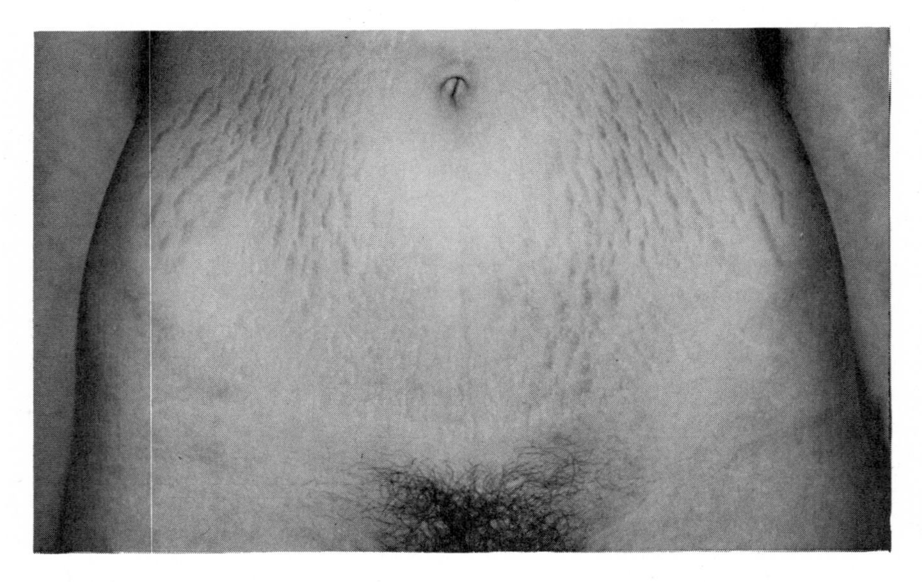

Figure 7–1. Permanent scars (stretch marks, or striae gravidarum) on the abdomen of a 23-year-old woman.

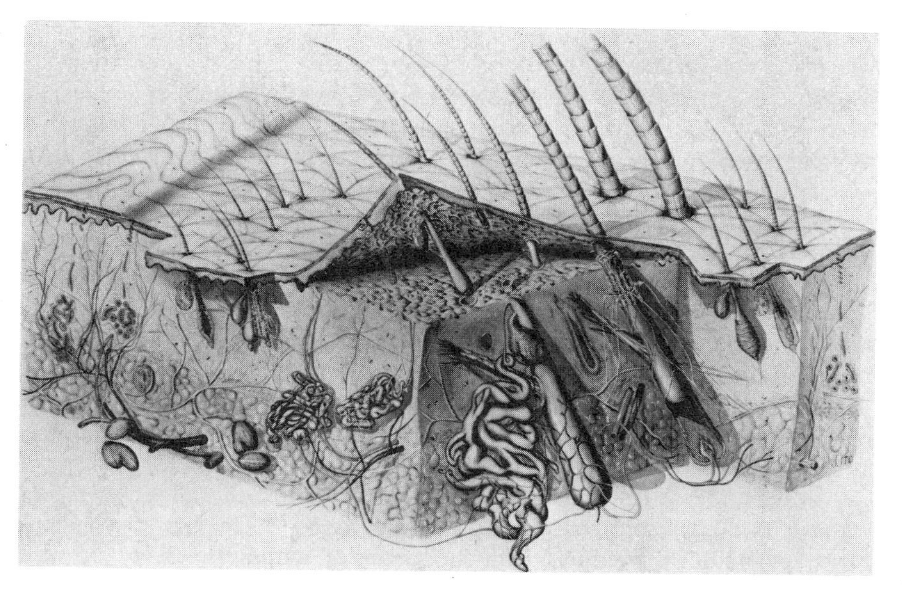

Figure 7–2. Diagram of human skin showing the various details referred to in the text.

Each of its numerous layers is oriented to complement the others structurally and functionally. Countless nerves, some modified as specialized receptor end-organs and others more or less structureless, come as close as possible to the surface layer to communicate with the environment, and nearly every skin organ is enwrapped by skeins of fine sensory nerves. In most areas skin is adapted for light touch sensibility rather than deep pressure.

Human skin is an apparently tangled and intricate network of arteries, veins, and capillaries. The consequent supply of blood, far in excess of the maximum biological needs of the skin itself, is evidence that man's skin is at the service of the blood vascular system, functioning as a cooling device. To aid in this function, sweat glands pour water upon its surface, the evaporation of which absorbs heat from the skin. Thus, when the environment is warm, the flow of blood is greater, and when great quantities of heat generated by muscle action must be dissipated, the blood flow through the skin is maximal. On the other hand, if the environment is cold and body heat must be conserved, the cutaneous blood vessels contract and allow only a small amount of blood to flow through. Besides controlling body temperature, skin also helps to regulate blood pressure by opening and closing certain sphincterlike vessels.

The lymph system of skin is almost as intricate as its vascular system. In the more superficial parts of the dermis, minute lymph vessels, which appear to terminate in blind sacs, are affluents of a superficial lymphatic net, which, in turn, opens into vessels that become progressively larger in the deeper portions of the dermis. Since the walls of these lymph vessels are very thin, the circulation of lymph is sluggish and controlled mostly by pressure, skeletal muscle action, and heat. Any external pressure interferes with its flow. Skin is so important in the immunological responses of the body that its lymphatic drainage is even more significant than its blood vascular system.

Grossly, skin is comparable to plywood, which is much stronger than a single board of the same thickness. Each of its various layers has its own properties. There are two principal layers: the superficial, entirely cellular *epidermis,* and the deep fibrous *dermis.* The former is composed of four or five variably distinct layers and is relatively thinner than the dermis which is much thicker and composed of two or more layers.

Endowed with remarkable lability, skin undergoes profound changes in response to stress. A restless tissue, it renews itself constantly, the rate depending upon factors from within the body as well as from without. It is almost constantly sustaining some more or less severe injury, which is repaired with dispatch. An English biologist, A. E. Needham, estimates that if a man sustained one minor wound a week, in a lifetime of 70 years he would have received 3,500 of them. If these wounds did not heal expeditiously, the consequent stress, loss of body fluid, or infection would cause his demise long before the end of his reproductive age. But, in spite of its restlessness and apparent fragility, skin is an unusually durable and sturdy organ, able to withstand and adjust to an infinite number of circumstances and abuses.

The characteristic features of the skin change constantly during a person's lifetime. Being a long-lived animal, man has prolonged periods of childhood, adolescence, adulthood, and finally old age and senescence. This means that from its formation *in utero* to senility his skin progressively undergoes uninterrupted changes. Sexual dimorphic differences are also apparent; the cheek of a woman, basically similar to that of a man, is qualitatively different. Furthermore, even a cursory appraisal of the characteristics of normal skin indicates numerous racial and individual differences.

Early in uterine life, skin begins to acquire distinct properties that protect it from the amniotic fluid. Its various suborgans—hair follicles, sebaceous glands, sweat glands—develop at different times. Because the head forms first, the growth sequences occurring in it are considerably ahead of those of the rest of the body. Hair follicles, for example, are first discerned on the scalp and face at about 3 months of fetal age and do not evolve over the trunk and limbs until about 4½ to 5 months and even later. Sweat glands in the hairy skin develop last.

In the palms and soles and in the axillae, they are formed about the same time as the hair follicles on the head. Nails, hair follicles, and sweat glands first appear during embryonic life and are not formed after late fetal life. Thus, at birth an individual has as many pores (orifices of glands) and hair follicles on the surface of his skin as he will ever have. The belief that new hairs are formed on certain body surfaces as one matures or becomes old is unfounded. The primordia that produce these "new" hairs are already present at birth. The dry skin of infants and children is regarded as the epitome of skin beauty because it is free of blemishes and superficially similar to the skin of young women. Children less than two years old sweat poorly and irregularly, and their sebaceous glands are small and nearly functionless.

It is natural that an organ constantly exposed to the vagaries of both external and internal environments should often be beset with problems. Thus, skin is afflicted with minor, often nameless, disorders most of which are repaired in due course without aids. Many of the subtle functions of skin and its varied responses to stress and disease still remain inexplicable.

The Surface of the Skin

The skin over the human body is pitted by the orifices of sweat glands and hair follicles and furrowed by intersecting lines that delineate geometric patterns in relatively smooth surfaces and form pebbled and cobblestone patterns over joints (Figures 7–3, 7–4). The designs formed by these lines in any one part of the body are roughly similar, but not identical, in all individuals. The lines are oriented to indicate the general direction of elastic tension. Countless numbers of them, deep and shallow, together with the pores give each area of the body a characteristic topography, which, like the deeper furrows and ridges on the palms and soles, is pretty well established before birth. The accompanying photographs give a general idea of the precision of these patterns. The fine details of each body surface are peculiar to every individual, but because those of the fingers have a high relief with more precise patternings and are easily reproduced, fingerprints are customarily used for identification.

Not all the lines on the skin surface are congenital; some are acquired after birth as a result of use or damage. The furrows on the forehead and face and those from the wings of the nose to the corners of the mouth are acquired from years of wrinkling the brow or from certain fixed facial expressions such as smiling and squinting in bright light or against the wind. These and other similar lines accentuate or

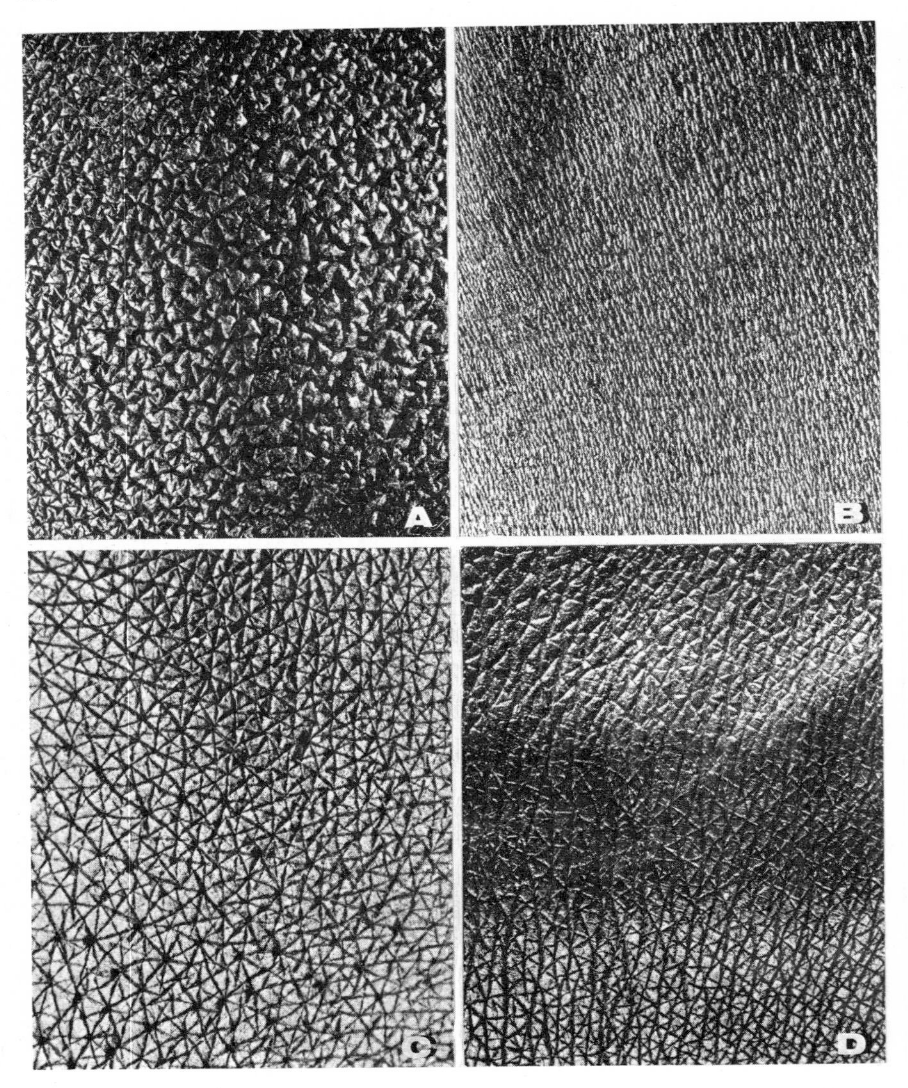

Figure 7–3. The surface of living human skin from the same subject, enlarged to show details of natural creases. A: elbow; B: antecubital fossa; C: back of hand; D: wrist. These markings were barely visible with the unaided eye.

exaggerate preexisting congenital lines and become more and more strongly delineated as the person ages. The direct pull of muscles on the skin, such as on the face and scrotum, result in the formation of permanent wrinkles. Certain occupations also leave skin marks which, depending upon their duration and severity, are transient or permanent.

Lines, furrows, and wrinkles, whether congenital or acquired, are

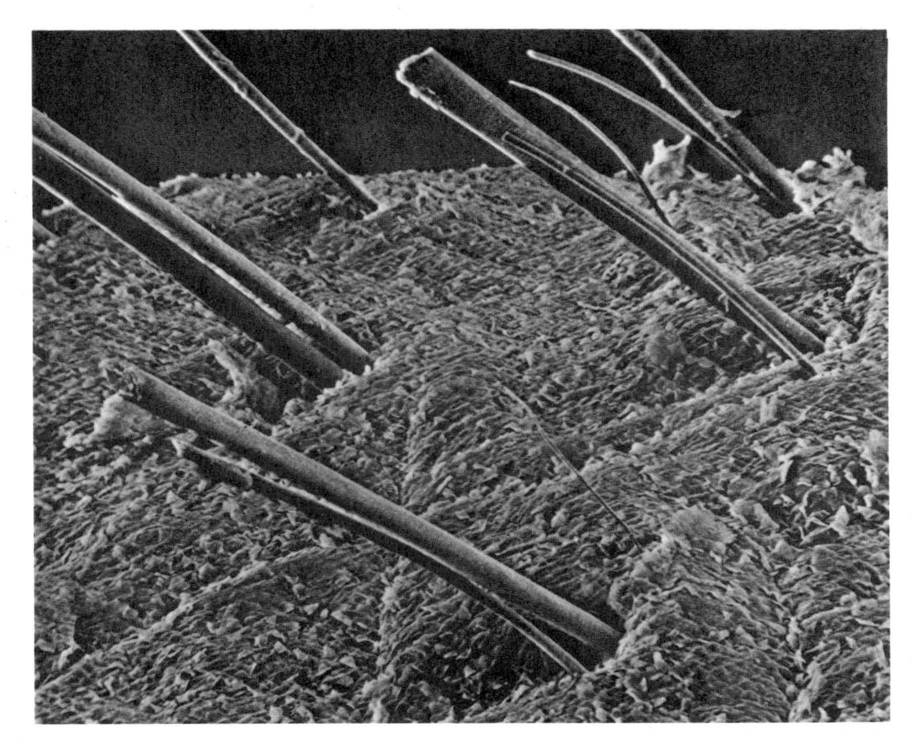

Figure 7–4. Scanning electron micrograph of a normal human scalp showing the different-sized hairs emerging from the hair canals. Note particularly the very fine vellous hairs. The scalelike structures are dead horny cells. (Magnification 135 X.)

influenced by the thickness of the epidermis and dermis, the particular orientation of the fibers in the dermis, the thickness of the underlying fat, muscle pull, and external factors. The best-known and best-developed markings on the skin are those on the ventral surfaces of the hands and feet. These are etched by distinct alternating ridges and grooves, which together constitute the dermatoglyphics. Close inspection of a fingerprint or of one's own finger or palm under a magnifying lens shows that the ridges are pockmarked at spaced intervals by small conical craters, the orifices of sweat glands. The ridges pursue a variable course, but their arrangements in specific areas follow a consistent structural plan. On the pulp of the farthest segment of the fingers, a specific configuration is always patterned on one of the three basic types—whorl, loop, and arch—or on a combination of them (Figure 7–5). Although apparently continuous, the ridges have many interruptions; they branch, vary in length, and have many other irregularities that give character and individuality to the prints.

Although the general configuration pattern of the ridges can be

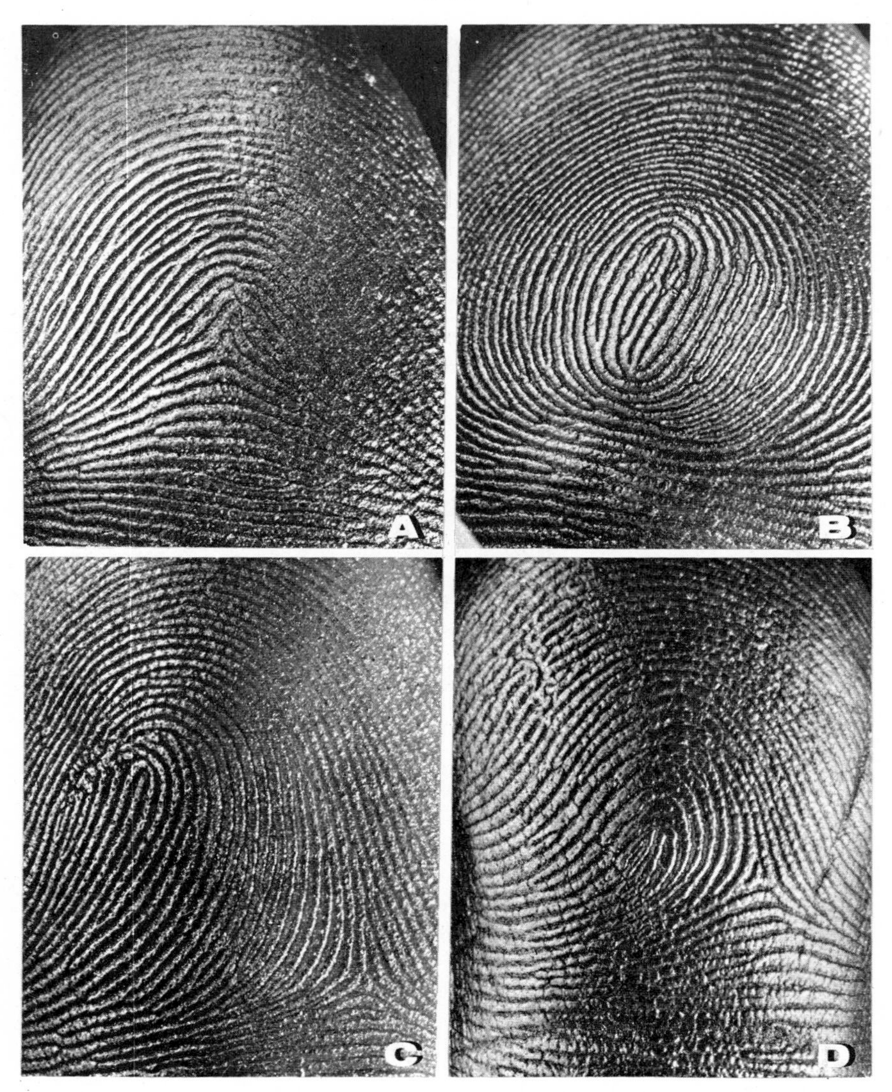

Figure 7–5. Dermatoglyphic patterns in the digits. A: arch; B: whorl; C: loop; D: combined form.

grouped according to the characters they have in common, some ridge details are not matched anywhere in the same individual or in any other individual. This infallible "signature" makes dermatoglyphics the most reliable criterion for identification. Although similar in identical twins, the details are never exactly alike. The development of these patterns is genetically controlled and established during the third and fourth fetal months. Major differences exist between the markings of the right and the left hand of the same person, and women have fewer whorls and more arches than men. The finger markings of some mentally deficient, epileptic, and insane persons apparently show consistent deviations from normal trends. These abnormalities may indicate that the factors responsible for, or predisposing these individuals to, these conditions must have been at work during the third and fourth fetal months to have so influenced their development.

Dermatoglyphics also have the obvious purpose of giving the hand or foot a better grip and of enhancing their tactile sensitivity. If the surfaces of the hands and feet were perfectly smooth, much of their effectiveness would be lost.

The Epidermis

The entirely cellular epidermis on the surface is relatively thin and is itself veneered, being composed of four or five variably distinct layers.

The epidermis is thicker on the palms and soles than it is anywhere else. If the fine details are omitted, it can be described as divisible into a lower layer of living cells and a superficial layer of compact dead cells (Figure 7–6). All of these cells, living or dead, are attached to one another by a series of specialized surfaces called attachment plaques, or desmosomes. Thus, instead of being completely fused, the cell membranes of adjacent cells make a zipperlike contact, with fluid-filled spaces between the contact areas. This structural pattern insures a concatenation, or linking together, of the cells to one another so that they cannot be easily sloughed off. Epidermal cells, which generally multiply at the basal layer in contact with the dermis, gradually ascend to the surface, manufacturing keratin as they go. Once they reach the upper part of the epidermis, the cells die.

The human body and that of most other vertebrates is completely encased in a thin layer of flattened dead cells called the horny layer. Although it is difficult to imagine, the velvety, supple skin of young persons is likewise covered by such a dead casing. When you scratch your own skin with your nails, you scrape off tiny but visible opaque flakes of the horny layer.

Human epidermis has some unique features. Peeled off intact from the dermis and viewed underside up, it reveals an intricate structure of branching ridges and valleys, columns and pits that establish an intimate interdigitation with the subjacent layer of the dermis (Figure 7–7). Striking regional differences are found in the architectural patterns of this structure, varying from almost a total absence in the eyelids and certain

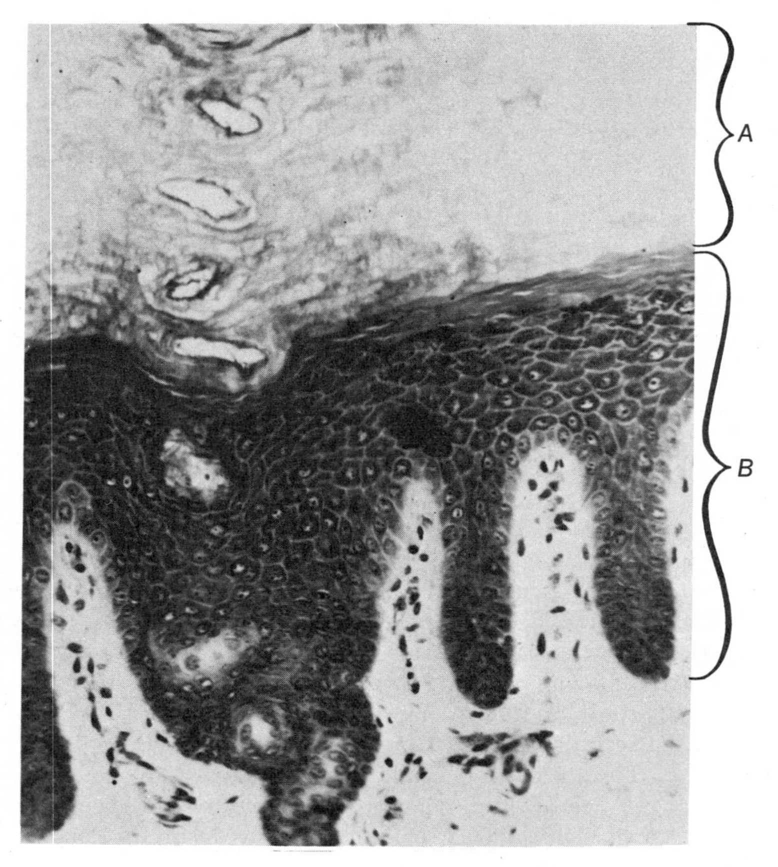

Figure 7–6. Photomicrograph of the epidermis from the palm. A: the lower third of the dead corneal layer; B: the living epidermis. This figure also shows the winding pattern of the duct of a sweat gland.

parts of the face to a very elaborate sculpture in the fingers and nipples. When viewed under the scanning electron microscope, the underside is even more complex (Figure 7–8). The ridges are generally more complex on the dorsal than on the ventral areas of the body. The thinner the epidermis and the denser the hair, the simpler is the pattern of the epidermal ridges. They are multiform in regions of much wear and tear, such as the friction surfaces of the foot and hand and the nipple and areola of the breast (Figure 7–7). Except in the scalp, which has an abundant hair population and elaborate epidermal ridges, the richness of hair cover and the complexity of the underside of the epidermis are in inverse relation.

During a lifetime, profound alterations occur in the epidermal ridges. In the scalp of the newborn, for example, the epidermal ridges are

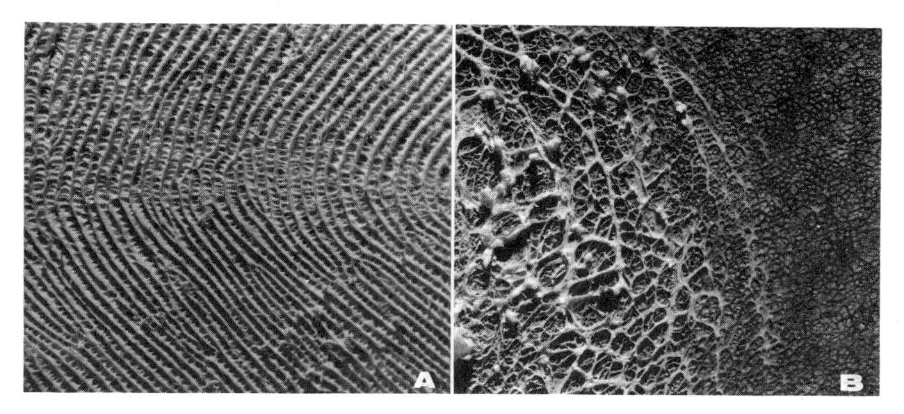

Figure 7–7. Photomicrograph of the undersurface of intact epidermis peeled off the underlying dermis. A: thumb of a six-year-old girl; B: nipple and areola of a woman's breast.

more or less rectilinear around the hair follicles. In adults they form a characteristic complex framework around the follicles and in old people the underside of the epidermis generally flattens out. Because of the variations in the number and depth of these ridges, the thickness of the epidermis cannot be measured exactly.

Each of the two distinct layers of the epidermis is in turn layered: a lower malpighian layer growing upon the dermis and the dead, horny layer on the surface. The lowest cells of the malpighian layer compose

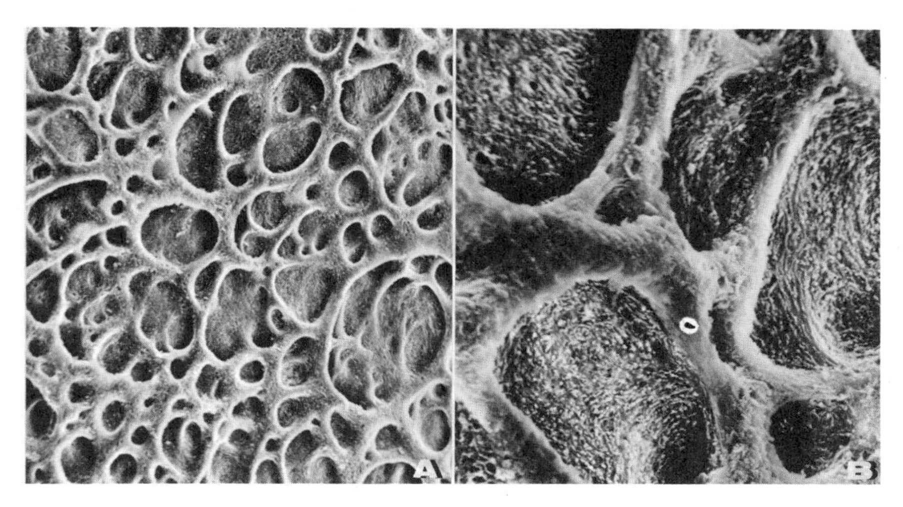

Figure 7–8. Scanning electron micrograph of the undersurface of the epidermis of the areola in Figure 7–7. A: magnification about 50 X; B: magnification about 1,000 X.

the basal layer, above which lies the so-called spinous layer. In the upper part of this layer, the cells gradually blend into a granular layer, the cells of which contain angular particles called keratohyalin granules. In the friction surfaces, the backs of the hands, and a few other surfaces, a conspicuous hyalinized layer, called the stratum lucidum, is apparent.

Unless epidermal cells reproduce themselves constantly, they are depleted. Cell division (mitosis) occurs chiefly in the basal layer, but in very thick epidermis it occurs higher up as well. Epidermal mitosis is rhythmic and cyclic, about twenty-seven days being required for epidermal cells to travel from the basal layer to the surface. Under certain conditions such as injury, less time may be involved, but when this happens the outer horny layer is generally faulty.

Situated between external and internal environments, the epidermis is sensitive to the changes of both. Such sensitivity is largely guided by the highly developed enzyme system within the cells. Having no intrinsic vascular supply, this system depends for its integrity upon its relation to the dermis, from which nutritive and other essential substances seep through the intercellular spaces of the epidermis to guide its growth and differentiation.

Since man forfeited most of the protection afforded by a hairy coat, he has developed many protective adaptations against his environment. Singular among them is the horny layer, which, though remaining thin, somewhat transparent, and flexible, has remarkable biological sturdiness. It is smooth and supple over the protected areas and joints and rough and stiff over the exposed areas. It is thicker on the back than on the ventral parts of the trunk and very thick and rugged over the friction surfaces. Although friction, abrasion, and chemical agents constantly wear away the horny layer, the lost cells are steadily replaced from below. When injured, the epidermis is replaced more rapidly than usual.

The flattened dead cells of the horny layer originate in the deeper layers of the epidermis. As the cells ascend to the surface, they manufacture keratin, a relatively inert, fibrous protein that remains inside the cells. Death, the fate of every epidermal cell, is orderly, each cell adding itself to the effective, dead, outer barrier. As the horny layer becomes desiccated at the surface, the cells exposed to air become less cohesive and are shed. The cells farther down are progressively more closely bound, forming a gradient that is less stable at the surface than below.

Although the horny layer is an effective barrier against the penetration of injurious substances, it is not completely impervious. It is a bulwark against most substances, but many do get through. The importance of this barrier is rarely appreciated until some noxious substance penetrates the epidermis. Only infinitesimally small amounts of some agents need to pass through to be injurious. Trace amounts of the two most common sensitizers of man, metallic nickel and the oleoresins of poison ivy, cause distress in hypersensitive skin. Even a casual brushing of the skin against the leaves of poison ivy can bring about severe dermatitis. It seems almost incredible that alloys containing nickel could discharge enough molecules of this metal on contact with the skin to penetrate the outer dead horny layer and cause the sensitive living tissues to react violently.

Many of the adaptive modifications in human skin are incomplete. A thicker and more impervious horny layer, such as that of the elephant and rhinoceros, might have been more efficient. Encased in such armature, however, skin would have lost its suppleness and much of its use in thermoregulation. To a body that must remain agile, such a cuirass would have been an impediment rather than an asset.

The Color of Skin

The pigmentation of the skin is designed primarily to protect it from the injurious rays of the sun. Melanin (Gk. *melas,* black), a yellow-to-black pigment, is the skin's major source of color and has the property of absorbing ultraviolet light that is noxious to living cells. The skin of some human races inhabiting the torrid regions of the world is more heavily pigmented than that of most people living in temperate zones.

Melanin pigment is formed by spiderlike cells, the melanocytes, which are found primarily between the basal cells of the epidermis (Figure 7–9). From the cell body of melanocytes, many long cytoplasmic processes called dendrites extend for some distance, worming their way between the epidermal cells. Melanocytes produce pigment granules and inject them into the surrounding epidermal cells where they arrange themselves like a protective shroud above the nucleus. Apparently each melanocyte has a more or less definite number of "satellite" epidermal cells into which it injects melanin. The melanocyte and its satellites constitute a "constellation." Regardless of race, differences in skin color result not so much from the number of melanocytes as from the amount of pigment manufactured by the melanocytes and from the depth of this pigment from the surface. The darkest Negro has about as many melanocytes in his skin as does the blondest Caucasian. In Negro skin, however, the melanocytes are genetically far more active and responsive to the stimulation of ultraviolet light. The action of light upon the melanocytes of the skin is especially evident when it is exposed to the sun. The resulting tan represents an accelerated production of melanin by the melanocytes. An increase in peripheral blood circulation also enhances the tan, e.g., after a hot shower. This is the result of the peripheral vasodilatory effect on the darkening of the skin.

Not all the body areas have an equal number of melanocytes;

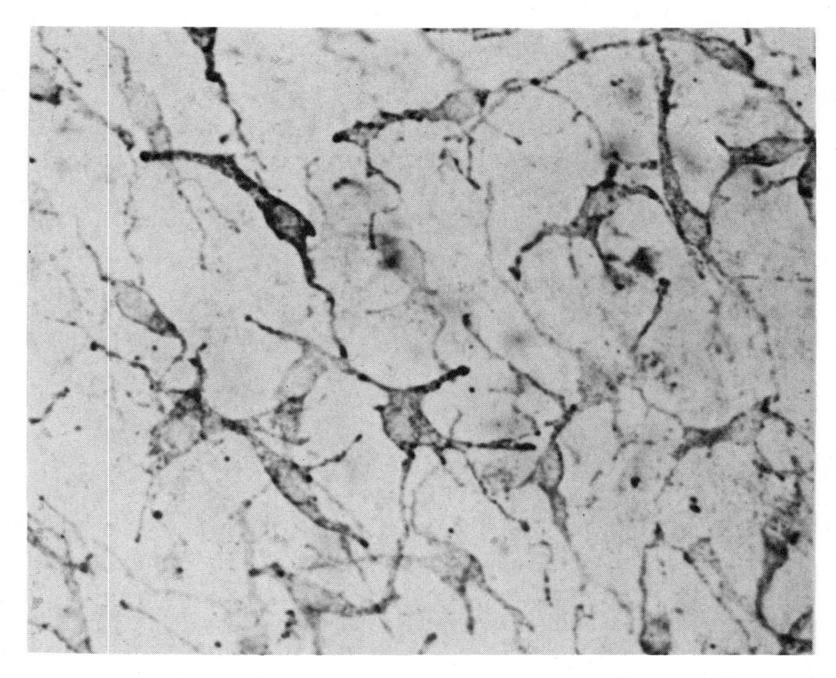

Figure 7–9. Photomicrograph of melanocytes at the base of the epidermis, seen from the underside of epidermis that has been peeled off the dermis.

there may be as few as 1,000 per square millimeter in the thighs and arms and as many as 4,000 in the face and neck. Assuming that all melanocytes have equal potentiality to form melanin, this should mean that the face tans more and sooner. Such an assumption, however, cannot be made without reservation.

Everyone is familiar with the partial or total failure of melanocyte activity that results in different degrees of albinism, either all over the body or localized in certain spots, as in piebaldism. Although albino skin usually has a normal complement of melanocytes in the epidermis, they are defective, lacking the major enzyme tyrosinase, which converts tyrosine to melanin. In a common skin disorder known as vitiligo, the melanocytes in certain areas of the skin gradually become permanently inactive and the areas are left a white or pink color. It is well to remember that even under the most normal conditions melanocytes may be active, partially active, or inactive.

Vast differences in skin color are found among nonhuman primates. Many of those with a heavy fur have no pigment at all except in the hair and hair follicles. In some the epidermis is heavily pigmented and the dermis is practically nonpigmented whereas in others just the opposite is true. In no case, however, does the epidermis lack melanocytes, even though these may be completely inactive.

The Dermis

A variably thick, fibrous dermis nourishes the epidermis and all the cutaneous organs and at least partially controls their growth and differentiation. It gives substance to the skin, the epithelial elements being a relatively small part of it. Pinching the skin over the various parts of the body shows it to be thicker in some areas than in others, more elastic, firmer, or looser. These differences are almost entirely due to the dermis. Leather is nothing more than a tanned dermis with the epidermis shaved off. Actually, human skin, with its relatively thick dermis, makes quite adequate leather.

The dermis is thicker in men than in women and on the dorsal rather than on the ventral side of the same individual. Quite thick on the nape and upper back of both men and women and on the palms and soles, it is naturally thinnest over and around joints and in areas of great mobility. Since all the vascular and nerve supplies of the skin are embedded in the dermis, it is literally a bed of tissue designed to hold and maintain these vascular and nerve mechanisms (Figure 7–2). Everywhere in the dermis blood vessels and both sensory and effector nerves abound.

The dermis is composed of two layers. The superficial, or papillary, layer is the most distinctive feature of human dermis. The understructure of the overlying epidermis grows upon this layer, which forms a negative image of that understructure. This papillary layer is rich in mucopolysaccharides, blood vessels, nerves, and very delicate connective tissue fibers. Beneath it, much thicker and coarser, lies the reticular layer, composed largely of a mesh of coarse collagenous and elastic fibers, the large twigs of cutaneous vessels, and the preterminal branches of the nerves. In certain parts of the body, such as the scrotum and around the nipples, under certain physiological conditions, muscle fibers are capable of contracting and causing a wrinkling of the skin.

Although the connective tissue fibers in the dermis seem to be heaped together without any particular order, they are actually arranged in horizontal layers with the fibers running alternately at right angles to each other. This pattern is somewhat clearer in the skin of infants than in that of adults. Some fibers are oriented perpendicular to the surface of the skin and help to anchor it to the connective tissue around the underlying structures. Collagenous fibers tend to be smaller in diameter and denser where the skin folds and creases or where it is tightly bound to underlying structures. The dermis is particularly adherent to the cartilages of the nose and ears.

Human dermis usually has no pigment-containing cells except in isolated spots such as moles; most of the pigment is found in the epidermis and hair follicles. In other primates, however, a few to numerous pigment-containing cells are found either in the papillary or in the reticular layer. In some monkeys, the dermis is sometimes so rich in pigment-containing cells that it appears to be abnormal.

The Hairy Coat

The quantity and quality of the hairy coat, which vary with different animals, probably reflect adaptations to the environment, though at times the adaptive function seems obscure. The heavy fur of animals in cold regions obviously protects them from exposure and prevents the loss of body heat. But can the luxuriant pelage of mammals that inhabit the torrid areas of the earth be explained in terms of a similar adaptation? Some marine mammals, like the whale and walrus, are naked; others, such as the seal and sea otter, have a rich fur. Adaptation to environment, then, is not the only explanation for the presence or absence of hair.

Hair is only one element in the adaptive patterns of the structure and function of skin. The total thickness of the skin, the type of horny layer on its surface, the amount and quality of glandular secretion, the quantity of stored subcutaneous fat, and the abundance of vascularity in the skin proper—all help an animal to adapt to specific environmental conditions. The heavy coat of hair that prevents heat loss in animals that inhabit a cold environment also protects the skin from the torrid sun and other hazards in animals that live in the tropics. Pachyderms—elephants and rhinoceroses—whose pelage is sparse, have an excessively thick and hardened skin. Man, a nearly naked animal, has a skin that is evidently adapted to a temperate or tropical environment; he surely would not have survived in cold climates without clothing and shelter.

The most outstanding superficial characteristic of man is his apparent nakedness. To be accurate, man is not hairless; even those surfaces of his body that appear to be naked—e.g., the forehead, the inner surfaces of his thighs, and parts of his abdomen—have hairs similar to those of the obviously furry surfaces, except that they are minuscule and colorless and therefore hard to see. The number of hairs on man's body, then, is not appreciably different from that of his furry relatives, such as the gorilla, orang, and chimpanzee. Moreover, in spite of apparent differences in hirsutism, men and women have about the same number of hairs on their bodies (Figure 7–10). With very few exceptions, the hair on the human body is vestigial and serves rather poorly to protect it.

Large or vestigial hairs participate in the mechanism of tactile sensibility. Around the hair follicles a collar of sensory nerves records disturbances in the position of the hair (Figure 7–11), e.g., pressure on the hair shaft. The most highly innervated hair follicles are on the face and the anogenital areas. Highly sensitive hair follicles are found strategically where they will be most effective. Squirrels, for example, have rows of such tactile hair follicles on the abdominal surface, where they make contact with tree trunks whereas lemurs have patches of sensory hairs on the inner surface of the wrist, and nearly all mammals have tufts of sensory hairs growing from molelike elevations around their muzzles.

These specialized tactile hairs, often referred to as vibrissae, or whiskers, are very large in nocturnal mammals (Figure 1–9) and in seals (Figure 1–10) and walruses grow extremely large. They are different from other hairs in that they emerge from follicles that are surrounded by a sinus filled with blood. Thus, besides providing an exquisite nerve supply, the blood-filled sinus also supplies the follicles with a delicate

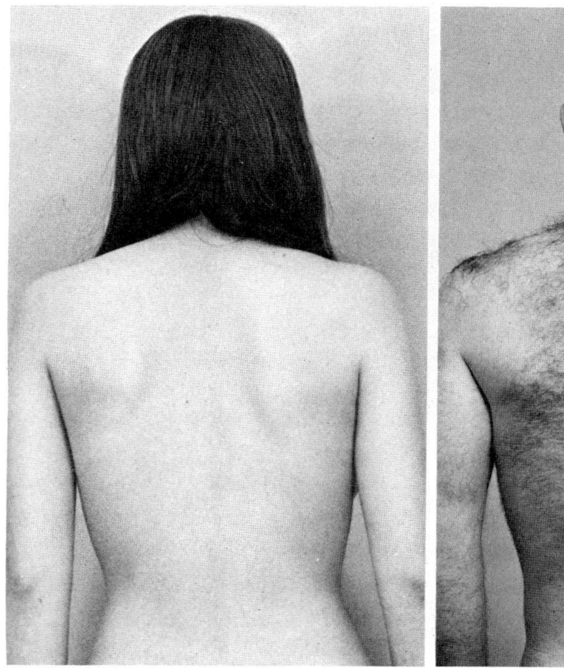
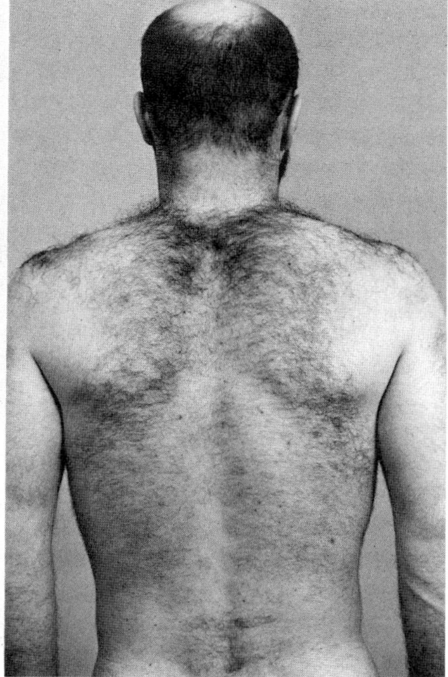

Figure 7–10. The amount of hair per surface area of body of these two individuals, including their heads, is about the same.

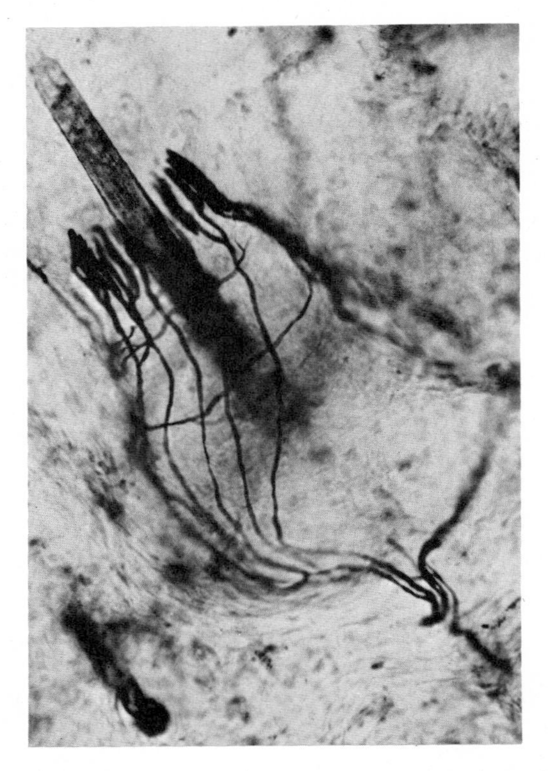

Figure 7–11. Nerve basket (follicle end-organ) around a vellous hair follicle in the pinna of the ear.

mechanism for recording pressure on the hair shaft. Despite the fact that many of his hair follicles are richly innervated, man is the only mammal with no sinus hair. His facial hairs, however, are equipped with highly organized nerve end-organs that render this area, particularly around the mouth, extremely sensitive.

Every observant person knows that, except for the scalp hairs and the beard, all other hairs grow to a specific length, are shed periodically, and are subsequently replaced by other hairs that attain the same length. Hairs do not grow continuously but are produced by follicles that have precisely controlled periods of growth and rest; otherwise, like all animals, man would require periodic shearing over the entire body to avoid becoming entangled in his own hair. Control of growth is the most important single factor of the biology of hair.

Hairs are manufactured by tubelike organs, or hair follicles, attached to and continuous with the epidermis and, depending upon the size, extending to various depths in the skin (Figures 7–2, 7–11). The shallot-shaped follicles grow at an angle to the surface, the bulbous part representing the growing end. Two-thirds of the way up, on the obtuse angle, is a bulge to which is attached a strap of smooth muscle fibers. When these muscles contract, they pull the follicles to a more or less perpendicular position and cause the hair to "stand." This action also puckers the skin at the base of the hair and forms a mound on the surface that is known as a goose pimple. In the center of the follicle is the hair, with a root extending to the bulbous end where it is being formed.

After a period of growth, when the hair has reached its characteristic length, the follicle does something different. The basal part of the hair becomes clubbed rather than cylindrical, with fibrous rootlets from the club anchoring the hair to the surrounding tissue of the follicle. While forming the club, the follicle shrivels up, and the lower part, or bulb, is almost dissipated. A resting follicle can be recognized at once by the clubbed hair, its short size, and its uniquely different structure (Figures 7–2, 7–11). This vestigial follicle remains dormant for variable periods of time. When it becomes active again, it must first form the bulb of a new follicle that will produce a new hair below the clubbed hair. As the new hair works its way to the surface, it loosens the club hair from its moorings and causes it to be shed.

In many mammals nearly all of the body hair grows and rests in seasonal phases and is shed at the same time. In man and in some other mammals, however, growth is mosaic; at any given time, some follicles in an area are growing, others resting. Pelts of animals are useless as fur pieces if they are collected when the hair follicles are quiescent and the clubbed hairs remain firmly attached to the tanned skin.

Hairs, even those of man, vary in color, diameter, and contour. The different colors are produced by the amounts and distribution of melanin pigment in them and by surface structure, which reflects the light in different ways. Hairs can be extremely coarse or so thin and colorless as to be nearly invisible. They are round when straight or alternately oval and round when waved at intervals; when curly and kinky, hairs are shaped like ribbons.

Human hair grows about one-third of a millimeter per day. This means that if the color or shape of a hair is altered while it is forming, several days elapse before the defect is visible on the surface. Melodramatic tales of the hair becoming entirely gray overnight are unfounded. A fact, however, that may serve to explain the "phenomenon" of "turning white in a single night" is that when the hair first begins to turn gray, only the coarser hairs are blanched. The latter are usually masked by the other thinner, normally pigmented hairs, which after a particularly traumatic experience are the ones that fall, whereas the coarse white hairs, being more resistant to stress, remain.

One more point should be clarified. Since hairs are dead elements, clipping, shaving, and singeing have no direct effect upon the follicles that produce them. Those who insist that shaving coarsens the hair are bad observers. Body hairs are tapered and therefore soft to the touch. When they are cut at the base, the soft tapered end is destroyed and the resulting sharp-ended stubble feels rough. One more item of misinformation is the growth of the beard in a dead person. Even after a clean shave, one or more millimeters of hair remain underneath the surface. After death, the whole skin shrivels and thus exposes these hairs lying beneath.

So much foolishness has been written and uttered on the subject of baldness that it seems advisable to explain a few facts. Baldness, a phenomenon of unusual cosmetic and biological interest, especially affects the scalp, which in man supports the growth of the longest hairs. However, since the human scalp and forehead are really one continuous and indistinguishable type of skin, any consideration of man's scalp must include the forehead, which, being nearly naked, gives the scalp its characteristic hair line. Androgenic hormones, which are responsible for the growth of all baby hairs, are also responsible for the demise of scalp hairs.

The hair follicles of the scalp of man are distinctive for their ability to grow uninterruptedly for years, producing hairs of considerable length. In most other mammals, scalp hairs are generally only slightly longer than body hairs. A few baboons, the lion-tailed macaque *(M. silenus)* and Celebes ape, some marmosets, and a few others have scalp hairs of variable length. Therefore, somewhat surprisingly, the human scalp, in which the hair follicles grow very dense and vigorously, is the one that so often becomes bald. Baldness is not a catastrophic destruction of hair follicles but a deliberate, systematic involution of them, culminating in organs similar to the primitive embryonic follicles, without an appreciable diminution in absolute numbers.

All human beings are, to a degree, bald. In early fetal life, the forehead is covered with hairs that are as long as those elsewhere on the scalp; in fact, the forehead is part of the scalp since there is no line of demarcation between the two. After the fifth month, the follicles on the forehead as well as elsewhere on the body undergo a gradual involution; witness the fact that newborn infants are often more hirsute than they are months or even years later. In rare cases of excessive hirsutism, the forehead may be as heavily clothed with hairs as the scalp.

By birth, the follicles are so small that the forehead has a naked appearance although numerically they remain relatively unchanged. During early postnatal months, these hairs become even more diminutive and nearly invisible, and the hair line is indistinct and often fairly low. During infancy, it retreats as the familial pattern of the hair line becomes established. The point is that the development of a naked forehead is actually part of the process of baldness. When male pattern baldness begins in the late twenties or earlier, the follicles behave exactly like those on the forehead of the fetus and infant.

Man is not the only mammal who becomes bald. Among the subhuman primates, the uakari (*Cacajao,* family Cebidae), the stump-tailed macaque *(M. speciosa),* and the orangutan have a "naked" forehead. In nearly all other adults, the scalp hair extends down to the eyebrows. In preadolescents, both the forehead and the scalp are hairy. The preadolescent uakaris from South America have a respectable hair growth on the scalp and forehead, which they lose as they attain maturity, first on the forehead and then over the entire scalp. Since this process is entirely comparable to that in balding men, much can be learned about baldness by studying these animals.

Baldness should be regarded, not as an affliction, but as a natural tendency of the scalp follicles to become very small as the individual matures. A society that is bombarded daily by the advertising media about the value and beauty of scalp hair is not receptive to the real fact: that this ornament is fated to be replaced by the more recent ornamental phase of total nakedness. Whether we like it or not, adult man is becoming progressively balder and, like the uakari of South America, will become entirely so in time.

Glands of the Skin

Skin glands manufacture a limitless variety of substances. Fish and amphibians produce slime, and toads have "poison glands," which secrete complex substances that make the animals unpalatable to potential predators. The skin of reptiles lacks glands, except in and around the cloaca where glands secrete malodorous fluids. Birds have practically no skin glands except over the dorsal part of the tail, where the uropygial (preen) glands secrete semifluid fatty substances that contain precursors of vitamin D. Birds rub their heads against these glands or squeeze the nipple with their beaks, distributing the oil over their feathers as they preen and perhaps swallowing some of it for the essential vitamin D without which they become ricketic. Mammals have the greatest assortment of skin glands. All modifications of these glands are based upon three standard patterns: mucoid glands, which secrete slime; sebaceous glands, which produce fatty materials; and tubular glands, which secrete watery or viscous materials. Though fashioned after these basic patterns, the structural modifications in different animals are too numerous to discuss here.

The secretions manufactured by these glands are remarkably varied in physical properties and composition. The oily secretions of the sebaceous glands are odorless or appallingly malodorous. Mammary glands secrete milky emulsions, and sweat glands pour upon the surface secretions that are watery or viscid, colorless or strikingly pigmented, odorless or odorous. Since animals find one another by tracking the scent, secretions with a specific odor seem to serve a social function. They also function in sexual attraction and in the establishment of territory. The many types of musks of hooved mammals and the secretions from the anal glands of carnivores are used as sexual attractants and for marking territory. These substances, called pheromones, will be discussed in Chapter 14. Other than debris and fecal matter, the specific odor of most animals emanates from the secretion of their skin glands. Even man has a characteristic odor that stems from the complex secretions of his skin glands.

Sebaceous Glands

Sebaceous, or oil, glands are usually attached to hair follicles and pour their secretion, the sebum, inside the canal of the follicle (Figures 7–2, 7–11). Hairs emerge upon the surface of the skin coated with sebum. In a few areas of the body, disproportionately large sebaceous glands are associated with very small hair follicles whereas in other areas the glands are free of hair follicles.

The outstanding structural feature of sebaceous glands is their mode of forming sebum. The undifferentiated glandular cells gradually synthesize and accumulate fat globules in their cytoplasm and become progressively larger and distorted until, when they have finally utilized all of their vital resources, they die and disintegrate. Sebum, which is the accumulation of fragmented sebaceous cells, is formed in a manner analogous to that of the horny layer on the surface of the epidermis. The pressure created by the sebum in the body of the glands, capillary traction through the ducts, and movements and compression of the skin—all aid in bringing about a secretion of sebum upon the surface.

Very little is known about the specific functions of sebum. This semiliquid mixture of bizarre fatty acids, triglycerides, waxes, cholesterol esters, and cellular detritus does not emulsify readily, is toxic to living tissues, and seems to serve little purpose except as an emollient. Some investigators have dismissed sebum as a useless substance and the sebaceous glands as archaic organs that do greater harm than good. Nonetheless, there is a specific plan in the distribution of these glands both in man and in other mammals. They are largest and most numerous on the face and around the anogenital surfaces. The skin of the forehead, around the nose and mouth, and over the cheekbones has beds of gigantic glands, the secretions of which keep these surfaces constantly oily. The rows of evenly spaced sebaceous glands at the border of the eyelids, called Meibomian glands, are so large that they are easily seen with the naked eye when one inverts the eyelids before a mirror. The glands on the genitalia produce copious and continuous amounts of sebum, which has a distinct odor and therefore may well be a human pheromone (see chapter 14).

Among the Primates only some of the lemurs have as many and as active sebaceous glands as man. In addition to their great numbers, their distribution in certain areas is singularly peculiar to man. Only man has rich populations of them on the edge of the rosy borders of the lips. The glands, few or lacking in infants and children, increase in size and number as individuals mature. A similar distinction in man is the presence of glands inside the cheeks and occasionally even on the gums and tongue. Diligent search for sebaceous glands in similar sites in other primates, including the great apes, has yielded practically no results.

Human sebaceous glands contain glycogen in those cells that either are indifferent or are undergoing differentiation. The disappearance of glycogen at the same pace that fatty globules are stored in the cytoplasm suggests that this carbohydrate is in some way utilized in the synthesis of sebum precursors. The sebaceous glands of some other primates contain very little glycogen, and most have none.

Among the distinctive characteristics of the sebaceous glands of man are their common disorders. Nearly all adolescents and even some adults are plagued with plugs of sebum impacted within the duct, a condition commonly known as blackheads. At the onset of adolescence, young people go through the distress of acne, a disease of the skin characterized by pustular inflammation of sebaceous glands. Neither blackheads nor acne lesions develop in the skin of other animals. In spite of constant efforts to ascertain the cause, little is known of the etiology of acne and of effective ways of treating or preventing it. Androgenic hormones, together with hereditary predisposition and diet, are indubitably the major inciting factors for acne. Of its provocative aspects, none is more paradoxical than its precise localization on the face, neck, shoulders, chest, and upper back. Although the sebaceous glands are usually very large in these areas, the skin has no common peculiarity and there is no apparent reason for the disorder to be restricted there. The large size and number of the glands can be ruled out. The skin of the anogenital areas and of the bald scalp, both of which abound in large glands, rarely develops acne lesions.

The size and function of sebaceous glands are mostly controlled by androgenic hormones. These glands, large in the newborn infant, become smaller during childhood. They gradually enlarge during puberty and attain full size in adults, being larger and more active in men than in women. The level of the total output of sebum, which is the only valid index of glandular activity, closely parallels the levels of androgenic hormones in individuals. The minimally developed glands of eunuchs when treated with androgens attain considerable size.

Sweat Glands

Many organs in mammals, most of which have nothing to do with sweating, are vaguely lumped together as sweat glands. Some of these produce scent or musk; some, mucoid substances; and others, colored secretions. Only a few actually sweat. The single common feature of all these organs is that they are tubular glands, some simple, others branched or complex. They have been divided into two general groups: apocrine, which are usually associated with hair follicles, and eccrine, which are not (Figure 7–12). Although these two types share some biological properties, they may have a distinctly different origin, a different structure, and usually a different function.

Most mammals have apocrine glands over their hairy skin. Eccrine glands are often absent or limited to a few restricted, glabrous areas. In the hairy skin of the progressively more advanced primate, the number of eccrine sweat glands tends to increase at the same time the number of apocrine glands diminishes. The prosimians have only apocrine glands in the hairy skin; eccrine glands begin to appear in some of the higher forms. The great apes have equal numbers of them, or more eccrine than apocrine glands; and man has the most eccrine glands, with apocrine glands restricted to specific body areas.

Apocrine glands in man have long been assumed to be primitive organs, the relics of a waning organ system, whereas eccrine glands are thought to be more recent and advanced. However, this assumption is challenged by the fact that the platypus, the most primitive extant mammal, has apocrine glands in the hairy skin and well-developed eccrine glands on its lips. The tree shrew also, believed by some to be the most primitive primate, has more eccrine than apocrine sweat glands in its hairy skin. Furthermore, the apocrine glands in the axilla of man are more numerous, larger, and more active than those found in any other primate. Although the comparative study of organs can give us some information on the phylogenetic history, such data are not sufficient to enable us to reconstruct their evolutionary pathway. Probably the evolution of sweat glands has not followed a single path from a common origin but has gone along parallel or convergent paths. Perhaps neither gland is more ancient or primitive than the other since both types have appeared independently in different animals.

Apocrine glands In man, some primates, and some domestic animals, apocrine glands develop embryonically in close association with hair

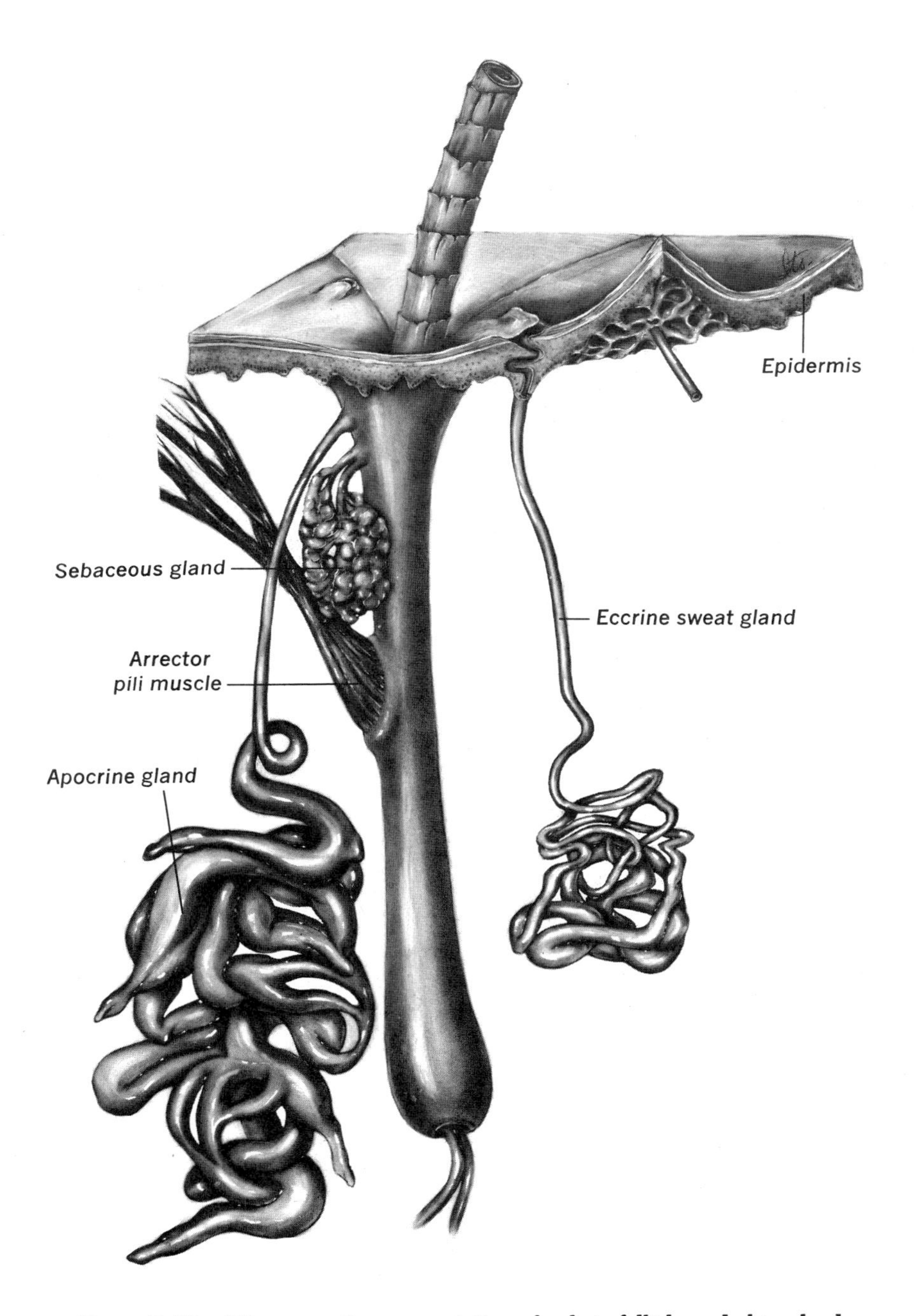

Figure 7–12. Diagrammatic representation of a hair follicle and skin glands.

follicles and in adults open inside the canal of the follicle. Nevertheless, in many adult animals, including most subhuman primates, the glands open directly upon the surface—near to, but not inside, the follicle orifices. Apocrine glands may even open onto completely glabrous surfaces, as in the ring-tailed lemur. Thus, apocrine sweat glands do not necessarily develop from and remain associated with hair follicles. Perhaps the fact that in man and in domestic animals the glands open inside the pilary canal is neither advanced nor primitive.

In a human fetus at 5 to 5½ months, rudiments of apocrine sweat glands appear nearly everywhere on the body. After a few weeks most of these rudiments disappear except in the external ear canal, the nipples of the breasts, the axilla, the navel, and the anogenital surfaces. These are the areas where the apocrine glands normally abound in adult human beings, but single glands can be found anywhere. The ancestors of man may have had apocrine glands widely distributed over the body, and the embryonic rudiments may be reminders of the history of a once widespread organ system. In the specific areas mentioned above, the glands are very large and numerous. In the axilla they are so large that the coils press upon each other forming adhesions and cross-shunts of such complexity that the glands are more spongy than tubular (Figures 7–13, 7–14). The complex of these large apocrine glands, together with an equal number of eccrine glands in the axilla, forms what is known as the axillary organ, which is found only in man, the gorilla, and the chimpanzee. It is less well developed in the orang and absent in all other primates studied.

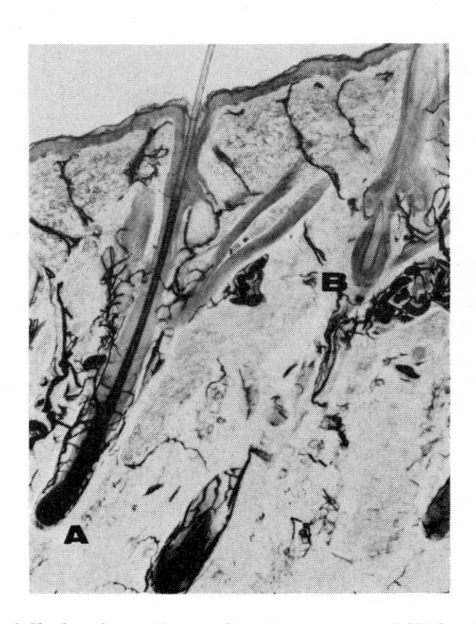

Figure 7–13. Hair follicles from the scalp. A: growing follicle; B: quiescent follicle.

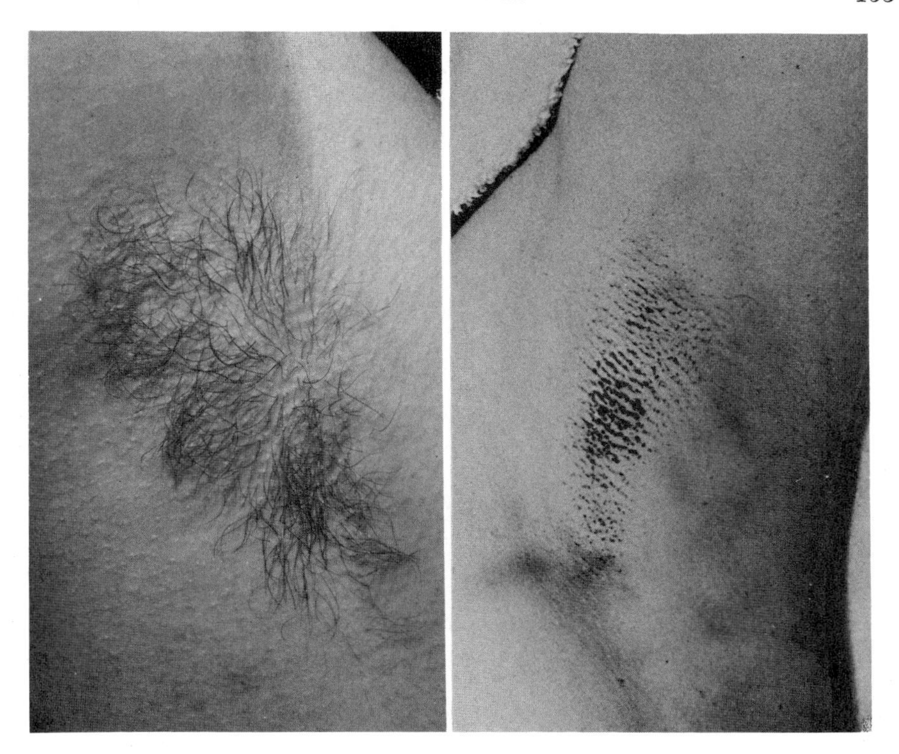

Figure 7–14. Left: the unshaven axilla of a 20-year-old woman; right: the shaven axilla with beads of sweat made visible with a dye.

In spite of their large size, apocrine sweat glands secrete only small amounts of a milky, viscid, pale gray, whitish, yellow, or reddish fluid. The secretion dries in glistening droplets with a gluelike consistency. Since in most individuals the axilla has profuse eccrine sweating, the contribution of the apocrine glands to the total sweat is small. If eccrine glands were not there, the axilla would be dry.

The unwashed axilla has a characteristic odor in man and the gorilla even though the sterile fresh secretion of both glands is relatively odorless. Left standing in contact with the microflora of the axilla, apocrine sweat decomposes rapidly and attains the characteristic odor of the unclean axilla. The odor of individual human beings comes mostly from apocrine sweat and from the sebum. Since the body odors of all other animals have a social or sexual significance, we must assume that this is the archetypical purpose of apocrine secretion even in man.

Apocrine glands are dispensable organs, unnecessary to the biological economy of modern man. They are primarily scent glands that secrete in response to stress and to sexual stimulation. The presence of eccrine sweat in the same areas provides a vehicle in which the viscid, concentrated apocrine sweat is diluted, brought into contact with microorganisms that decompose it, and distributed over larger surfaces to allow a better aeration and thus a stronger odor. Though now useless, apocrine glands must have played some role in human society before the cosmetic industry flourished.

Eccrine glands In most mammals eccrine glands are lacking or limited to friction surfaces. They are relatively similar in structure in most mammals, unlike the apocrine glands, which sometimes vary in structure and function. Cats, dogs, and rodents have eccrine glands only in the digital pads; elsewhere on the body they have only apocrine glands. In the pig, organs resembling eccrine glands are found only on the flat snout and in the peculiar carpal organs. Eccrine glands appear in the hairy skin only in the more advanced primates; the chimpanzee and gorilla have more eccrine than apocrine glands, and man has nearly all eccrine glands on his body skin. All primates have eccrine glands on the friction surfaces. In the prehensile tail of some South American monkeys, the lower surface is differentiated like the palms and is equipped with sweat glands like the palm. Similarly, the gorilla and the chimpanzee, which rest the weight of the anterior part of their bodies on their knuckles as they walk, have the skin there specialized like friction surfaces and abounding in eccrine glands.

Man has 2 to 5 million sweat glands over his body, with an average distribution ranging from about 150 to 340 per square centimeter. In the adult body, sweat glands are most numerous on the palms and soles, then, in decreasing order, on the head, the trunk, and the extremities. Some individuals have more glands than others, but the difference is not sex related. Since all sweat glands are formed before birth, the density of their population is greatest at that time and gradually decreases. The individual differences in the population of sweat glands could be due to differences in body size. Some individuals sweat more profusely than others, but this is not strictly related to the number of glands they have, since a relatively large number of glands, although apparently normal, are functionless.

The single function of sweat glands in man is to secrete water upon the surface in order to cool the skin as it evaporates. Sweating on the palms and soles, however, keeps these surfaces damp and prevents flaking of the horny layer, maintains tactile sensibility, and enhances grip. A dry hand neither grips well nor is very sensitive. Eccrine glands, then, can be divided into two groups: those on the hairy skin, which respond to thermal stimulation and help to regulate body temperatures, and those on the palms and soles, which respond to psychogenic stimuli and keep the friction surfaces moist. In the axilla, a hairy surface, there are numerous glands, some of which respond to thermal and some to psychogenic stimuli. The same ones, however, apparently do not respond to both stimuli. In some individuals, some glands on the face and scalp also sweat in response to certain drugs and hot spices. The glands on the friction surfaces and those on the hairy skin have not only a different function but also a different developmental sequence.

In human embryos the glands on the palms and soles begin to develop around 3½ months whereas those on the hairy skin appear at 5 to 5½ months when all of the other cutaneous structures are already formed. This dyschronicity in development may also represent a fundamental difference in the history of these glands. Those on palms and soles, which appear first, are also present in most other mammals except the hooved ones. Perhaps phylogenetically they are the most ancient. If this assumption is correct, then responding to psychogenic stimuli would be the original function of sweat glands. Since the glands on the hairy skin have a more recent evolutionary history and are designed for a different purpose, they may not have yet become capable of responding effectively to heat stimulation.

Significant in this regard is the fact that the sweat glands on the hairy skin of subhuman primates, although apparently perfect structurally, are nearly always nonfunctional. The skin of monkeys remains dry even in a hot environment. The only exception we have seen is in a rhesus monkey which, dying of cardiac failure, perspired profusely only on the face.

The glands in the hairy skin respond to a number of drugs that have antagonistic properties. Under nearly identical conditions, they often have varied responses in different individuals and even in the same person. Together with water, the glands excrete variable quantities of sodium and other electrolytes, which after heavy periods of sweating must be replenished. Since recent studies suggest that sweating involves the active transport of water, the excretion of sodium is essential. Interestingly, however, the duct of these glands is very long and well supplied with blood vessels and reabsorbs much of the sodium from the sweat as it flows through the duct. The apparent vagaries in physiology could reflect the newness of these glands and their still unsettled physiological performance.

Eccrine sweat glands are indispensable to man, and, though insensible, function constantly under the control of the autonomic nervous system. Sweating is essential for preventing overheating of the body. Notwithstanding the daily campaigns against it by the manufacturers of antiperspirants and deodorants, sweating is normal, even essential.

8

CHARACTERISTICS OF TEETH

Thy teeth are like a flock of sheep
that are even shorn, which came up
from the washing; whereof every one
bears twins, and none is barren
among them.
(Song of Solomon)

Because of the special as well as the diversified roles they play, teeth reveal a great deal about an animal. Examination quickly shows whether they belong to a carnivore or herbivore, and the amount of wear and shearing discloses not only the age of the owner but the mode of chewing as well. Furthermore, in man the shape of the face and even the expression of an individual are affected by the plane of occlusion, that is, by the biting surface or the way teeth from the upper and lower jaws come together. Teeth are functionally versatile: they can chew, gnaw, grind, and tear food as well as rip, slash, gnash, and snarl. As such, they are not limited to procuring and preparing food for digestion but are important adjuncts in communication. Whether prehistoric or modern, set in rows in the jaw or rattling loose, they are nearly indestructible and always interesting. Their shape and organization were recorded millions of years ago, imprinted or embedded on rocks as fossil remains that have been one of our primary sources of information about prehistoric animals.

Because teeth are often the only fossils of extinct animals, they have been the subject of extraordinary controversy, mistakes, and even scientific fraud. In 1922 a careful study of a single, water-worn molar tooth from the Snake Creek of Nebraska "disclosed" that it belonged to a fossil manlike anthropoid from the Pliocene, which was named *Hesperopithecus haroldcooki.* Imaginary complete male and female figures were reconstructed from this one tooth! Five years later, however, this same tooth was shown to have belonged to an early American peccary. This was obviously a mistake. However, an outright fraud was perpetrated in England that centered around a supposedly prehistoric portion of a brain case and lower jaw. These remains, allegedly found at Piltdown in Sussex between 1908 and 1913, excited enormous interest; despite the doubts and reservations expressed by some anatomists, many judged the remains to be those of the most primitive type of man yet discovered. Named the "dawn-man," or *Eoanthropus dawsoni,* these remains were later exposed as fraudulent.

It would be wrong, however, to emphasize the mistakes and the fraud because the careful study of teeth has provided much basic information on evolutionary trends and adaptations.

A tooth is a curiously difficult object to define, particularly if one is faced with the whole array of toothlike structures found in animals. Teeth are mineralized, mainly ectodermal appendages found at the margins of the jaws and used chiefly in eating. In mammals they are set in sockets and held in place by fibrous tissue. Such peg-and-socket joints are called *gomphoses,* a barbarous word that the ancient physician Galen (p. 88) borrowed from the Greek name for a bolt used in ancient shipbuilding. Some fish and amphibians have teeth almost anywhere around the mouth: inside the jaws, in the throat, or even, like the sawfish, outside the mouth. Teeth are not always equally mineralized; some animals have only horny protuberances, and others, like the baleen whales, have unerupted teeth only during fetal life. Baleen plates are hardened palatal ridges. Teeth are used for eating, for attack, for defense, for carrying objects, for digging up roots, and even for locomotion. Bull walruses use their long, massive tusks to pull themselves on to the ice. Teeth are also a sine qua non for most articulate speech (a toothless person cannot say the word *teeth*) and are useful in expressing emotion, such as signaling cooperation and understanding, snarling, expressing pain, or in grinning or subtly smiling.

Types of Teeth

From an evolutionary point of view, anthropologist John Buettner-Janusch considers teeth to have maintained "conservative traits" since they have not changed as much as other structures. One can therefore expect to find the teeth of related forms separated by great periods of time relatively more alike than some of their other organs. This would indicate that teeth reached some degree of stability early in evolution. We must also note that when, in evolutionary history, a tooth is lost, it will not appear again. Thus the number of teeth of any specific animal cannot be a criterion for determining the ancestral species of a descendant.

Different functions determine the diversity of form in the teeth of animals. Sharks and other carnivorous fish have pointed, cutting teeth arranged in numerous rows. Those fish that crush hard molluscs have evolved flattened oral plates made of several teeth fused together. Fish and reptiles generally have more teeth than mammals. Ancient birds had reptilianlike teeth, which are no longer found in modern forms. Except for highly specialized species, mammals have four types of teeth in each jaw, which are from front to back: (1) *incisors* (*L. incidere,* to cut into), sharp, chisel-like anterior teeth used for nipping, cutting, and combing; (2) *canines* (*L. canis,* a dog), a single tooth in each half jaw behind and lateral to the incisors, usually long, pointed, and well marked in carnivores (the upper canine, or "eye-tooth," is longer than the lower one); (3) *premolars,* behind the canine and replacing the deciduous molars; and (4) *molars* (*L. molare,* to grind), usually with an expanded crown and a complicated pattern of little projections, or cusps, on the surface. Premolars and molars together are known as the "cheek" teeth. In some mammals they are separated from the incisors or canines by a gap called a diastema. The size, shape, and presence or absence of any of these four types of teeth are characteristics of each genus or species, and the probable evolution and significance of the types have provoked much study.

The number and pattern of mammalian teeth can be expressed in a simple formula provided one is looking at the upper and lower half jaws with the incisors in front. Thus the primitive mammalian dentition can be expressed,

$$\frac{3.1.4.3.}{3.1.4.3.} = \frac{11}{11} = 22 = 44 \text{ teeth.}$$

Such an animal has 3 incisors, 1 canine, 4 premolars, and 3 molars in each half jaw. Since man has both deciduous and permanent dentitions, another formula is necessary: man's deciduous dentition is $\frac{2.1.2.}{2.1.2.}$, indicating that there are 2 incisors, 1 canine, and 2 molars in the upper and lower halves of the jaws in the deciduous dentition.

Man's permanent dentition can be expressed as,

$$\frac{2.1.2.3.}{2.1.2.3.} = 16 = 32 \text{ total},$$

showing that there has been a reduction from the primitive mammalian type of dentition by 1 incisor and 2 premolars.

Notwithstanding their long evolutionary history and the wide variety in form, size, habits, behavior, and nutrition, nonhuman primates have for the most part retained relatively stable teeth with little change in their dental formula. Only the aye-aye *(Daubentonia)*, a curious, highly specialized prosimian, has drastically departed from the basic primate pattern with a dental formula of

$$\frac{1.0.1.3.}{1.0.0.3.} = 9 = 18 \text{ total}.$$

The following examples of dental formulae show that the number of permanent teeth is somewhat consistent:

Tarsiers	$\frac{2.1.3.3.}{1.1.3.3.} = 17 = 34$ total
Lemurs	$\frac{2.1.3.3.}{2.1.3.3.} = 18 = 36$ total
Spider monkeys	$\frac{2.1.3.3.}{2.1.3.3.} = 18 = 36$ total
Marmosets	$\frac{2.1.3.2.}{2.1.3.2.} = 16 = 32$ total
Rhesus monkeys	$\frac{2.1.2.3.}{2.1.2.3.} = 16 = 32$ total
Gorillas	$\frac{2.1.2.3.}{2.1.2.3.} = 16 = 32$ total

These formulae indicate that primates have usually retained a somewhat primitive molar pattern and variable specialization of the incisors and canines. Incisors have changed little, except in the prosimians, where they are variable and highly specialized. In the Lorisidae, Indriidae, and Lemuridae, the lower incisors, together with the canines (except in *Indri*), are horizontally directed as tightly packed, slender teeth forming a "dental comb." As a rule, canines are pointed and large (more so in males) and are used for various purposes, not the least of which is display. The upper canine of man has a particularly long root extending upward in the maxilla (thus *eye-tooth*), which is a remnant of an ancestral feature.

Replacement of Teeth

The simple, pointed teeth of many lower vertebrates are replaced by eversion (turning or rolling outward) as new ones develop on the inside of the jaw. The functional teeth of reptiles alternate with neighboring tooth germs; as the mature teeth are lost, tooth germs grow into functional ones. The replacement of these latter by successive tooth germs at their base insures an almost continuous replacement.

In mammals teeth are replaced vertically, as in man and other primates, or horizontally, as in elephants. Not all six cheek teeth in each half jaw of elephants are present at once. As the first deciduous teeth are lost in front, unerupted ones behind them move forward to replace them as permanent molars. Some contend that if elephants lose one of these molars, they become so exasperated at being unable to chew properly and so maddened by the pain of the others grinding against a raw gum that they degenerate into "rogue" elephants. Kangaroos, manatees, pigs, and to a limited degree apes and men also show some horizontal type of replacement.

When a child sheds his two deciduous molars, not all the resulting space is filled by the two smaller premolars that replace them. This allows a shift forward of the first permanent molar. If for any reason one of the premolars is lost or removed, the molars move slowly forward, as in the horizontal type of replacement. The forward horizontal shift of teeth is brought about by the continuous to and fro working of the jaws, by the way the lower jaw is slung, and by the pressure of masticatory movements. A dentition like that of fish and reptiles—generations of teeth succeeding one another—is called polyphyodont, whereas that of mammals is called diphyodont, i.e., some of the first acquired teeth are shed. The eruption of the first dentition in the young child appears in the following order from front to back:

		Roots completed
Deciduous incisors	6–9 months	18–24 months
1st deciduous molars	12–14 months	30–33 months
Deciduous canines	16–18 months	38–42 months
2nd deciduous molars	2–2½ years	36–40 months

After an interval of three to four years, the series continues with the eruption at age six to seven of the first permanent molars behind the deciduous molars. As the permanent molars continue to erupt, replacement starts at the front in somewhat the following order:

		Roots completed
Permanent incisors	6–9 years	9–11 years
Permanent premolars	11–12 years	12–14 years
Permanent canines	9–12 years	12–15 years

Finally the two other permanent molars erupt:

		Roots completed
Permanent 2nd molars	11–13 years	14–16 years
Permanent 3rd molars	17–25 years (or later)	18–25 years

This order is a uniquely human feature (Figure 8–1).

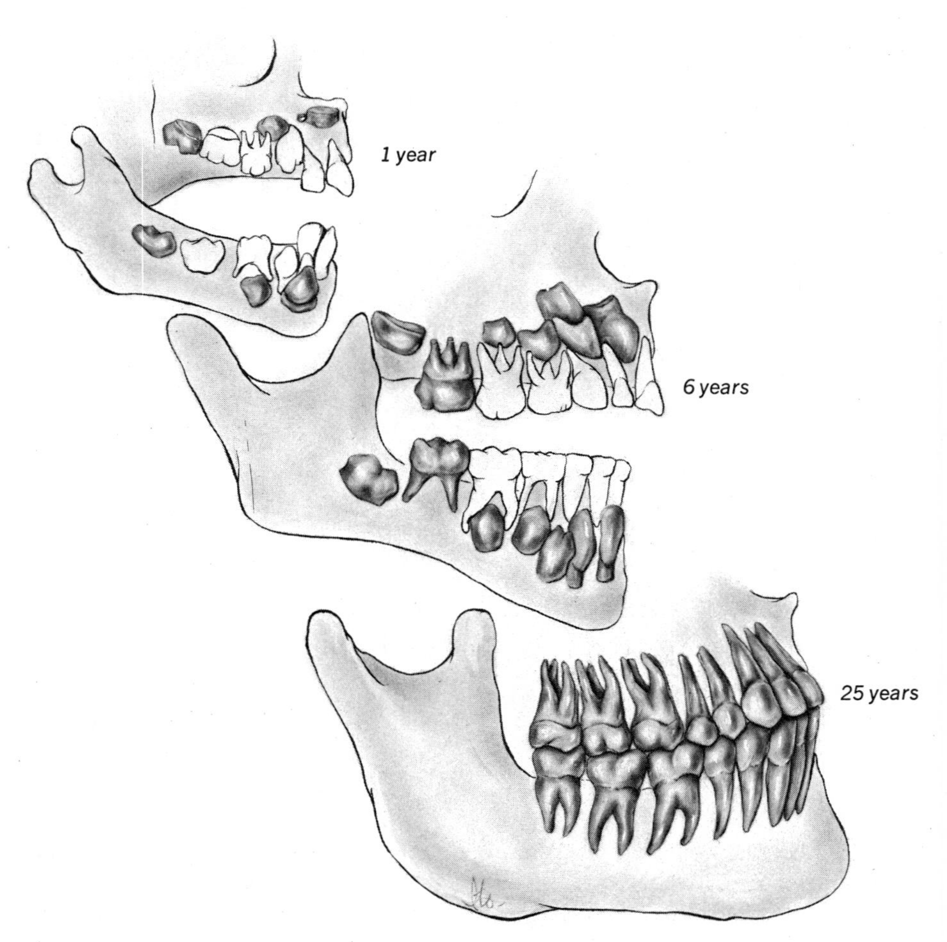

Figure 8–1. The eruption and replacement of primary and secondary dentition. The permanent (secondary) teeth are shaded.

In early nineteenth-century England, the second permanent molar was known as the "factory tooth," since its eruption coincided with the child's introduction to factory work. Since our permanent molars resemble deciduous molars more than permanent premolars, some consider them to be latent members of the first dentition. Such a theory naturally delights those who argue that man carries fetal characteristics into adult life.

Structure of Man's Teeth

A tooth has a crown above the gum line and a root below it (Figure 8–2). The hard mineralized material of a human tooth is hollowed out at the base by a pulp cavity, which contains soft connective tissue, nerves, and blood vessels. The tooth itself consists of dentine on the inside, a hard, bonelike tissue containing about seventy-five percent inorganic salts (mainly calcium phosphate in the form of calcium hydroxyapatite), covered by an outer layer of enamel, the hardest and most nearly indestructible substance in the human body. The enamel covers only the crown. Another very hard tissue (cement) surrounds the root of the tooth. The composition of the cement is similar to that of bone and can be considered homologous to the so-called bone of attachment in lower vertebrates, which unites the teeth and the jaws.

Enamel

In the developing tooth, enamel is laid down by tooth germ cells called ameloblasts, and once deposited it becomes permanent. It is dead tissue and if damaged or chipped cannot be repaired since the cells that formed it disappear. Enamel is made up of crystalline prisms oriented in characteristic ways to toughen the substance to resist the many abuses it must suffer. Alterations in the rate of enamel formation and changes in nutrition at birth may cause faults in it called neonatal lines. Variations in the closeness with which the enamel prisms are packed and the characteristics of the underlying dentine produce differences in enamel color. In older people the enamel becomes worn and shows more obviously than before the yellow dentine beneath.

Since the ameloblasts that form the enamel are derived from ectoderm, that part of the tooth can be considered to be homologous to other skin derivatives. Rodent incisor teeth grow throughout life, an indication that ameloblasts and other tooth-forming cells remain active. To prevent an overgrowth, these animals must keep their teeth ground down at all times. Although dead and mainly inorganic, enamel is per-

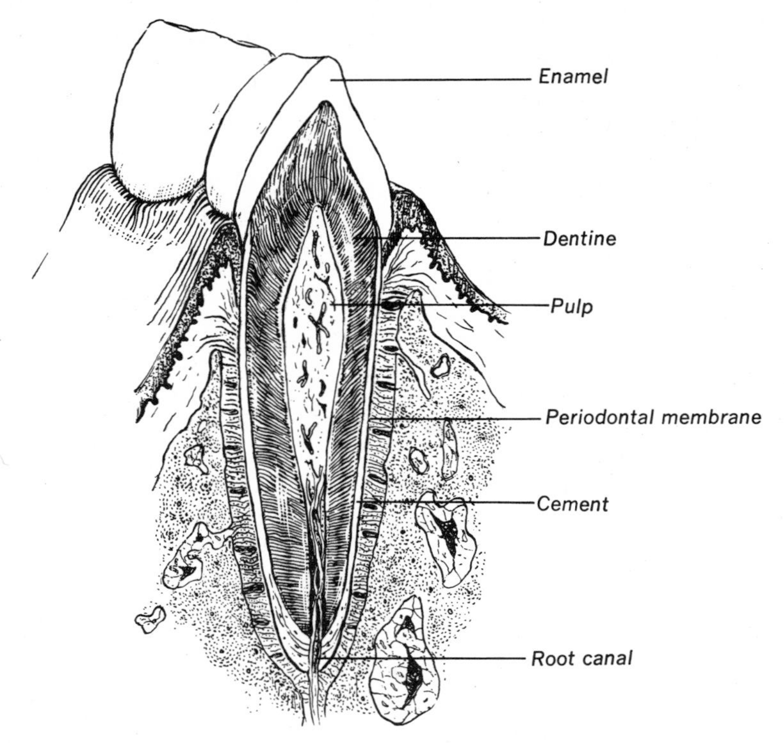

Figure 8–2. The main elements of a human tooth.

meable to water, to substances of low molecular weight, and to some large dye molecules. Experiments with radioactive isotopes have shown that some ions penetrate enamel with remarkable rapidity, perhaps passing through it by way of the keratinlike protein that cements enamel prisms together or through minute cracks, fissures, and fibrous lamellae.

Dentine

The bulk of the tooth is made up of dentine, a hard yellowish material (but not so hard as enamel) with a high degree of elasticity. It is laid down by cells called odontoblasts on its inner surface surrounding the pulp cavity. As dentine is formed, the odontoblasts leave behind them a long protoplasmic strand, the dentinal fiber of Tomes, named after Sir John Tomes (1815–1895). As more dentine is laid down, the processes become longer and the dentinal substance is traversed by millions of narrow dentinal tubules (the early microscopist Leeuwenhoek, in 1678, patiently counted 4,822,500 of what he called "pipes" in a human molar). Since dentine is laid down in rhythmical phases of activity and rest, the increments give it a lined appearance. The number of incremental lines in the teeth of children can give a somewhat accurate gauge of their age. This tubular nature of dentine, which is present in most mammalian teeth, is really an ancient type of tissue first found in fish. In some fish the dentine is full of blood vessels that form a network of capillaries within it. On rare occasions a capillary becomes incorporated in developing human dentine.

Although a full complement of dentine is achieved after completion of root formation, under certain circumstances, including normal aging of the owner, an additional (secondary) dentine can be laid down. Mild stimuli that affect odontoblasts, thinning of surface coverings, attrition, or even caries can encourage the deposition of secondary dentine. Dentists sometimes attempt to stimulate a deposition of secondary dentine by placing mild irritants at the bottom of a cavity that has extended close to the pulp. The degree of deposition of secondary dentine, therefore, can also be a guide to the age of a tooth.

The largest mass of dentine with no enamel over it is the elephant's tusk. Ivory is fine-grained, almost solid dentine. The tusks of elephants are modified incisors that grow continuously throughout the life of these animals. What enamel was present at the tip of the tusk is lost soon after eruption. The continuously growing tusks of walruses are the upper canines, present in both sexes. Enamel is present, but there is a central core of granular dentine, which often identifies walrus ivory. The single tusk of a narwhal (a toothed whale) is an incisor, characteristically spiraled, usually present only on the left side of males.

Cement and Periodontal Membrane

Cement tissue attaches the human tooth to the jaw. This tissue surrounds the tooth at its root except at the opening of the root canal, and extends upward to the junction with enamel. Each human tooth is set into an individual socket in the jaws (alveolus), lined by a vascular periodontal membrane. Fibers of connective tissue that pass from the socket bone into the cement hold the tooth firmly in place and are the principal component of the periodontal membrane.

The periodontal membrane is richly supplied with blood vessels. When a tooth is extracted, the vessels are ruptured and blood escapes, but it soon clots in the socket and is replaced first by granulation tissue and later by bone. When the periodontal membrane becomes inflamed, it swells and forces the tooth upward, projecting its crown above the level of its neighbors. In biting or chewing, this tooth is struck first: the contents of the sockets are compressed, the sensory nerves in the periodontal membrane are stimulated, and a painful sensation results.

The membrane is well supplied by nerves. The slightest degree of contact between teeth or the gentlest pressure exerted by something wedged between two teeth causes an immediate sensation. A new filling that rises ever so slightly above the level of the usual occlusion plane can soon cause acute discomfort.

Pulp

The pulp cavity of a tooth extending from the base of the crown to the root is filled with a soft tissue (pulp). At the tip of each root is an opening or openings through which pass the blood vessels, lymphatics, and nerve fibers that supply the tooth. Excessive wear, caused by eating certain foods, the habit of overchewing food, or nervous grinding, can result in so much loss of tooth substance that the pulp cavity is exposed; this may become infected with eventual formation of periapical abscesses. Although deprived of its blood and nerve supply, a tooth can remain usefully in place for many years. Teeth that have been knocked out can sometimes be made to stay in position for years if they are quickly replaced in their sockets and splinted. Such teeth are not re-innervated and are dead objects. In such cases, the root surface is invaded by the bone of the socket, which holds the tooth tightly enough in place. Attempts to replace a lost permanent tooth in a young person by transplanting a tooth germ from a more posterior position of the same jaw have been only moderately successful. Tooth germs have not been transplanted successfully from one individual to another.

Table 8–1. Chemical Composition of Tissues of Permanent Teeth

	Enamel	*Dentine*	*Cement*
Inorganic	96–97 percent (Various apatites mainly hydroxy-apatite and some fluoroapatite)	70–75 percent (Hydroxyapatite)	55–60 percent (Apatite deposited with a collagenous matrix)
Organic	3–4 percent (Mainly a protein with a resemblance to keratin)	25–30 percent (Collagenous material)	40–45 percent (A collagenous fibrillar network)
Calcium	35–36 percent	34–35 percent	35–36 percent
Phosphorus	17–18 percent of 100 gm Ash.	17–18 percent of 100 gm Ash.	17–18 percent of 100 gm Ash.

Development and Eruption of Teeth

Teeth begin to develop in human embryos in the seventh week. At first only an ectodermal dental lamina is present from which tooth germs grow inward into the tissues of the developing jaws (Figure 8–3). Deciduous incisors begin calcification about the fourth month of fetal life and the second deciduous molar about the sixth month. Calcification of the crown of the incisors, which of course are still unerupted, is completed about two months after birth and of all deciduous teeth by the first year after birth. Man has twenty deciduous teeth that provide a functional dentition up to the age of about seven. Most mammals have a deciduous dentition, but in some it is suppressed, and in others it is shed during fetal life. In guinea pigs and seals such shed crowns of the first dentition are occasionally found on the gums, in the amniotic fluid, or in the stomach of the newborn animal.

From an evolutionary point of view, deciduous and permanent dentitions are probably relics of the multiple generations of teeth (poly-

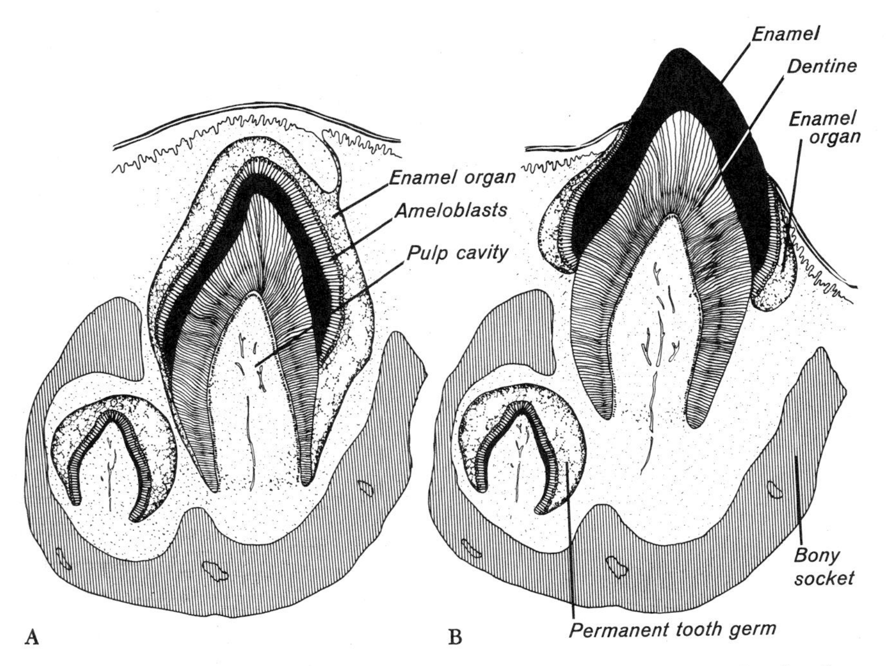

Figure 8–3. A: development of tooth germs. The enamel organ of the deciduous tooth is still attached to the gum surface by a strand of tissue. B: the eruption of a deciduous tooth with the loss of the enamel organ.

phyodont) in lower animals. Continuous successive generations of teeth seem to fulfill two important functions: replacement and the addition of teeth in an animal that grows larger as it grows older. Theoretically, under optimal conditions a marine animal such as a shark could live for an indefinitely long time, and the continual replacement of its teeth should be an advantage. This would not be the case in animals that reach a fixed genotypic size at a certain age and in animals with a limited life span. The deciduous dentition of man (the "foundation" dentition) is a functional pioneering dentition that prepares for the next generation. Although of short duration, the deciduous dentition greatly influences the succeeding one. Modern dentistry is increasingly aware of the importance of attending to the proper arrangement (spacing) and functioning (chewing) of the first dentition. Although not erupted, all deciduous teeth are well developed at the time of birth and have been exposed to the same environment as that of the rest of the embryo and fetus.

Except for the molars, man's deciduous teeth are smaller than the permanent ones. Anyone who has been bitten by a two-year-old can tell that the cusps of first erupted teeth are quite pointed. Moreover, their crowns are more bulbous, the neck more constricted, and the shorter roots have relatively larger pulp cavities. The resorption of roots commences some eighteen months before a tooth is shed. Since they are covered by a softer enamel, deciduous teeth undergo attrition earlier than permanent teeth.

In general, the deciduous teeth of man resemble those of the great apes. Man, however, lacks a diastema between upper canine and lateral incisor and has smaller canines and different arrangements of the molar cusps. The order of eruption of man's deciduous teeth follows that of Old World monkeys with the canine erupting after the first molar and before the second. In apes the deciduous canine erupts last. The second permanent molar of apes appears much earlier than that of man and interrupts the replacement of the anterior deciduous teeth. The second molar of man appears after all deciduous teeth have been replaced. This order of eruption is almost certainly associated with man's longer, drawn-out period of childhood, during half of which most deciduous teeth remain functional. The delay in fusion of the ossification centers of the human skull can also be correlated with the late appearance of the second molar, which does not start to erupt until certain ossification centers in the base of the skull have become fused. The third molar does not erupt until the whole bony base of the skull has undergone advanced fusion.

Eruption is an actual movement of the tooth relative to the jaw and is brought about mostly by the longitudinal growth of the root. Additional force is generated by appositional growth of bone near the erupting tooth. As the crown grows upward, the root forms downward from its edges, gradually surrounding the pulp with dentine and cement. The pulp enlarges by active growth, and this creates a tissue pressure involved in the process of eruption. The tissue in front of the erupting tooth is loosened until the cellular cap over the enamel (Nasmyth's membrane) comes up into contact with the gum epithelium and fuses with it. The crown has only to break through this bloodless barrier to erupt and therefore no bleeding usually occurs. Eruption is an active process with some force behind it. If an erupting tooth hits the edge of another already erupted (impaction), it will tilt on its axis. Teeth move or can be moved by extensive treatment within the jaw substance even after their roots have been formed.

Paleontologists and anthropologists have attempted to explain the evolutionary pattern of cusps in mammalian molars. Some of the ingenious theories that have been propounded, however, have been exercises in morphology that largely disregard morphogenetic processes and sound embryological or paleontological evidence.

For example, the cusp pattern of the mandibular molars of *Dryopithecus* (tree-ape) of the Pliocene, has been considered to be a primitive feature of hominid dentitions. These molars have three cusps disposed buccally and two lingually, with a small sixth cusp commonly placed distally. The hominid first mandibular molar (and that of modern man) has a not dissimilar array of cusps and intervening fissures. There are five cusps and the main fissures form a Y. Contemporary man has second mandibular molars with four cusps in a cruciform pattern, but patterns intermediate between this and the dryopithecoid pattern have also been found.

The dentition of the australopithecines (especially the deciduous) is remarkably human in pattern. The dental arches are evenly curved (parabolic) as opposed to the more parallel arrangement of the ape's postcanine teeth. No precanine diastema exists in the upper jaw, the incisors are fairly small, and the canines hardly project beyond the level of the neighboring teeth. The cheek teeth are large compared with the incisors, but are more like human teeth than those of apes. Attrition of these teeth is more pronounced than in apes, again probably an indication of a greater degree of grinding movements than that in apes. The cusp pattern of the lower molars is dryopithecoid. From comparisons such as these, some have argued that some australopithecines (*Paranthropus*) were herbivores, used their hands for feeding, and probably made tools. Just how correct these deductions are is still not known, but they indicate the fascination of trying to reconstruct the sort of creatures these fossil forms were.

The jaws and teeth of typical neanderthaloids were large, and the crowns of the teeth were similar to those of modern man except that the incisors were shovel-shaped with prominent marginal ridges on the lingual side. A feature of many molars is their taurodontism, i.e., the presence of a long "trunk" before the bifurcation of the roots. In its most marked form, the root consists of a thick trunk terminating in a concave surface surrounded by two or three very short, webbed roots. Typical taurodont molars are sometimes seen in modern man (Mongoliforms). These could be the expression of a genetic inheritance from Neanderthal man or a character established in earlier forms and retained even by living man.

Occlusion

Opposing teeth from each jaw erupt at about the same time so that they come to bite against each other as they rise from the jaw. When all deciduous, and later, all permanent teeth have erupted and are in their final position, the jaws can be closed and held together in the resting position so that all the cusps, depressions (fossae), and the cutting edges of the two dental arcades are brought into articulation. The way projecting cusps fit into the depressions of opposing teeth is called the occlusion pattern, and each dentition has its own pattern. The occlusal pattern of the more primitive forms of man showed the incisors biting together edge to edge. That of modern man, particularly in the western hemisphere, shows the upper incisors occluding in front of the lower ones and giving rise to an "overbite." In the cheek region, the overbite decreases and the third molars occlude almost crown to crown with the upper third molar extending slightly over the back of the lower. The plane of occlusion (the biting surfaces of teeth) is not always horizontal and has often a distinctly downward curve.

Various types of malocclusions develop when teeth erupt at the wrong time and in the wrong order. Orthodontists correct malocclusion with mechanical devices that force teeth to move into a correct occlusion. This is done as much for practical as for esthetic reasons since malocclusion can result in uneven biting that later causes painful and disabling changes in the temporo-mandibular joint. Malocclusion may well be one of the prices man pays for bipedalism and erect posture. In the evolution of primates, as jaws receded, one incisor and two premolars (from the generalized mammalian complement) were lost. Man, however, did not lose enough teeth. The upper lateral incisors, the lower second premolars, and the third molars are the best candidates for elimination to avoid faulty eruption. The eruption of the first two out of the plane of occlusion may be due to insufficient widening of the jaws at the age of six or seven. At this age the brain of most children is approaching adult size with a relatively large cranium, but as yet no well-developed air sinuses are present in the facial skeleton, and the jaws are often too small and inadequate to accommodate all the teeth that erupt. The eruption of the third molar occurs when the jaws are about fully elongated, generally at the end of bone growth. About fifteen percent of the population shows faulty eruption of the third molar, which is ironically called the "wisdom" tooth (dens serotinus).

Caries

The most common disorder of man's teeth is caries. This seldom occurs in other animals, although close inspection of the teeth of many wild-caught monkeys shows carious lesions. This often painful, unsightly, and exasperating condition is not often found in primitive societies and is infrequent in hardy native populations. Caries is rare in Australian aborigines who eat a native diet but soon appears when European food is ingested. Those who hold that animals do not live long enough to suffer caries should remember that it can occur in the deciduous dentition of children two to three years old. Apart from the contributing factors of overcrowding, malocclusion, and the fact that the arrangement of teeth in the jaw leaves numerous interstices and crevices for the accumulation of bacteria-ridden particles of food, little explanation can be given for the prevalence of this condition. Differences in hardness and composition of individual teeth, the composition of saliva, excessive contact with carbohydrates, the presence of particular bacteria that flourish under appropriate oral conditions, and genetic and other factors doubtless affect its incidence. Many puzzling questions remain unanswered. We do not know why, in two individuals eating the same type of food and living in the same locality with apparently the same "type" of teeth, caries develops in one and not in the other. Nor do we know whether man would develop caries if his teeth were spaced in such a way that no food particles could collect between them. Now that dental research is attaining full scientific stature, answers to these questions and solutions to these problems should be forthcoming.

9

MAN'S INTERNAL ENVIRONMENT

By an apparent contradiction the living being maintains its stability only if it is excitable and capable of modifying itself according to external stimuli and adjusting its response to the stimulation.
(Charles Richet)

The word *environment* is used to denote surroundings or surrounding conditions or forces that influence or modify. The atmosphere in which an organism lives, with all it contains and encloses, constitutes one important aspect of its external environment. The conditions of this environment are constantly changing, and for survival the organism must be able to adjust to them. It accomplishes this either by changing with the environment or by making appropriate adjustments so that its inner substances remain unchanged. The temperature of cold-blooded creatures, for example, is roughly the same as their outer environment. Warm-blooded organisms, however, must make changes so that, regardless of temperature alterations outside, they are able to maintain the status quo. Man can be exposed for long periods to temperatures of 180° F or higher and arctic foxes to temperatures −40° F and colder without showing alterations in body temperature. The body also has adjustment mechanisms that enable it to cope with changes or disturbances occurring within it. Vigorous muscular action, for example, causes such quantities of heat that if it were not dissipated through the skin and lungs we would soon perish.

Numerous agents in the mammalian body have as their main function the maintenance of a stable environment in that organism. This means that the substances of the organism must themselves be so sensitive to changes outside or within that they are always ready to trigger off the proper responses to maintain the equilibrium of the body. This phenomenon is called homeostasis. In other words, the body is equipped with numberless instabilities that, in the totality of their function, bring about stability. The brain and all the elements of the nervous system, the total cardiovascular system including the blood itself, lungs and kidneys, spleen and other lymphoid tissues, the glands and endocrine organs are all engaged in this mission of stability. The coordinated function of all these organs and tissues is to keep the living being safely stable.

In 1855 a distinguished French physiologist, Claude Bernard, concluded that the body has a *milieu interne* and that important "internal secretions" produced by various organs profoundly affect the specific organs that respond to them. Ernest Henry Starling later coined the word *hormone* (Gk. *hormánein,* to excite) to designate those chemical substances liberated directly into the bloodstream by certain ductless, or endocrine, glands. The effects of different hormones are frequently interrelated, several being able to affect the same target organ in different ways. Most endocrine glands are under the control of the specific secretions of the pituitary (hypophysis) at the base of the brain. These secretions are in turn controlled by the amounts of hormones that the individual endocrine organs secrete. In this way many delicate control systems under the overall control of the hypophysis are established to maintain homeostasis.

Very small quantities of some hormones can produce explosive responses. Physiological amounts of some hormones are secreted in micrograms (one microgram = 1/1000 milligram), more powerful ones in nanograms (one nanogram = 1/1000 microgram). Endocrine glands that need to maintain their action constant secrete some hormone at all times, whereas those that function in rhythmical cycles secrete intermittently. One of the most significant aspects of some, mostly the steroid hormones, is that to a large extent they are chemically and functionally interspecific. Protein hormones, on the other hand, with the possible exception of thyroxin, are more species specific. A number of hormones are as effective in fish as they are in man. All biological processes are at least in part under hormonal control.

Hormones are part of an internal communication or signaling system. Endocrine glands are comparable to a series of transmitters that emit specific chemical signals, which, once distributed throughout the body by the vascular system, influence the activities of sets of cells sensitive to them, regardless of the distance from the emitting organ. Thus, certain cells are sensitive to one or more hormones; those that respond to a specific hormone are the target organs of that hormone. The endocrine system communicates much more slowly than the nervous system, but it sustains the message over a longer period of time. The quantity of the hormone arriving at a target can be increased or decreased, and its effect is correspondingly enhanced or reduced.

As soon as a hormone has begun to accomplish its physiological mission its rate of secretion usually begins to decrease, that is, information is transferred back to the endocrine organ to check its activity. On the other hand, if an endocrine organ is undersecreting, such information decreases and the gland continues to secrete more hormone. Thus, the rate of secretion of a hormone is controlled by the body's need for it.

Combinations of hormones from two or more endocrine glands bring about rhythmical or cyclical changes in the prevailing hormonal climate, altering the type of hormone acting at the target area and the response of the target organ. Thus, if one supposes that endocrine organ B is under the control of endocrine organ A through mediation of hormone H1, the latter causes B to secrete a second hormone, H2, which will continue to be produced as long as B is stimulated by H1 from organ A. But sometimes certain concentrations of H2 can influence organ A to inhibit the production of H1, and a fall in the production ensues. This feedback mechanism reaches a certain level of activity and then cuts itself off, somewhat like a thermostat controlling a heater. As soon as the concentration of H2 falls below the cutoff threshold, organ A produces hormone H1 again, and the process is repeated.

A third organ, C, would experience these changes through its vascular supply in the form of successive waves, or rising concentrations, of first H1 and then H2. If organ C, however, were sensitive only to the hormone H2, then its response would reflect only variations in the concentration of H2 and exhibit a cycle of stimulation, regression, and rest. It could also happen that during the period of rest yet another endocrine organ, D, perhaps also under the control of A, would exert an effect on organ C and thereby produce quite different results. A mechanism similar to this brings about the rhythmic series of changes in many organs during the reproductive cycle. Some endocrine substances can exert positive feedback control, which has been observed during the reproductive cycles in primates.

Hormones of one type or another have potent effects at different

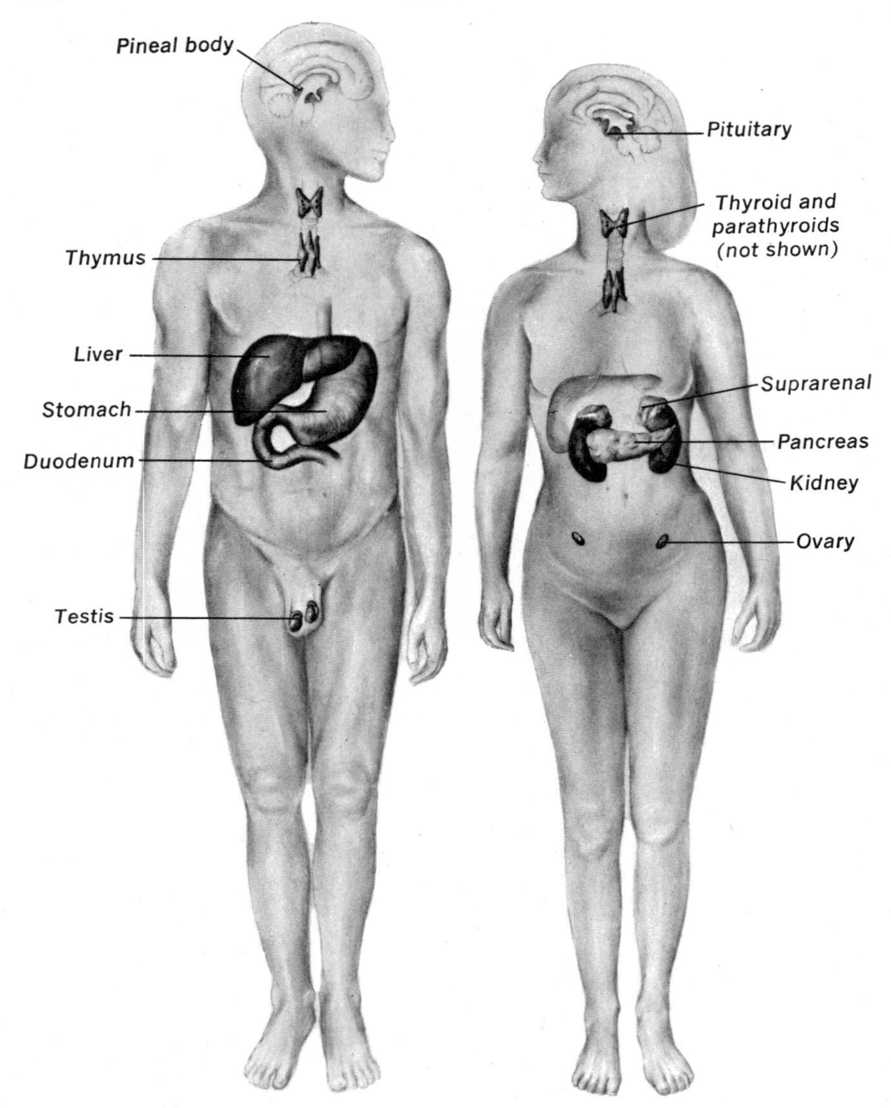

Figure 9–1. Distribution of the major endocrine organs.

periods of an animal's life—embryogenesis and fetal life, growth, and full maturity. As an animal ages, its endocrine activity often decreases. Each period has its own variables that involve the type and number of hormones, the activity of the glands that produce them, and integrative mechanisms of the system at large, the vascularity and sensitivity of target tissues, and possibly the genetic constitution of particular species. Years ago, when endocrinology was still in its infancy, the noted Scottish anthropologist Sir Arthur Keith suggested that morphological differences, particularly in the skeleton, which characterize the various subgroups of man, could be caused by inherited instructions that control the endocrine system. Adult maternal hormones and hormones of placental and fetal origin can affect the fetus; and alterations in the concentrations of some hormones, such as cortisone from adrenal glands, can sometimes produce havoc in organogenesis.

Hormones are produced by easily identifiable organs with a compact and distinctive form and structure, as well as by groups of cells incorporated in organs that have other functions (Figure 9–1). The lining of the gastrointestinal tract has hormonal functions, and islets of endocrine tissue are disseminated throughout the pancreas, the exocrine functions of which are controlled by secretions from cells in the lining of the small intestine. Several types of cells with an endocrine function are found in the placenta of many mammals, and the uterus of pregnant mares produces a substance that stimulates the ovaries.

Endocrine glands have been classified according to their position in the body, their component tissues, their embryological origin, and the chemical nature of their secretions. Through their secretions they all have detectable effects on the cells of other organs and ultimately on the appearance and shape of man. His size, musculature and obesity, strength and vigor, his hairiness and pigmentation, the external texture of his skin, the appearances of the reproductive organs and mammary glands are all influenced by hormones.

The precise way in which hormones act on target cells is not yet understood. They may influence nuclear instructions to the cytoplasm or interfere directly in the activity of enzyme systems in the secretory regions of cells. Why one type of cell should respond to a hormone when another does not remains unknown. In some instances, however, a precise and repeatable response with a known quantity of a particular hormone, such as insulin, can be produced. Some hormones exert what may be called a "permissive" action in that small quantities are needed at all times to enable the animal to respond to particular stimuli and to prepare it or its organs to respond, as with adrenal cortical steroids. When these hormones are absent, numerous and widespread changes occur.

The administration of some few hormones, especially those of the anterior pituitary, can bring about the formation of antihormones. This means that subsequent administrations fail to elicit as marked a response as the initial one. An antihormone thus neutralizes the effect of the injected hormone and is occasionally effective against similar hormone preparations from different species. This is really an antigen-antibody response to a foreign protein.

The Pituitary

Considering its importance, the pituitary gland (hypophysis cerebri) is remarkably small. It resides in a bony closet in the middle of the base of the skull, attached by a short stalk to the undersurface of the diencephalon of the brain. It is intimately associated with the two internal carotid arteries, with a collection of venous sinuses known as the cavernous sinus, and with several cranial nerves that pass through the sinus on their way to the eyes and face. Since the hypophysis lies above one of the paranasal air sinuses of the skull and close to the roof of the nasal cavity, ancient anatomists, even Vesalius, believed that the gland produced nasal mucus. (The word *pituitary* is derived from the Latin *pituita,* slime.)

The gland is roughly oval in shape, about half an inch long, and in man weighs just over 0.5 gm. It is composed of an anterior lobe (adenohypophysis), a posterior lobe (neurohypophysis), and an intermediate lobe and pars tuberalis. The embryological origin of the anterior part is an upgrowth from the roof of the primitive pharynx of the embryo called Rathke's pouch, after a nineteenth-century German anatomist, M. H. Rathke. The posterior lobe develops as a down-growth from the floor of the third ventricle of the brain.

The anterior lobe, by far the largest part of the gland, contains three distinct types of cells, which can be distinguished in histological sections by various staining techniques and in electron micrographs by the type of granules present in each cell. The secretory cells produce a number of hormones, several of which control the action of other endocrine glands. For this reason the pituitary has been referred to as the master endocrine gland. Only some of the hormones from the anterior lobe have been identified and their properties fully characterized. Other hormones are not yet clearly defined, and their presence has been inferred only from their biological action.

Of the hormones that are well known, the growth-promoting principle, or somatotrophic hormone (STH), affects the growth of the skeleton and is responsible for its final individual shaping. An overproduction of this hormone during the period of skeletal development results in gigantism; conversely, its underactivity during childhood results in diminished body growth, or dwarfism. If excessive secretion of the somatotrophic hormone should occur during adolescence or after normal bone growth has ceased, the bones of the hands and feet lengthen disproportionately, the jaws grow, the teeth become widely separated, and the lower jaw projects forward (prognathism); some visceral changes also occur. This condition (Figure 9–2), known as acromegaly (Gk.

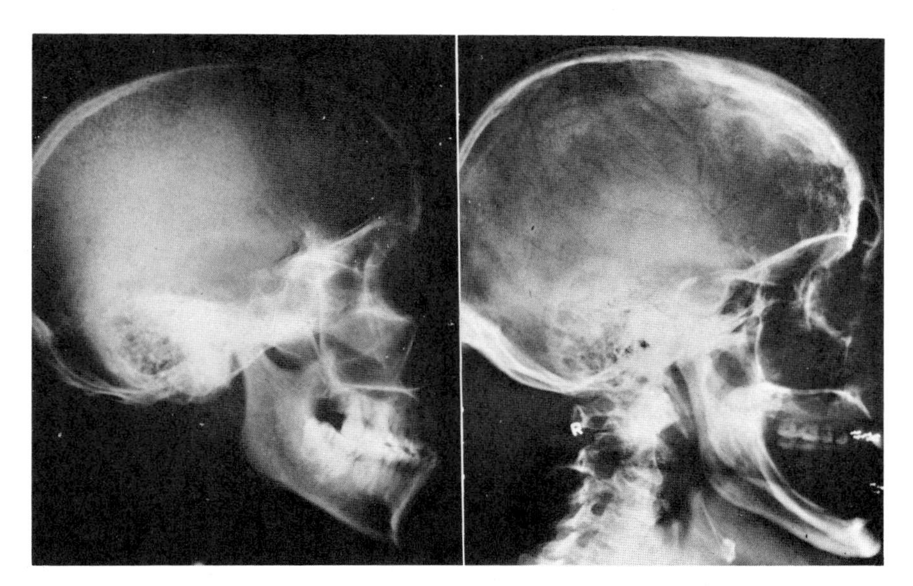

Figure 9–2. X-ray photographs of the human skull. Right: normal person; left: acromegalic person. Note especially the elongation and forward projection of the lower jaw (prognathism). *(Courtesy of Dr. Monte Greer, University of Oregon Medical School.)*

acro, extremity; *mega*, large), is found in individuals who are often strong but become flabby and weak later in life. Many apparently normal but large, big-boned people as well as certain racial subdivisions display acromegalic features; one must assume that some inherited mechanism produces a slight overactivity of the anterior lobe.

Hormones concerned with reproduction also arise from the anterior lobe. The follicle-stimulating hormone, FSH, stimulates the growth of ovarian follicles and of the testicular seminiferous tubules; the luteinizing hormone, LH, stimulates follicular activity and the formation of the corpus luteum and causes the testicular interstitial cells to produce androgens.

The anterior lobe also produces the adrenocorticotrophic hormone, ACTH, which stimulates the adrenal cortex to secrete corticoid hormones. Thus, through the mediation of the adrenocortical hormones, the pituitary has control over metabolic processes. Some evidence exists that it also produces hormones that directly affect metabolism. The thyrotrophic hormone of the pituitary, TSH, which affects only the thyroid gland, has been isolated in a relatively pure state. The removal of the pituitary causes rapid involution of the thyroid, adrenal glands, and gonads; thus it is possible to physiologically thyroidectomize, adrenalectomize, and gonadectomize animals (or man) merely by removing the pituitary. The intermediate lobe secretes a substance that stimulates the melanocytes in the skin of frogs and perhaps other vertebrates, including man. Some doubt remains, however, about the purity of this melanin-stimulating hormone.

The posterior lobe of the pituitary has a uniform structure consisting of modifed nervous connective tissue cells (neuroglial cells) called pituicytes. Many nerve fibers from the hypothalamus also run through the posterior lobe. Its chief hormone is believed to be the antidiuretic principle, which is formed by the cells of the hypothalamic nuclei, passed down the axons of these cells through the hypothalamo-hypophysial tract as neuro-secretory material, and stored or liberated in the posterior lobe. This hormone induces the cells of the renal tubules to reabsorb a certain amount of the water originally filtered through the glomeruli. Should the hypothalamus, its tract to the posterior lobe, or a greater part of the posterior lobe be damaged, the animal develops a condition called diabetes insipidus. Without the action of the posterior lobe principle, thirty percent of the water is not reabsorbed by the renal tubules and excessive amounts of urine are produced.

The Pineal Body

The pineal body is an organ on the back of the roof of the third ventricle of the brain. It is a small ovoid structure (0.2 gm) that has always been something of a puzzle. Its purpose was by no means clarified when the seventeenth-century French philosopher René Descartes suggested that it is the seat of the soul. Far from being vestigial, it seethes with enzymes, especially those involved in the synthesis of monoamines, apparently important in brain function.

The claim of the pineal body to be a full member of the endocrine family is best substantiated by its role in reproductive physiology, that is, the development of sexual activity and other rhythmic phenomena. An indisputable endocrine function of the pineal is its secretion of a hormone, melatonin, which inhibits the action of a melanophore-stimulating hormone in fish and amphibians. In these species the pineal body acts as a photoreceptor. Melatonin has been reported to have questionable results on human melanocytes. In man, tumors in the pineal region are sometimes associated with sexual precocity. The gland has a complex structure, its various parts being characteristically modified in different vertebrates. In certain primitive vertebrates the parapineal is differentiated as a photoreceptor, and in the tuatara *(Sphenodon),* a primitive reptile from New Zealand, it lies in the middle of the head beneath the skin as a parietal eye to which the human pineal gland is in no way related. It becomes calcified later in life, and the deposits can be detected in X-ray films of the skulls of elderly people.

The Thyroid

The thyroid has an interesting evolutionary history. Very primitive chordates have in its stead rows of mucus-secreting cells on the floor of the pharynx. In somewhat more advanced chordates, the gland became tubular and then lost its connection with the pharynx altogether; only at this stage did it exhibit any endocrine activity.

The thyroid of man consists of two large lobes connected by a thin band that crosses the second to fourth rings of the trachea. It weighs between 20 and 30 gm and is heavier in women, particularly during menstruation. It is relatively large in children and smaller in the undernourished. The gland consists of numerous follicles enclosed in an envelope of connective tissue, the lobes lying one on each side of the trachea. Even more than the other endocrine organs, the thyroid has a very rich blood supply and is probably the best vascularized gland in the human body. Estimates suggest that some nine pints of blood (equivalent to the total blood volume) percolate through its vascular bed in about an hour.

The follicular cells of the thyroid secrete a hormone, once thought to be thyroxine, but now said to be triiodothyronine, a substance similar to thyroxine but with one less iodine atom. The gland possesses large amounts of iodine since it "takes up" iodine from the bloodstream with great facility.

The thyroid hormone influences the general metabolic activities of the body. Its failure to function properly during childhood results in defective and inhibited growth and mental retardation, a condition called cretinism. Lowered metabolism, changes in the skin, loss of hair, and accumulation of water in the tissues (myxedema) are all due to a similar failure of thyroid function in adults. On the other hand, overactivity of the thyroid brings about an increased metabolic rate, loss of weight, increased nervousness, and other changes that sometimes culminate in a toxic condition. These disorders clearly illustrate that normally just the right amounts of hormones bring about physiological responses that are essential for efficient everyday life, but too large or too small quantities can produce profound functional and even anatomical damage (Figure 9–3). Fortunately, many of these abnormalities can be reversed with appropriate treatment. The activity of the thyroid is closely linked with that of the anterior lobe of the pituitary through its thyrotrophic hormone. Curiously, other pituitary hormones, such as those that promote growth

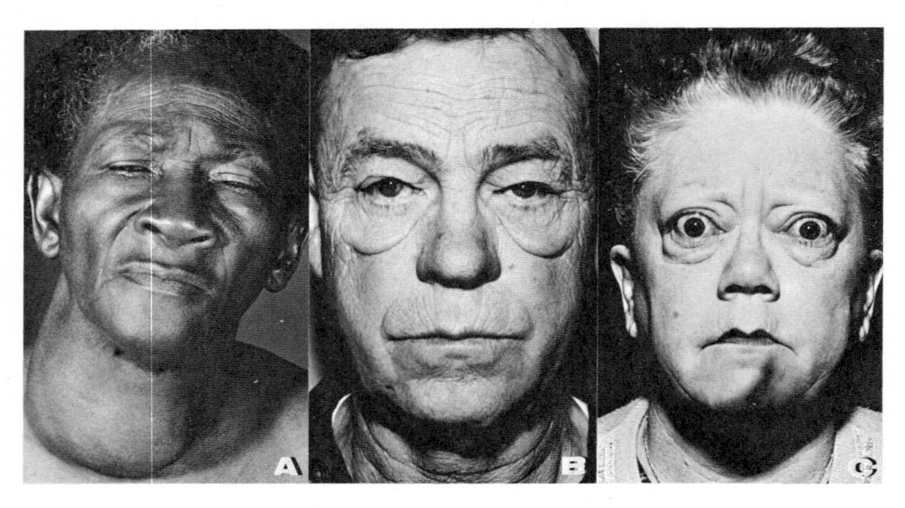

Figure 9–3. Three cases of malfunction of the thyroid. A: nontoxic goiter, an enlargement of the thyroid caused by insufficient secretion of triiodothyronine (thyroxine); B: myxedema, failure of thyroid function, which results in the accumulation of water in tissues; C: exophthalmia, with bulging eyes, a condition brought on by overactive thyroids. *(Courtesy of Dr. Monte Greer, University of Oregon Medical School.)*

and gonadal activity, do not seem to be fully effective without the collaboration of the thyroid hormone.

A second type of epithelial cell exists in the thyroid gland of most mammals including man. Called parafollicular or light cells, they appear to function independently of follicle cells. After hypophysectomy they increase in number. These cells are now considered to be the source of a second thyroid hormone, thyrocalcitonin, whose action seems to influence the calcium level of systemic blood.

The Parathyroids

Four—sometimes fewer, sometimes more—small parathyroid glands, weighing up to 0.1 gm each and found on the back surface of the thyroid within its capsule, produce a proteinous hormone, parathormone, which controls the metabolism of calcium and phosphorus. Total removal of the glands results in a fall of calcium ions in the blood to such an extent that the ensuing muscle tetany causes respiratory and cardiac failure and death in a few days. This hormone, therefore, not only ensures the proper relation between the calcium in the blood and that in the skeleton, but also is a vital factor in bone development and growth. It also affects the density of bone in later life, and, since the bone calcium is "turned over" at a rate of up to twenty percent of the total calcium per month, overactivity of the organ soon results in decalcification and underactivity in the deposition of extra amounts of calcium. If the thyroid is removed, some parathyroid tissue must be left behind. Whether the anterior lobe of the pituitary exerts some control over the parathyroid is not absolutely certain.

The Thymus

The thymus is a mass of lymphatic tissue, arranged in two main thyme-shaped lobes situated in the thorax anterior to the great vessels above the heart. Whether the thymus secretes a hormone is not certain, but its size and life history are closely related to the maturation of the body, and it responds to certain experimental conditions. In general, its size is inversely related to the level of circulating steroid hormones. At birth the organ, which develops from the pharynx, is a flattened, pinkish mass between the sternum and the pericardium weighing about 13 gm. It enlarges steadily, and by puberty has tripled its weight. After this it gradually and variably involutes. Some thymic tissue remains throughout life and is present even in quite elderly individuals. The removal of the thymus is not known to have any endocrinological results in mammals. Although it plays a key role in the establishment and maintenance of some of the body's immunity mechanisms, its significance in terms of hormone secretion remains unknown. Regardless of what endocrinological properties it may have, the thymus can no longer be dismissed as unimportant. All vertebrates, except certain primitive fishes, have one either in relation to the gills, the neck, or within the thorax. Furthermore, more and more attention is being focused on the thymus because of its role in the control of cell-mediated immunity.

The Pancreas

The pancreas is a long, lobulated gland lying transversely on the back wall of the abdominal cavity, mostly behind the stomach, at the level of the first and second lumbar vertebrae. Most of the pancreas is a digestive organ that secretes a complex of enzymes known as pancreatic juice into the first segment of the small intestine. This aids in the digestion of fats, proteins, and carbohydrates. Scattered throughout the gland are some half million compact islets of about a hundred cells each, which secrete the hormone insulin and glucagon. First described in 1869 by a German pathologist, Paul Langerhans, who called them "clumps of epithelial-like protoplasmic cells," they are now called islets of Langerhans. Insulin is essential for carbohydrate metabolism because it controls the synthesis and storage of glycogen in the liver. Glucagon has the opposite action of insulin; by mobilizing and depleting glycogen from the liver, it induces a rise in blood sugar. Other tissues of the body also secrete some glucagon, but the pancreas is its main source. Thus, the pancreas, whose key exocrine function is the breakdown of fats, proteins, and carbohydrates in the intestine to aid absorption, also contributes two essential hormones that control the synthesis, storage, release, and utilization of tissue carbohydrates.

The discovery of insulin is intimately associated with the disease diabetes mellitus, which has been known since the time of Hippocrates. But the description "mellitus" (L. *mel,* honey) is attributable to Thomas Willis, who noticed in the late seventeenth century the sweet, honeylike taste of the urine from diabetic patients. J. von Mering and O. Minkowski showed in 1889 that when the pancreas is removed from a dog, the animal develops diabetes mellitus and soon dies. Not until 1921, however, were F. Banting and C. Best able to extract from the pancreas a complex proteinous material, the essential regulator of carbohydrate metabolism, which they named insulin. In the absence of insulin, there is such an increase in the sugar concentration of the blood that it is eliminated by the kidneys. Seemingly, the hormone stimulates the utilization of sugar by the tissues, with glucagon regulating its release from the liver into the blood and inhibiting sugar formation from body protein.

Other hormones are clearly involved in this process, particularly those of the anterior pituitary. The diabetogenic hormone, or growth hormone (STH), is one. When experimentally injected into animals, it causes a diabetic state and the destruction of some of the cells in the islets. Twenty-five percent of those afflicted with diabetes mellitus have a family history of the disease; its inheritance may be due to a recessive gene (m) together with certain secondary factors.

The Adrenals (Suprarenals)

These two essential endocrine organs, each weighing from 5 to 10 gm, are enclosed within the renal fascia on the upper surface of each kidney. The right gland is shaped somewhat like a pyramid, the left like a crescent. Each gland is composed of a cortex, whose secretion is essential to life, and a medulla, whose secretion is important but dispensable. Most vertebrates possess recognizable suprarenals, but fish have clusters of separate isolated cortical and medullary tissue alongside the large blood vessels between the kidneys. This tissue is derived from the neural crest of the embryo, and the action of its secretion is closely related to that of the sympathetic division of the autonomic nervous system. The cortical tissue develops from localized thickenings of epithelium near the site of origin of the gonads. This close embryological origin of the adrenal cortex and gonads has often been cited as evidence that some structural and functional relation exists between them. Moreover, accessory cortical tissue is often found near the gonads.

There are differences in the adrenal/body weight ratio among mammals: a high ratio, like that in predatory carnivores, has been related to the short, sudden dashes after prey and indicates "sprinters" rather than "stayers," such as cows. Wild rats, for example, have considerably larger adrenals than the docile, inbred, laboratory ones. In man and most active primates, the ratio is closer to that of the "sprinters."

The cortex of the adrenal consists of three zones, one within the other, composed of cells rich in lipid substances, steroids, and vitamin C. The suprarenal is very well vascularized, being supplied by vessels from the aorta, the renal arteries, and those supplying the diaphragm, a further indication that the gland is of crucial importance. Of the many different hormones that have been extracted from the cortical tissue, the biological properties of cortisol and aldosterone are the best known. Cortisol participates in carbohydrate metabolism, and aldosterone controls salt and water metabolism. The removal of the adrenal cortex or its destruction by disease (described by Thomas Addison in 1855) is fatal unless cortical hormones are administered. Insufficient cortical secretion renders the body more sensitive to stress, such as exposure to cold and physical injury. An overactive adrenal cortex or tumor of this tissue causes a condition known as Cushing's syndrome, characterized by abnormal carbohydrate metabolism and fatness of the face, neck, and trunk (Figure 9–4).

The adrenal medulla is an entirely different organ and its propinquity to the cortex that envelops it is only circumstantial. It is brown and about one-tenth the total mass of the gland. Its large granular cells are surrounded by numerous blood-filled sinusoids. The rich nerve supply to the medulla consists of fibers from the sympathetic division of the autonomic system that arise from the lower thoracic and upper lumbar levels of the spinal cord. These nerves stimulate the activity of the medulla, which secretes the excitatory hormones, adrenaline and noradrenaline. When adrenaline is injected experimentally, it causes a transient rise in blood pressure from the constriction of some but not all of the peripheral blood vessels, a rise in the heart rate, an increase in the metabolic rate, and a rise in the blood sugar level. Thus, the secretion of the adrenal medulla gears the body to react advantageously to sudden demands or crises. There is also some evidence that the medullary hormones control the secretory activity of the cortex.

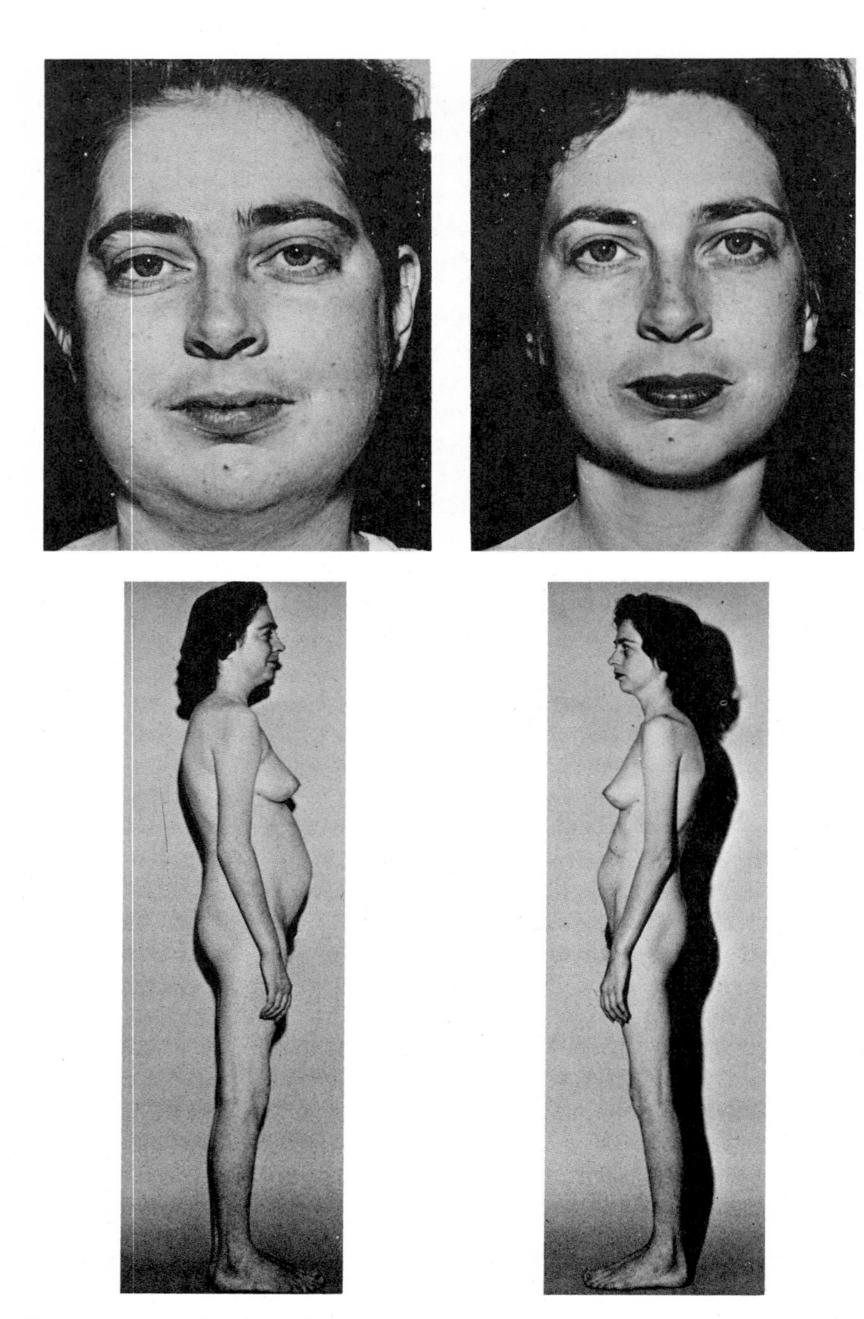

Figure 9–4. Cushing's syndrome. Left column: a patient suffering from the disease; right column: the same patient after surgical removal of the adrenal cortical tumor. *(Courtesy of Dr. Monte Greer, University of Oregon Medical School.)*

The Kidney

Under certain conditions, for example, constriction of the renal artery or loss of a great deal of blood, the kidneys secrete renin into the blood which acts upon a blood protein called angiotensinogen and converts it into angiotensin II. This new substance is a vasoconstrictor, which physiologically acts differently from the adrenalines. The concentrations of renin and angiotensinogen vary in different species, and the action of renin shows some species specificity. Renins from primates, for example, act upon the angiotensinogen of other mammals, but the action is not reciprocal. The renin-angiotensin system is a safeguard mechanism that maintains blood pressure even after hemorrhage or severe wounding.

Other Aspects of the Internal Environment

Many other factors besides the hormonal ones already described affect man's internal environment. So that every cell can survive and function properly, provision must be made to supply its needs and remove its waste products lest these disturb its equilibrium. Reserves must be made available when activity is increased or when sudden demands for extra effort strain the internal environment. Just as the whole man functions because of numerous control systems equipped with feedback mechanisms, so each living cell is a microcosm, subject to similar controls. The vascular system transports essential gases and nutriments to the cells and draws waste products away from them. The concentration of substances carried by the blood and its corpuscular components is critical for the proper functioning and even survival of every cell. Even too much or too little water can be dangerous to cells.

When animals left their watery habitats to become terrestrial, they evolved control mechanisms that eventually affected the stabilization of each constituent. Some of these mechanisms were present in primitive animals and were modified only when their descendants became terrestrial and warm-blooded. Just as the size of the brain increased and nerve activities became more refined, so did the sensitivity and efficiency of the control mechanisms. To realize the importance of keeping the brain well supplied with oxygenated blood, one need only reflect on the consequences of cutting off the oxygen supply—irreversible damage and unconsciousness in about two minutes unless breathing is resumed.

Indeed, the power of certain control mechanisms is so strong that we are unable to stop their operation by our own efforts. For example, if we try to hold our breath, sooner or later we must expel the air and take another breath. Start running: soon both your respiration and your heart rates increase. Drink several bottles of beer: eventually and inevitably the excessive water has to be excreted.

Cells depend for what they need on their environment—on tissue fluids, on an efficient blood supply, and on lymphatic drainage. Indefatigable injector of blood vessels and lymphatics, Frederik Ruysch declared in 1696 that the body consisted of "vessels variously arranged," emphasizing the importance of William Harvey's demonstration of the circulation of blood in 1628. Ruysch's enthusiasm for injections was his undoing and led him to declare that the cerebral cortex was nothing more than a complex mass of blood vessels. Neither Ruysch nor Harvey, as far as we know, was aware of the complexity of the interrelated mechanisms

Table 9–1. Some Records of Human Performances

Type of performance	*Time/ Distance*	*Performer*	*Date*
Deepest dive in water	217 ft., 6 in.	R. Croft	1967 (Navy Record)
In water without breathing	6 min., 29 & 4/5 sec.	M. Pauliquen	1912
In water without breathing *	13 min., 42.5 sec.	R. Foster	1959
Swimming:			
100 yd. sprint	46.8 sec.	S. Clark	1961 (World Record)
Continuous swimming in a tank	87 hr., 27 min.	M. Huddleston	1931
Swimming long distance **—2,300 mi.	30 days, 2 hrs.	F. Newton	1931
Running:			
100 yd. sprint	9.1 sec.	R. Hayes	1963 (4 times) (World Record)
220 yd. sprint	20.0 sec.	T. Smith	1966 (World Record)
Jogging:			
159 mi., 562 yds.	24 hrs.	W. Hayward	1953

* *Before submerging breathed pure oxygen for 30 minutes.*

** *Swam the length of the Mississippi River from Clinton, Oklahoma to New Orleans, Louisiana.*

Table 9–1. Some Records of Human Performances—*Continued*

Type of performance	*Time/ Distance*	*Performer*	*Date*
Walking:			
Speed–¼ mi.	1 min., 22.5 sec.	F. H. Creamer	1897 (World Record)
Distance–133 mi., 21 yds.	24 hrs.	H. D. Neilson	1960 (World Record)
High jump	7 ft., 6¼ in.	P. Matzdorf	1972 (World Record)
Broad jump	29 ft., 2½ in.	R. Beamon	1968 (World Record)
Throwing			
Baseball	443 ft., 3½ in.	D. Grate	1953 (World Record)
Football	300 ft.	R. Waterfield	1945 (World Record)
Basketball	92 ft. field goal	J. Harkness	1967 (World Record)
Javelin	307 ft., 9 in.	J. Lusis	1972 (World Record)
Shotput	71 ft., 5½ in.	R. Matson	1967 (World Record)
Hammer	250 ft., 8 in.	W. Schmidt	1972 (World Record)
Weight lifting	1,339¼ lb.	V. Alekseev	1970 (Olympic Record)

that control respiration, heart action, and blood pressure. However, equipped with hindsight, we can detect in their writings numerous clues to these mechanisms, which later became apparent when anatomical and physiological investigation had revealed respiratory centers in the hindbrain, vagal and sympathetic innervation of the heart, and the heart's own intrinsic atrioventricular bundle and "pacemaker." We now know about chemoreceptors in carotid and aortic bodies, pressure receptors within vessels, and other pathways that conduct impulses to the brain. Experiments with many species of mammals have given us a clear insight into the neural, chemical, and physical control of respiration and blood flow and, more recently, of the mechanism of cellular respiration.

All mammals appear to be equipped with such mechanisms. Among diving mammals, the modifications of certain reflexes and responses enable them to perform under water in a way that surpasses the ability of any human swimmer (see Table 9–1). Michael Wenden of Australia has covered 100 meters, free style, in 52.2 seconds, and Stephen Clark of the United States has reached a record speed of 4.89 miles per hour through water.

Some dolphins, however, have traveled over timed distances at speeds of 15 (*Lagenorhynchus*) to over 21 knots (*Stenella*), and trained *Tursiops* have dived to over 1,000 feet. Even though man resorts to artificial aids to help him underwater, he is still surpassed by porpoises and dolphins. The anatomy of man and other primates, however, reflects none of the numerous structural modifications of truly aquatic mammals. On the contrary, man's structure and the mechanisms controlling his internal environment seem fashioned specifically for a terrestrial existence. They appear to have undergone adaptation to sudden, rapid, and repeated demands on bodily stamina caused by continued physical endeavor. His long legs and arms and relatively mobile vertebral column and head have equipped him to be the sporting, exploring, widely distributed, and competitive creature he is. Even a cursory reading of Table 9–1 will indicate man's astonishing versatility. With the advanced training techniques now available, any one of these records will sooner or later be broken.

10

MAN'S REPRODUCTIVE PATTERNS

. . . research has carried mankind to the very threshold of willfully directed reproduction.
(Emil Witschi)

Except for those organisms that reproduce vegetatively (asexually), most living beings are equipped with special reproductive organs, male testes and female ovaries, called gonads in animals. These produce, nurture, and liberate germ cells, sperm (spermatozoa) and eggs (ova), which at the time of fertilization unite to form a single cell, the zygote. Once fertilized, the egg divides and continues to divide and differentiate until a new individual is formed, itself usually possessing gonads, which, in their turn, will become functional and repeat the process. Before this process could happen, mammals had to undergo a number of important and interrelated adaptations involving all parts of the reproductive apparatus and the organs that control them.

A potentially prolific creature, man is characterized by a high reproductive rate. His proclivity to reproduce has become especially apparent in the last hundred years during which the total population of the world has more than doubled. Several factors have contributed to this success: (1) his life expectancy has considerably increased; (2) the mortality of women during and immediately after pregnancy has lessened; (3) the survival rate of infants has greatly increased; (4) the young family is much more mobile than before and can move with ease to a better environment; and (5) disease, famine, war, neglect, poverty, and failure to develop resources have all lessened in various degrees in many parts of the world. Thus, human reproduction has become highly successful because of increasing gains in fertility (ability to reproduce), low mortality, high fecundity (quantity produced) and natality (successful births), marked longevity, and extraordinary mobility. Man is the most widely spread of all terrestrial species, and his reproductive prowess has been so successful that the problem now is to control rather than to increase his numbers.

In mid-1971, there were 3.7 billion people on the earth, 7.4 million more than the previous year. With the current world's growth rate of about 2 percent, the world population increases by 200,000 people daily and at this pace should double itself in 35 years. The highest growth rate in the world, 2.9 percent, is in South America; but Asia, with a growth rate of 2.3 percent and 2.1 billion inhabitants, added the greatest number of people to the world last year: 48 million. The United States has a 1.1 rate, up from 1.0 percent in 1970; by the middle of 1971 there were 2,300,000 more American citizens than at the same time the previous year.

Several of man's anatomical features decisively affect his reproductive capacity. His brain is large at birth, about a quarter of its adult weight, a fact that makes the infantile period more hazardous and precarious. With so large a head for women to deliver, there has evolved a female pelvis large enough to allow the birth of viable young. Again, human growth proceeds at a relatively slow rate; hence the reproductive activities remain potential until a certain maturity has been reached. The child's reproductive organs keep pace with his general growth and become functional only when he and they reach a certain size and state of development.

Puberty

Man spends about one-sixth of his total expected life growing from childhood to puberty. The attainment of puberty is gradual; it is not a particular day in an individual's life but rather a period during which the development of reproductive activity is speeded up and finally established. The process is mediated through the activity of the anterior lobe of the pituitary. From fragmentary information we infer that this organ, or that part of the brain that controls it (the hypothalamus), is programmed to regulate the development of reproductive organs at a particular age in each species of mammal.

Climate was once believed to influence the onset of puberty, and children in the tropics were thought to reach puberty earlier than those in cooler areas. If such variations do exist, they are probably due to other causes, such as genetic factors and the quantity and type of food available. In any event, the onset of puberty is gradual in man and is delayed longer (10 to 16 years) than in any other mammal. During recent decades the average time of onset has, however, been reached earlier in girls of western Europe and the United States; some evidence suggests that this is correlated with the attainment of a critical weight.

At puberty an accumulation of fat in the skin over the pubes elevates the area into the *mons Veneris* in women and the *mons Jovis* in men. Hair also appears there, frequently differing in color and quality from that found elsewhere. The physical and behavioral changes characteristic of puberty are considered in chapters 11 and 12. Discussed here are only the changes in the reproductive organs.

The first menstrual discharge (menarche) signals the onset of puberty in girls. This biological experience terminates the period of preparation of the reproductive mechanism and indeed of the whole body and inaugurates the years of reproductive activity. The first menstruation marks the point at which reproduction becomes a physical possibility, though not necessarily a desirable experience. Girls still need several years before they mature to a state when reproduction can occur optimally for both the mother and young, and perhaps even for the father.

Progressive testicular function accompanies puberty in boys. Androgenic hormones (androgens are substances that conduce to masculinity), produced by certain cells in the testes, bring about the physical changes that signal the transition from childhood to adolescence. The tubules of the testes commence to produce spermatozoa, and the accessory male reproductive organs—prostate and seminal vesicles—begin to secrete seminal fluid. Boys become capable of ejaculating semen at puberty, but, like girls, they need several more years of maturation to be fully effective.

The activity of the gonads is controlled by gonadotrophic hormones from the anterior lobe of the pituitary. These stimulate the ovary and the testis to secrete the steroid hormones that guide the development and maturation of the reproductive organs and other sex-related organs. If ovaries or testes fail to develop, or if they are removed before puberty, these effects fail to materialize. The removal of the testes in boys before puberty results in eunuchs, individuals in whom all typical male characteristics have failed to develop. Such persons may grow to full size but retain the build, distribution of hair, voice, and temperament of boys and are, of course, sterile. The effects of castration performed after puberty are different and depend largely on the age and maturity of the individual at the time, but it nearly always leads to obesity.

Puberty can be delayed experimentally in animals by severe alterations in their diet and sometimes by alterations in the altitude where they are reared. It is also delayed in premature babies and in twins. It can be advanced by injecting gonadotrophins. Massive administration of sex hormones can also affect the physique of young human beings (Figure 10–1) and other animals. This practice has been used in cattle raising and has been tried, unwisely, by certain types of athletes to improve their performance. Massive doses of sex hormones often result in gonadal damage and reduced reproductive vigor and may even shorten life.

Puberty appears at different times in different species of mammals (Table 10–1). Some correlation exists between the size or weight of adult animals and the length of time that precedes its onset. With some exceptions, such as bats, small mammals reach puberty when only a few months old whereas the largest mammals reach it after several years. Primates have an absolutely longer period from birth to puberty, the larger ones reaching puberty at an older age than the smaller ones and man latest

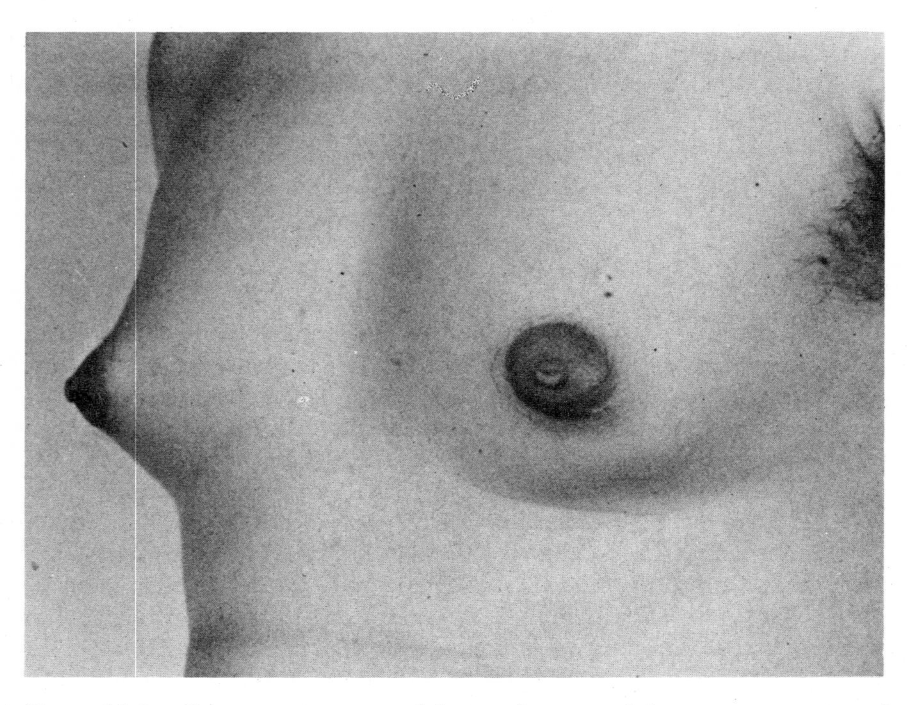

Figure 10–1. Gynecomastia in an adolescent boy caused by an oversecretion of estrogenic hormones, probably in his testes.

Table 10–1. Age at Reaching Puberty in Various Mammals and Primates

Mammals		*Primates*	
Mouse	5–7 weeks	Marmoset	14–15 months
Rat	6–9 weeks	Ring-tailed lemur	1½–2 years
Golden hamster	7–8 weeks	Bush baby	1½–2 years
Guinea pig	8–10 weeks	Rhesus monkey	2–4 years
Rabbit	6–9 months	Pig-tailed macaque	4–4½ years
Dog	6–8 months	Chacma baboon	2½–3½ years
Sheep	7–8 months	Gelada baboon	5 years
Pig	7–8 months	Langur	6–7 years
Cow	9–10 months	Gibbon	8–10 years
Mink	10–11 months	Chimpanzee	7–9 years
Horse	11–12 months	Man	10–16 years
Cat	12–15 months		
Common pipistrelle	15–18 months		
Sperm whale	18–24 months		
Blue whale	2 years		
Common seal	6–7 years		
Brown bear	6 years		
Indian elephant	9–14 years		

of all. Considering his size and physique, man reaches puberty later than any other animal except, perhaps, the elephant.

Adolescence and Sexual Maturity

Adolescence is the period extending from puberty to full maturity. In some mammals it is synchronous with sexual maturity. Females are capable of becoming pregnant and successfully giving birth to young and looking after them immediately after adolescence begins. In other mammals, however, a gradual but steady development of the gonads and secondary sexual characteristics marks adolescence. Man experiences the longest adolescence of all. He becomes in all ways best equipped for reproducing when he reaches full sexual maturity at the end of adolescence. Then his gonads and accessory organs are full grown and at peak function to allow successful conception. The most advantageous circumstances for the bearing and rearing of young also exist, including responsible maternal (and paternal) attitudes toward the well-being and future of the offspring.

Human sexual maturity is attained anywhere from 16 to 22 years, but a specific time is impossible to state since economic and social factors affect us directly and in many subtle and unknown ways. For example, the widespread use of oral contraceptives may increase the average age of primigravid women (first pregnancy) since they can deliberately postpone motherhood until their situation improves socially and financially. No one can tell how this will influence future generations. A girl can become pregnant in early adolescence or, rarely, even before her first menstrual discharge. There have been newspaper reports of pregnancy before ten years. Usually, however, a short period of "adolescent sterility" occurs during which ovulation is irregular or nonexistent. Pregnancy in adolescent girls carries risks to both mother and young; hence it is biologically sensible that our sexual mores should continue to protect the adolescent.

Sexual maturity is followed by full physical maturity when an individual attains adulthood in all anatomical and physiological ways. This usually coincides with the end of bone growth. It is not correct, however, to assume that human ontogeny stops with the cessation of bone growth. In men, for example, the beard, eyebrows, and body hairs become progressively stouter until about 50. Such hairs as those in the nostrils and ears, particularly conspicuous in middle-aged and old men, are relatively inconspicuous in the young. In many small mammals physical maturity occurs a few months or even weeks after sexual maturity. Bone growth in small primates ceases a few years or months after the onset of reproductive competence. In the great apes it ceases by 14 to 15 years, and in human beings some bones (such as the clavicle) continue to grow until 21 to 24 in women, and until 23 to 26 in men. The bones of the skull continue to grow for a year or two longer, even if only slightly. Some authorities maintain that men and women are best proportioned and most comely in the latter part of adolescence, before physical maturity has broadened their shoulders and thickened their hips.

Reproductive Patterns in Women

Men are "constant breeders," that is, from puberty until the decline of sexual activities in old age, they are, within optimal conditions, capable of effecting impregnation at any moment. The principal reproduction events, however, are reflected in women. At puberty the reproductive organs begin a succession of sexual cycles each lasting about 28 days. The start of each cycle is marked by a period of menstruation, during which the inner lining of the uterus breaks down and, together with some blood, is discharged through the cervix and vagina for from 2 to 8 days. The length of the cycle, the amount of the discharge, and the length of the menstruation period become regularly established a few years after puberty. Even then, some irregularity can be considered normal since the whole phenomenon is under a delicately balanced hormonal control. The critical point in each cycle is the 13th to 14th day after the start of menstruation when ovulation takes place, i.e., when an oocyte (precursor to an ovum) is liberated from the ovary.

During the first part of each sexual cycle, the cellular coverings of one or more oocytes within the ovary develop into fluid-filled follicles. At the end of this follicular phase, a follicle ruptures (ovulation), and the oocyte begins its journey through the uterine tube toward the uterus. Fertilization takes place in the outer part of the uterine tube when the oocyte undergoes its final maturation and becomes a fertilized ovum. During the follicular phase the cells of the follicle release steroid hormones, collectively called estrogens. Through the action of these powerful hormones the lining of the uterus (the endometrium) is prepared to receive the fertilized ovum, and other organs respond in such a way as to make fertilization and pregnancy possible. After ovulation, the cells of the ruptured follicle, which had been secreting estrogen, become very large and form a different endocrine gland, the corpus luteum, whose secretions, estrogen and progesterone, are essential for successful pregnancy. These two hormones act synergistically to further prepare and then maintain the endometrium for pregnancy.

Thus, each sexual cycle reflects the changes that take place in the ovary primarily and in the uterus secondarily. Each ovarian cycle has a follicular phase followed by ovulation and then by a luteal phase. During each uterine cycle menstruation is followed first by an estrogenic phase and then by a progestational phase, so called after the two principal hormones secreted by the cells of the ovarian follicle and by the corpus luteum. When pregnancy does not occur, the corpus luteum soon degenerates, and as a result the endometrium breaks down and is cast off in the menstrual flow. These cyclic events follow one another with some regularity. The ovarian cycle is controlled by gonadotrophins from the anterior lobe of the pituitary to be described later. The uterine cycle is controlled by the steroid hormones, estrogens and progesterone. In turn, the amounts of ovarian hormones secreted influence the output and the type and quantity of each gonadotrophin secreted from the anterior pituitary, establishing a negative feedback system.

The sexual cycle, then, represents an interplay and interaction of pituitary and ovarian hormones. Knowledge of the balance and control in this system led Gregory Pincus and his associates to formulate an oral contraceptive pill to inhibit ovulation and so to control fertility. The mechanism of the action of the "pill" will be discussed in chapter 12.

When all goes well and ovulation occurs within a very few hours before or after sexual intercourse, conception takes place at the very moment that sperm and oocyte come together. The oocyte once released is a senile cell, with only about 12 hours to live. Thus a woman is actually fertile for only a few hours during her cycle of about 28 days. Since sperm are viable for only 2 to 5 days inside the female reproductive tract, the total days of fertility can be calculated at a maximum of 5 under exceptionally ideal conditions. Fertilization, or conception, occurs high up in the first third of the uterine tube and not in the uterus. After being liberated into the vagina, the sperm make their way through the cervix, across the body of the uterus, and up the tortuous uterine tubes. In spite of such a long course, the sperm cover the distance with great facility.

The oocyte and sperm unite to form a single cell, the zygote, which within hours starts to divide, first into two cells, then four, eight, and so on until a solid cluster of cells, a morula, has formed. This division occurs while the zygote is slowly descending to the uterus. Four to 5 days after fertilization the morula becomes hollowed out into a small fluid-filled sphere of cells called a blastocyst. On the fifth day it reaches the uterus where it becomes attached to the surface of the endometrium and then is buried within it. This phenomenon of implantation marks the start of the true intrauterine life of the conceptus. It is also the point in time when the conceptus begins to become increasingly involved in the reproductive events within the mother. Shortly after implantation chemical signals liberated from the site of implantation act on the maternal pituitary, possibly also on her ovary, inhibiting further sexual cycles and suppressing the menstrual discharge that would have occurred 14 days after ovulation. In this way the conceptus, safely embedded in the succulent uterine lining, ensures its development.

The outer layer of cells of the blastocyst, known as the trophoblast, plays an important part in implantation. As pregnancy advances, the trophoblast assumes the role of an endocrine organ, producing several kinds of gonadotrophins as well as estrogens and progesterone. It also forms the placenta, an organ essential for transporting gases and chemical substances to and from the embryo and fetus while secreting hormones in its own right. Pregnancy can aptly be described as an endocrine syndrome, with its initiation, progress, and termination all under strict hormonal control.

The duration of pregnancy (gestation) lasts about 267 days from the time of implantation, or 280.2 ± 0.3 days, standard deviation 9.2, from the onset of the last menstrual period. This period is more variable in women than in any other mammal so that the day of birth cannot be infallibly predicted. The growth of the conceptus is usually divided into two main phases, embryonic and fetal, the former lasting eight weeks, during which the main organ systems of the embryo appear and differentiate. The fetal period is marked by increase in growth, changes in shape, and the completion of the differentiation of all systems except the brain and spinal cord, which continue to develop until some years after birth.

Birth (parturition) is usually divided into three stages: (1) the cervix of the uterus becomes dilated and the part of the fetus to be delivered first (normally the head) is presented; (2) the fetus is expelled from the mother; (3) the placenta is expelled. Immediately after the birth of her child, the mother passes through the puerperium (L. *puer,* a boy; L. *parere,* to bear), a period that in the past was extremely hazardous for the mother because the raw uterine surface to which the placenta had been attached easily became infected and caused puerperal fever. Today, antisepsis and antibiotics have largely obviated this danger. During the first three to four days of puerperium, the raw surface of the uterus exudes some blood (lochia), which by the tenth to twelfth day becomes brown and then yellowish white. Some exudate may continue for three or more weeks until the uterine lining is completely repaired.

Under normal circumstances a lactation period of about nine months follows during which the neonate is nursed by its mother. Man usually gives birth to one young at a time. Twins occur once in every 85 pregnancies, triplets once in every 85^2 pregnancies, quadruplets once in every 85^3 pregnancies, and so on, following what is sometimes known as Hellin's law of multiple births.

The end of the reproductive life in women (menopause) varies from 35 to 55 years of age (average 49.5 years) and may last for several years, which are marked by a more or less gradual cessation of menstruation. The various signs and symptoms accompanying and following the menopause can mostly be ascribed to a decline and cessation of ovarian activity, but there may be other, still unknown causes.

The Ovary and Uterus

The ovaries, which lie on each side of the true pelvic cavity (Figure 10–2), are suspended from the back of the broad ligaments that extend transversely across the pelvic cavity and wrap around the uterus. Classical anatomists likened these ligaments to the wings of a bat and called them alae vespertilioni. Along the top edges of the ligaments the variably coiled uterine tubes extend from near each ovary to the tubo-uterine junction. They are also called fallopian tubes, after the sixteenth-century Paduan Gabriello Fallopius, who described them. An industrious anatomist, he had no idea of their function and thought they were chimney pipes for the escape of "sooty humors" from the uterus.

The ovaries of a newborn girl contain all the eggs, or primary

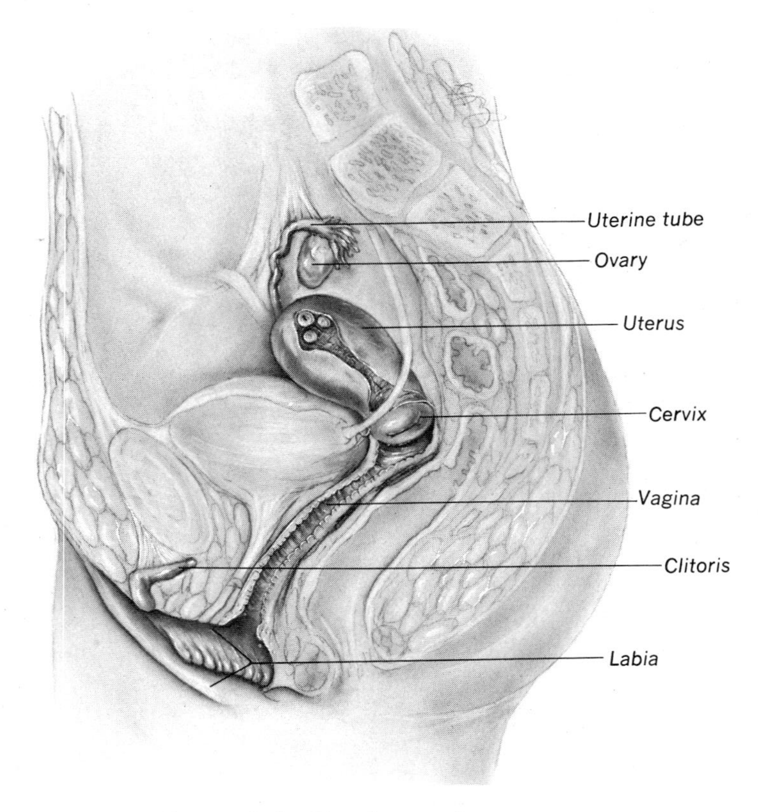

Figure 10–2. Female reproductive organs.

oocytes (up to 700,000), they will ever have. The oocytes are at the periphery of the ovary and are surrounded by a thin capsule of cells, which along with the oocytes form primordial follicles. The oocytes are formed during the division of germ cells that originate in the wall of the primitive gut or yolk sac of the early embryo, when the shape of most organs is only blocked out. Subsequently, somewhat haphazardly, they invade the dorsal body wall of the embryo, traveling along the stalk of the yolk sac. The germ cells that come to rest in the presumptive gonads survive and multiply rapidly until about the time of birth, when the full complement of primary oocytes is reached; those that wander to other presumptive organs seem to perish.

The number of oocytes declines rapidly over the first ten years of life with a steady fall during the reproductive period and complete annihilation at menopause. We do not know what brings about this early wholesale slaughter of oocytes. It almost seems as if an innate timing mechanism within some oocytes causes them to survive longer than others. Whatever the explanation, enough remain viable to provide one (rarely two or more) for each 28-day cycle during the reproductively competent period. Thus, a woman who began to ovulate at the age of 14 and continued at the rate of approximately once each lunar month until the menopause at age 45 (this is a generous estimate) would ovulate at most only 404 times. All the other oocytes would have perished by the time of the menopause. Furthermore, not all ovulated oocytes are of the best quality; many are faulty and die even if fertilized.

At the time of puberty a few of the oocytes and the capsule of cells around them start to grow in response to follicle-stimulating hormones (FSH) from the anterior pituitary. The follicle cells increase in number and size, but the oocyte enlarges only slightly (Figure 10–3). (The size of a mature oocyte is about 120 μ, just barely visible to a naked eye endowed with perfect vision.) Why only a few follicles respond to FSH stimulation is not known. What the selective factor is in those that do, or what growth factors are involved, also remains a mystery. It is difficult to explain why in man and a few other species of mammals only one follicle at a time eventually reaches maturity, whereas in others (the sow, for example) as many as 15 mature at one time. This law of constancy in the number of follicles reaching maturity at any one time is so firmly implanted that even when parts of an ovary are removed or when a number of follicles start to grow at once, only a characteristic number of them survive. For some reason the others degenerate, become atretic, and die. Again, the answer may lie in some internal timing mechanism within individual oocytes that makes them susceptible to hormonal stimulation.

A follicle that starts to grow is progressively stimulated by the pituitary gonadotrophin, FSH. In each human ovarian cycle, one follicle usually enlarges ahead of all others. Its cell capsule becomes two-layered: an inner membrana granulosa immediately develops about the oocyte, which presumably provides it with nutrition, and a surrounding theca interna, which rapidly develops into a highly vascularized endocrine tissue. This secretes estrogens when stimulated first by FSH and also, in some species, toward the time of ovulation by another pituitary gonadotrophin, LH (luteinizing hormone).

Three important events now occur. First, as the follicle continues to grow, especially through an increase in the number of granulosa cell layers, a fluid-filled cavity develops in that layer (Figure 10–3). The space progressively enlarges, increasing the total size of the follicle until it is many times larger than it was initially. In general, the final size of each follicle is related to the size of the adult mammal. (The size of ovulated mammalian oocytes is remarkably constant, 90 to 150 μ, whether of a blue whale or of the smallest shrew.) The mature follicle in a human ovary can reach 1.5 cm in diameter; in a large fin whale it can

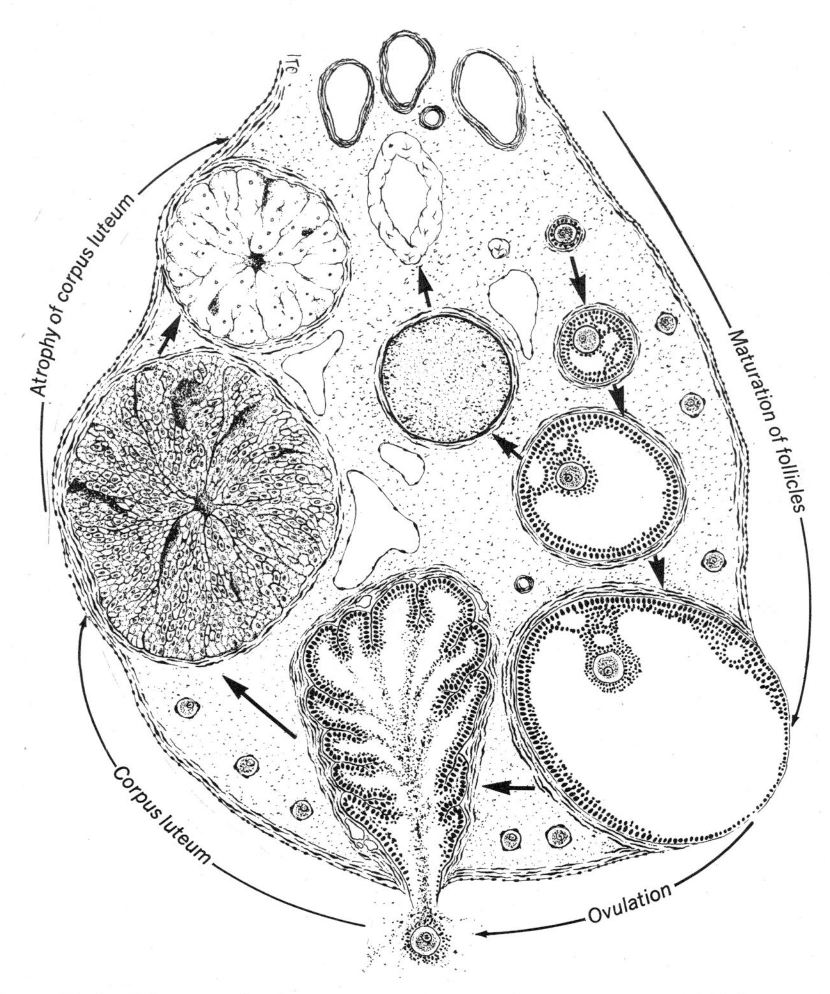

Figure 10–3. Diagram of a human ovary showing the maturation of follicles, ovulation, and corpus luteum.

reach 7 cm. The fluid it contains provides a watery medium in which the oocyte is more easily carried from the ruptured follicle to the open end of the oviduct. The first phase of each sexual cycle must, therefore, be long enough to allow adequate follicular growth. Since this is the more variable phase in the cycle, ovulation does not always occur on the thirteenth or fourteenth day, counting from the first day of the last menses. Ovulation has been reported in women as early as the sixth day and as late as the twenty-third day (see chapter 12).

Second, while the follicular space is developing, the oocyte gradually acquires a thick, translucent membrane called zona pellucida (Figure 10–4), the origin of which is still unknown. This is traversed by minute pores, which high resolution electron micrographs have shown to be occupied by slender processes from the surrounding granulosa cells as well as from the oocyte itself. No doubt, these processes aid in the transfer of materials to and from the oocyte across the zona pellucida. In spite of being porous and elastic, the zona seems to have a protective function. Certainly, in ferrets, goats, and some other mammals, the administration of steroid hormones causes those oocytes that lack a zona either to enlarge disproportionately or to divide prematurely and degenerate. Perhaps the real function of the zona becomes apparent later, when the fertilized egg is inside the uterine tube. Here it may act as a barrier between the dividing cells of the conceptus (indeed it may help hold them together in a cluster) and the maternal cells that line the tube and perhaps also prevent implantation until the blastocyst has safely reached the uterus.

The third event occurs in the oocyte itself. The primary oocyte contains the same number of chromosomes as the maternal somatic cells, namely 44 + 2X (diploid). At about the time the space (antrum) develops in the follicle, the primary oocyte divides unequally into a large secondary oocyte and a tiny, unneeded polar body that ultimately degenerates. This is the first of two maturation divisions (the second usually

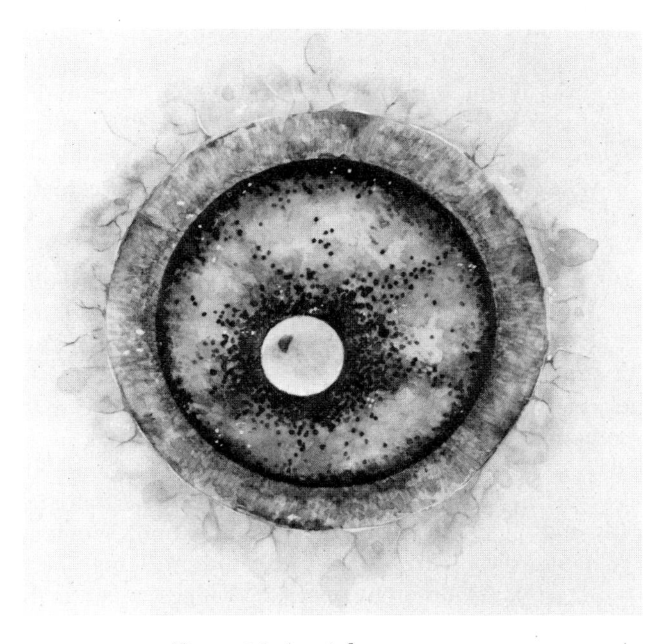

Figure 10–4. A human ovum.

occurs later, when the sperm enters the egg) that brings about a reduction in the number of chromosomes to 22 + X (haploid). At fertilization the diploid number is reconstituted by the chromosomes contained in the spermatozoon, which is also haploid. Should the number of chromosomes in the egg not be reduced exactly in half, the formation of the new individual could be abnormal (mongolism, Turner-Klinefelter syndrome, and so on).

A follicle nearing maturity is under the influence of both FSH and LH from the anterior pituitary. At this time, the theca interna attains such a size that in some mammals near the time of estrus it has been called a thecal gland and is considered to be the source of estrogens. The mature follicle was first described, though not understood, by a Dutch embryologist, Regnier de Graaf, in the seventeenth century and is often called a graafian follicle.

Ovulation

Ovulation (the release of the secondary oocyte from the follicle) is brought about primarily by the action of pituitary gonadotrophins on the follicle and in most mammals is related to manifestations of heat, or estrus. A change in the balance between FSH and LH causes a sudden enlargement of the follicle with a disproportionate secretion of follicular fluid. Under this pressure, the most superficial part of the follicle becomes thinned and eventually tears; the follicular fluid oozes out carrying with it the secondary oocyte. The cells lining the open ampulla of the uterine tubes are covered with long, hairlike cilia that beat in wavelike motion toward its lumen, causing a mild current that eventually draws the egg into its entrance. Unlike most other mammals, women give no particular indication that ovulation has occurred. Midway through the cycle some women experience variable degrees of abdominal pain *(mittelschmerz)* probably because of spasmodic contractions of the uterine tubes, and most women undergo a slight rise and fall in body temperature.

At this time distinct changes in the reproductive tract are reflected in certain behavioral patterns, including receptivity to the male. Vaginal discharges, sometimes blood-stained, originate from cervical glands; hyperemia of the genital tract and, in some primates, changes in the hue and swelling of perineal skin also occur. The female participates in courtship antics and may actively seek sexual gratification. Women, however, give no overt manifestations of estrus and show no apparent increase in the intensity of response, although distinct changes occur in the vaginal epithelium, the endometrium, uterine tubes, and even mammary glands and skin.

The Corpus Luteum

This endocrine gland develops from the cells of the ruptured follicle when the oocyte has been released. The cells of the membrana granulosa grow into large granulosa-luteal cells that secrete progesterone. Most of the theca interna cells remain at the periphery of the corpus, but some literally invade the luteal tissue. The theca-luteal cells almost certainly continue to secrete estrogens, but their fate varies in different mammalian types. The corpus luteum is fully vascularized and active by about the sixth day after ovulation. This significant point coincides with the time when the blastocyst (if fertilization has occurred) has reached the uterus and is about to become implanted. When implantation does occur, the corpus luteum remains large and active and continues to maintain the endometrium. The fate of the corpus luteum during pregnancy varies in each group of mammals; in some it persists throughout. In women the corpus luteum is active only during the first two months of pregnancy, after which it regresses as its endocrine functions are taken over by the placenta. Shortly after parturition, only a small fibrous scar remains in the ovary.

In a cycle that does not culminate in pregnancy, the corpus luteum remains active for 13 to 14 days, then degenerates, and menstruation follows almost immediately. The secretions from the corpus luteum determine the length of the sexual cycle from ovulation until menstruation. This is the phase in the reproductive pattern that is most constant in length. The formation of the corpus luteum is initiated mainly by LH and is maintained in some mammals, but probably not in women, by a third pituitary gonadotrophin, luteotrophin, or LTH. When the progesterone produced by the corpus luteum reaches a sufficiently high level, it depresses the production of pituitary gonadotrophins, without which the corpus luteum collapses. If progesterones are given *before* ovulation, they depress the production of LH and ovulation does not occur, no corpus luteum develops, and the endometrium is not prepared for pregnancy (see chapter 12).

If a corpus luteum of the ovarian cycle retrogresses, that cycle ends and a new one begins. In some mammals abortion follows the removal of a corpus luteum of pregnancy. In pregnant women the corpus luteum, or, for that matter, the whole ovary, may be removed without disturbing the pregnancy, provided that the placenta has become established, that is, at least two months after conception.

Various animals have accessory or aberrant forms of corpora lutea. In mares and elephants, for example, the first corpus luteum of pregnancy is replaced by a second formed from ruptured and unruptured follicles.

Rhesus monkeys may have both an aberrant form of corpus that does not seem to produce progesterone and accessory types derived from unruptured follicles. In badgers, in which the implantation of fertilized eggs is normally delayed by months, further ovulation takes place during that period of delay, and corpora are formed by these follicles. Curiously, these animals sometimes have about four still unimplanted blastocysts and twelve or more corpora.

The Uterus and Menstruation

The adult, nonpregnant human uterus is pear-shaped and as big as a clenched fist, about 10 cm in length and 6 cm thick. It consists of a single "horn," with a slitlike cavity continuous above with the uterine tubes and below with the vagina through the cervical canal (Figure 10–2). It is unlike the uteri of most other mammals (except most primates and bats), which are bicornuate with two horns usually leading to a small, common uterine chamber. The human uterus in nulliparous women lies above and at the back of the bladder within the folds of the broad ligament (Figure 10–2). It is turned forward in relation to the long axis of the vagina (anteverted) and is bent slightly forward on itself (anteflexed). It is supported in this position by condensations of pelvic fascia (cervical ligaments), folds of pelvic peritoneum, and by the two cordlike round ligaments that are homologues of the gubernaculum of the testis.

During pregnancy the uterus rises up in the abdominal cavity and enlarges naturally more than any other organ in the human body (Figures 10–5, 10–6). After parturition it eventually involutes to nearly its original size, but may not always return to its original position. This leads to its becoming retroverted in line with the vagina and raises the danger of gradual prolapse of the uterus down the vagina.

The uterus is derived embryologically by the partial fusion of two embryonic paramesonephric, or müllerian, ducts, first described by the German anatomist J. P. Müller in 1825, when he was twenty-four years old. Very occasionally the human uterus develops abnormally and has two horns. We have seen three such abnormalities in twenty-five years, one of which contained a conceptus that would have caused a disastrous rupture of the small single horn had it been allowed to progress. The criss-cross arrangement of muscle fibers in the unicornuate uterus may have an evolutionary significance associated with bipedalism.

A human fetus lies curled up inside the uterus, with head bent well forward, arms folded, and legs doubled up (Figure 10–7). It assumes an almost spherical form within its membranes, in contrast to quadrupeds, which are more elongated. Each of the two main types of uterus, unicornuate and bicornuate, with its special muscular arrange-

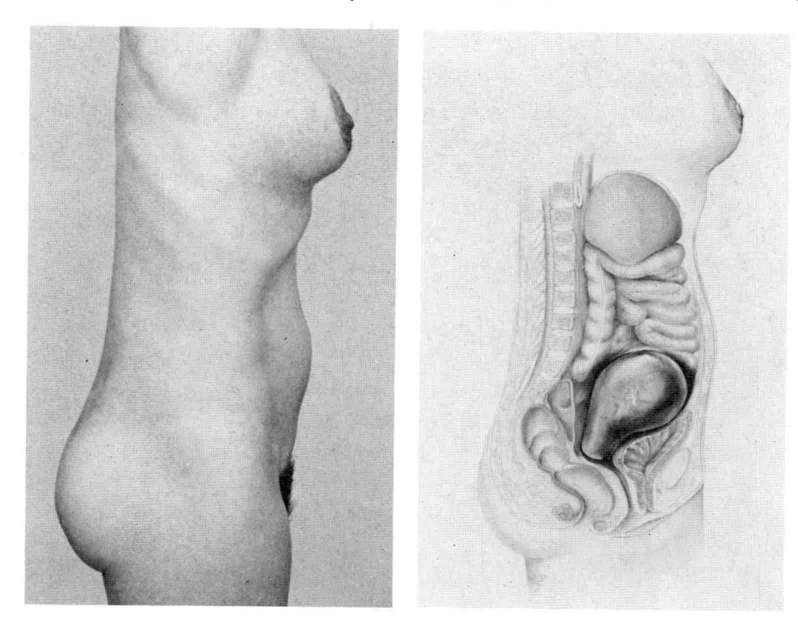

Figure 10–5. A woman in very early pregnancy. Diagram shows the size and position of the uterus.

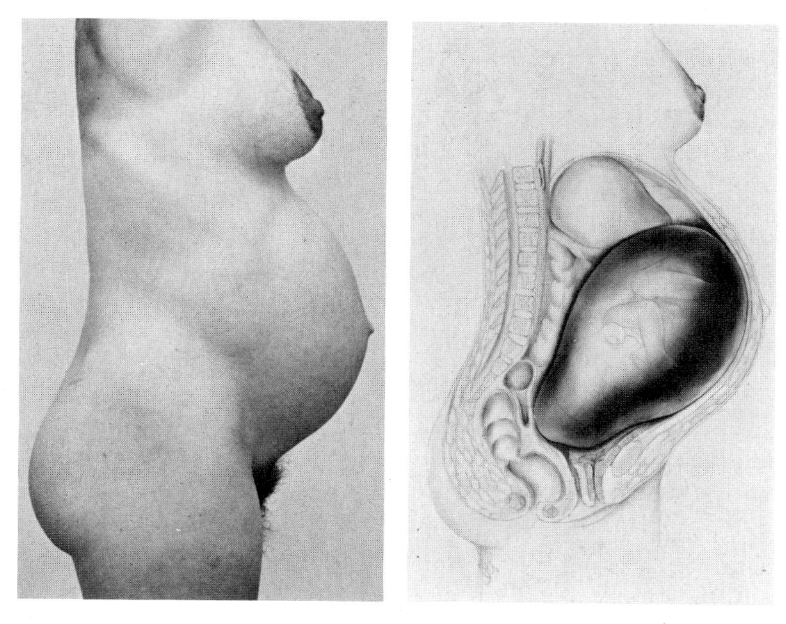

Figure 10–6. The same woman at eight months of pregnancy. Diagram shows the enlargement of the uterus and the crowding of the viscera.

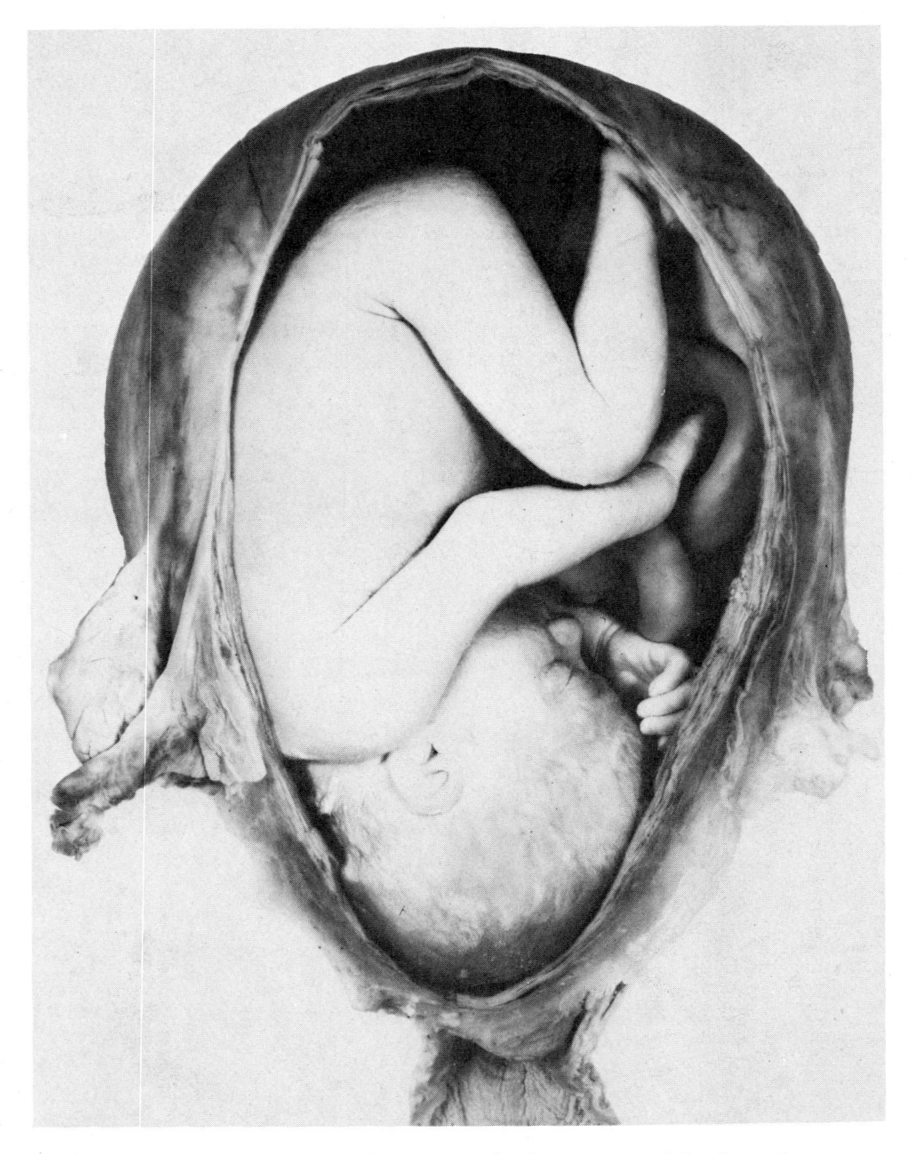

Figure 10–7. A five-month-old fetus inside the uterus and fetal membranes.

ments, is adapted for expelling differently shaped conceptuses. A unicornuate uterus is not well adapted for accommodating more than one fetus. In the bicornuate uterus of a rodent, where each horn has its own cervix, as many as eight fetuses may be lined up like peas in a pod, and successive births present no problem. When the human uterus bears more than one fetus, an obvious hazard is obstruction at birth. Even in twin pregnancies the danger of the fetal heads interlocking is always present.

During the early part of the ovarian cycle, a layer of endometrium hardly a millimeter thick lines the narrow, triangular, slitlike cavity of the human uterus. This progressively thickens under the influence of estrogens, and after ovulation, under the influence of estrogens and progesterone. There is first a proliferative phase during which uterine glands and blood vessels grow. After ovulation follows a secretory phase when the endometrium thickens further, the glands become coiled and active, and the blood vessels enlarge and become spiral. At this stage the endometrium has a thick compact superficial layer, a thicker spongy intermediate one, and a basal layer. Small specimens of endometrium can be removed under anesthesia to investigate its degree of development during any of these periods. (An abnormal endometrium can be stripped away by curettage to allow its replacement by healthy tissue.) Proper implantation of a blastocyst occurs only when the endometrium, under a balanced control of ovarian hormones, develops to an optimum receptive state, which lasts for only a few days or hours. During the secretory phase, the uterine glands secrete a sticky material known as uterine "milk," which provides an appropriate environment for the blastocyst during the critical hours before and during implantation.

At menstruation the two outermost layers of the endometrium become necrotic and slough away together with some blood, leaving the basal layer to repair the lining at the start of a new cycle. Since there are no sensory nerves in the endometrium, the process is relatively painless. The pain that some women experience at the onset of menstruation is as yet inexplicable. It may be due to spasms of uterine muscle or to uterine contractions or distension caused by the increased volume of fluids within the cavity. Whatever the explanation, the best and nearly always infallible cure for it is pregnancy. There are several theories about the cause of the menstrual flow. Clearly, the changes that occur in the endometrium are under hormonal control, and the regression of the corpus luteum precipitates menstruation. Perhaps a local slowing of blood flow to the endometrium causes anoxia with a resulting death of tissues.

The specific purpose of menstruation is obscure. Not least of its imponderables is why it should occur only in certain primates, since it seems to be a waste of tissue and essential substances such as iron. Several hundred milliliters of blood are lost each month, obviously a drain on a woman's reserves and a constant call on her blood-forming bone marrow. Apart from its social disadvantages and discomfort, it often leads to tiredness, bad temper, and anemia. From a biological point of view, menstruation should not occur at all!

Mammals that do not menstruate show a decline in the secretory phase of the endometrium but no tissue destruction or loss of blood. By contrast, the human endometrium seems "overdeveloped" at the end of the cycle. The dramatic changes it undergoes seem to indicate that the glands, blood vessels, and surrounding tissue are so greatly engorged that if implantation does not occur they will stagnate and die. The endometrium, then, cannot be "returned" to its presecretory condition, and the slightest change in the factors that maintain it causes its dissolution.

After implantation, all menstrual cycles cease and do not recommence until near the end of the suckling period (lactation). Lactation, however, does not altogether inhibit ovulation. If a child is not suckled, ovulation and menstruation are usually reestablished a few months after his birth; if he *is* suckled, ovulation sometimes resumes without menstruation. Proper cycles develop later when weaning begins. In some polyestrous and monestrous mammals, e.g., rodents, carnivores, and some primates, a postpartum ovulation occurs within a few days of birth, which is frequently, but not always, associated with delayed implantation. Greater bush babies *(Galago crassicaudatus)* will mate effectively only 2 or 3 days after parturition; the ensuing pregnancy does not interfere with the suckling of the other offspring. Moreover, since the gestation period of these animals is about 110 days, the next young may be born while the previous one is still nursing; when this happens, the older infant is rejected in favor of the neonate.

Menopause

Menopause marks the end of reproductive life in women. Usually it extends over several years during which menstruation occurs less and less regularly until it ceases altogether. Other changes often occur, e.g., in the vaginal epithelium, which indicate a withdrawal of ovarian hormones. Menopause seems to be accompanied by a relative exhaustion of oocytes from the ovary, since the few oocytes in the ovaries of postmenopausal women are markedly deficient in follicular cells. Since menopause often precedes follicular exhaustion, the anterior pituitary continues to secrete gonadotrophins, but to no avail.

Menopause is unique to human females; nothing comparable to it has been seen in wild mammals. Captive primates and rodents show a decline in, and even a cessation of, fertility, but aged rhesus monkeys and chimpanzees, which are no longer able to reproduce, continue to menstruate. Menopause was probably rare among primitive women since they seldom reached the age when it developed. Menopause itself does not seem to be the cause of senescence; many women live in the healthiest condition as long after the menopause as before it. However, menopause is often preceded and followed by certain inexplicable mental disturbances and depressions, which, fortunately, can be alleviated by ovarian hormonal therapy.

Reproductive Patterns in Primates and Other Mammals

Few mammals have reproductive patterns similar to those of women. A majority have only one or more breeding seasons per year, separated by periods of reproductive inactivity (anestrus). During a breeding season, a succession of sexual cycles may occur, a condition described as polyestrus, or only one long, sustained estrus occurs in the spring, with perhaps another later in the summer or early fall. This is the monestrous condition found in cats and ferrets (Table 10–2). In polyestrous mammals the cycles vary from 4 days (rat, mouse) to 35 days (chimpanzee), but ovulation always occurs at a particular time in relation to estrus, the time of maximum receptivity. Ovulation is said to be spontaneous in these mammals because it occurs whether or not mating takes place. A few common examples are given in Table 10–3.

In monestrous mammals ovulation is usually induced by the stimulus of mating. If, regardless of mating, conception does not occur, a condition known as pseudopregnancy develops. Corpora lutea form but last only for a short time; in some mammals there is evidence of a luteolytic factor originating in the uterus and exerting a direct effect on the corpus. Whether the polyestrous condition is more "primitive" or more "specialized" than the monestrous type is difficult to say. The former extends the reproductive pattern over a variably long period, but restricts opportunities for conception in one individual. The latter concentrates mating opportunities into a single period, but repeats them less often each year. Obstetric departments would be overcrowded if women were monestrous. This could happen, if, for example, women using oral contraceptives all decided to conceive during the summer to have their babies in early spring. This is a warning to family planners. Details of some mammals that exhibit monestrous cycles are given in Table 10–2.

Only some primates have a menstrual discharge; no other mammals have, with the possible exception of the elephant shrew *(Elephantulus)* of South Africa. In these insectivores a curious polyplike body develops as part of the endometrium that breaks down and discharges at the end of the cycle. Some have considered this curious phenomenon to be a forerunner of menstrual discharges in higher primates.

Some primates are more amenable to laboratory experimentation than others, and consequently more is known about their reproductive patterns. Outstanding among these are the macaques, of which the rhesus monkey is best known. Lorises, galagos, and several species of lemurs

Table 10–2. Details of Reproduction in Some Monestrous Mammals

Species	*Breeding season*	*Length of estrus*	*Ovulation*	*Remarks*
Hedgehog (*Erinaceus*)	May to September	Not known	Probably induced	Pseudopregnancy occurs.
Shrew mole (*Neurotrichus*)	February to April July to August	Up to 33 days	Induced, needs several matings	Pseudopregnancy lasts 9–10 days.
Mole (*Talpa*)	March to April Autumn	20–30 hours	Induced	
Rabbit (*Oryctolagus*)	May to July	Almost continuous in breeding season	Induced	Pseudopregnancy lasts 16–17 days.
Dog (*Canis*)	Spring and autumn	7–9 days	Probably spontaneous on second day of estrus	Pseudopregnancy lasts about 2 months.
Red fox (*Vulpes*)	December to March	6–10 days	Spontaneous	Pseudopregnancy occurs.
Cat (*Felis*)	Spring and autumn	9–10 days	Induced	Estrus recurs 15 days later in absence of male. Pseudopregnancy lasts 36 days.

Table 10–3. Details of Polyestrous Cycles in Some Mammals

Species	*Type of cycle*	*Length*	*Details of estrus*	*Ovulation*
Rat (*Rattus*)	Polyestrous all year	4–6 days	9–20 hours	8–10 hours after start of estrus
Mouse (*Mus*)	Polyestrous all year	4–6 days	1–2 days	2–3 hours after start of estrus
Guinea pig (*Cavia*)	Polyestrous all year	16–17 days	6–12 hours	10 hours after start of estrus
Pig (*Sus*)	Polyestrous all year	20–22 days	2–3 days	36 hours after start of estrus
Cow (*Bos*)	Polyestrous all year	17–23 days	12–22 hours	10 hours after start of estrus
Goat (*Capra*)	Polyestrous September to November	19–21 days	24–48 hours	During estrus
Sheep (*Ovis*)	Some anestrous in summer; others polyestrous all year	16–18 days	30–40 hours	24 hours after start of estrus
Horse (*Equus*)	Polyestrous March to October	21–23 days	6–7 days	Toward end of estrus

These statements refer to animals kept under laboratory or domestic conditions in the northern hemisphere.

can now be bred at will by controlling the time of light and darkness in their environment and by checking the time of their estrous cycle and thus the receptivity of the female and a host of other details. New information is available daily as the various primate centers in the United States and elsewhere learn to breed the lesser known and rarer species. This has become a vital part of the programs of these centers because the future of many species of existing primates is precarious. If active breeding policies are not developed soon, far too few primates will be available for investigation of the genetic, behavioral, and biochemical characteristics so important for understanding the order, and thus, man. Most primates are polyestrous; only some of the lower forms display a limited breeding season. Bleeding from the uterine cavity as a result of destruction of the endometrium is slight in many species and occurs only as overt menstruation in most Old World monkeys, apes, and man.

The details of reproductive events in those primates most assiduously studied are given in Table 10–4. Most Old World monkeys also display changes in their perineal skin during the cycle. This variously pigmented sex skin becomes more conspicuous and tumescent during estrus. Gelada baboons *(Theropithecus)*, for example, have sex skin on the chest (called a "necklace" of caruncles) as well as in the perineal region. Most macaques display striking color changes, and Celebes apes and chimpanzees have the most marked swelling. In chimpanzees the volume of the swollen sex skin can amount to as much as 1,400 cc and is greatest on the fifteenth day of the cycle. After ovulation both coloration and tumescence subside.

Reproduction in the Male

The male reproductive organs are shown in Figure 10–8, with the viscera immediately around them in their proper place. The testes start functioning the moment they begin to differentiate in the fetus. Although fetal testes contain only primordial germ cells, some endocrine cells in them secrete androgenic hormones that profoundly affect the development of male characteristics in the embryo. In fact, the effect of these hormones predisposes the developing central nervous system to maleness. The endocrine function of the testes continues uninterruptedly after birth and steadily accelerates near puberty. The secretions of the testes constantly fashion the animal in physique and behavior, from infancy to puberty, and at the same time prepare the gonad itself for effective reproduction.

In man the production of sperm is evident just before puberty, when the secretion of male sex hormones, androgens, also greatly in-

Table 10–4. Reproductive Patterns in Some Primates

Species	*Breeding season*	*Length of cycle*	*Ovulation*	*Menstruation*
Bush baby (*Galago senegalensis*)	December to July (all year in captivity)	36–42 days	Estrus lasts 5–6 days	
Spectral tarsier (*Tarsius spectrum*)	All year	23–24 days	Estrus lasts about 24 hours	
Spider monkey (*Ateles geoffroyi*)	All year	24–27 days	Spontaneous	
Crab-eating macaque (*M. irus*)	All year	25–29 days	Coitus mainly on day 7–11	2–6 days
Rhesus monkey (*M. mulatta*)	All year in exp. colonies	23–33 days (Mean 27.36 ± .17 S.D. = 5.7)	About 13th day	4–6 days (range 2–11 days)
Chacma baboon (*Papio ursinus*)	All year	29–42 days	Spontaneous	4–9 days
Gibbon (*Hylobates hoolock*)	All year	21–43 days	Spontaneous	2–5 days
Chimpanzee (*Pan satyrus*)	All year	34–35 days, longer in young females	16th day onward	2–3 days (range 1–7 days)
Man (*Homo*)	All year	21–34 days (Mean 28.32 ± 0.6 S.D. = 5.41)	6–23rd day (Mean 14 ± 2 days before menstruation)	2–8 days (usually 5 days)

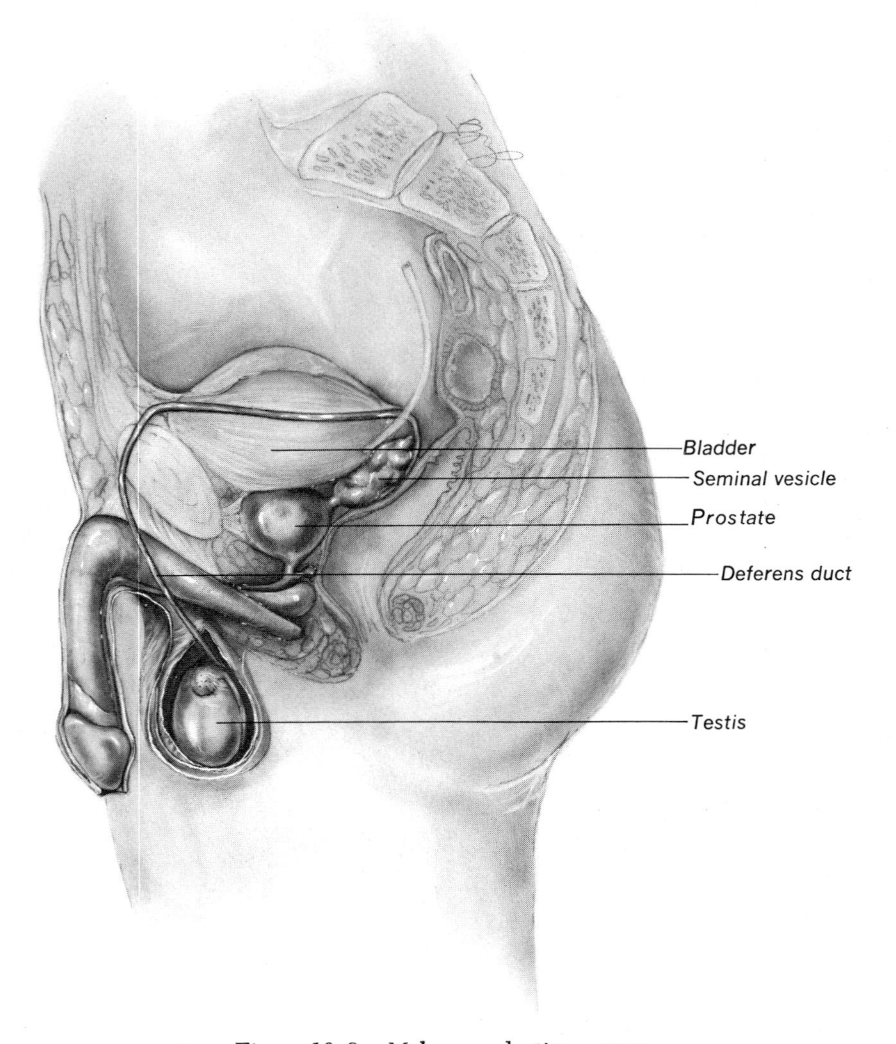

Figure 10–8. Male reproductive organs.

creases. Androgens, whose main function is to induce the expression of all male secondary sexual characteristics, are secreted by the interstitial, or Leydig, cells. Spermatozoa are manufactured in the long, coiled seminiferous tubules. Interstitial cells lie in clusters between the tubules and are under the control of pituitary gonadotrophins. Unlike most other mammals, which experience a rut, or male breeding season, man's seminiferous tubules, as far as we know, are active throughout the year, regardless of season. Under the most favorable conditions, man's testes may show some increased activity, probably as a result of increased sex drive with a concomitant increase in pituitary activity. Male rats, guinea pigs, and most primates are also constant breeders, but male hedgehogs, squirrels, most carnivores, and hooved animals have very precise breeding seasons when they frequently become strongly aggressive toward other males and often defend their territory with deadly pugnacity. Men may have sublimated this drive into other activities but are often unpleasant to each other about sexual matters. Men have no true menopause, although their testicular activity declines gradually. However, some atypical men have sired children when quite respectably aged; we do not know the world record but it is certainly in the eighties. Male readers may be relieved to know that their own sexual prowess will not reduce their longevity, libido, or power of reproduction as they age. Testicular activity is, however, reduced by excessive heat and cold, alcohol and other drugs, certain poisons such as cadmium salts, and maybe by certain diseases such as mumps.

The testes of boys descend into the scrotum just before or at birth, through the inguinal canal. The scrotum is a wrinkled sac of skin, haired, often pigmented, and one of the few regions of the skin (the eyelid is another) that accumulates very little subcutaneous fat. A midline seam indicates that the scrotum was formed embryologically from two elements on either side of the primitive cloaca, the common orifice of the reproductive, excretory, and alimentary systems. The testes of some mammals (whales and elephants) remain intraabdominal and can be very large. Relative to body weight, a dolphin of Japanese waters is said to have the largest testes (up to 8 kg), which are in addition considered a delectable gustatory delicacy. In adult rats and seals testes have an inguinal position, and in bats and some rodents they pass into the scrotum only during the breeding season.

The descent of the human testes into the scrotum is probably brought about, among other factors, by the pull of the gubernaculum (L. *gubernator,* a pilot), a fine strand of tissue extending in embryos from the testis across the abdominal wall into the inguinal canal. The scrotum seems to have a heat-regulating function; spermatozoa can be formed and mature only at a temperature lower than that in the abdomen. When in abnormal cases the testes do not descend into the scrotum (cryptorchidism), they atrophy soon after puberty, and the individual is effectively a eunuch.

Certain fishes, reptiles, and birds have copulatory organs that can be inserted into the female genital tract. All mammals have a penis with a passage (urethra) that conveys both urine and semen to the exterior. Monotremes have a penis on the ventral wall of their cloaca, and marsupials have it, of all places, behind the scrotum. In most mammals the penis is in front, along the ventral abdominal wall. In the bull, horse, seal, whale, dolphin, and others it is completely concealed within a sheath of abdominal skin, the preputial sac, from which it emerges during erection. In man, many primates, tree shrews, and some bats, the penis is partly concealed, pendulous, free from the abdominal wall, and only partly covered by a short prepuce. The penis is usually flaccid and becomes enlarged and erect during the various antics that precede mating. This is brought about by an engorgement and distension of spongelike blood vessels, known as cavernous tissue, within its shaft. Many carnivorous mammals have a bone in their penis, the os penis, or baculum, whose length can provide an estimate of age; in bull walruses this attains a stupendous size. Many primates, particularly the prosimians, have a middle cartilaginous rod in the penis; man has none. Some seals have an os clitoridis.

Spermatozoa, formed in the seminiferous tubules at a rate of about 50,000 per minute, leave the testis by way of the long, coiled, narrow tube, the epididymis, amassed on one side of the testis (Figure 10–8). The unraveled human epididymis measures several yards. Spermatozoa become motile and finally mature during the long journey in this tube. From the epididymis spermatozoa pass along the vas deferens to the ejaculatory duct that enters the relatively short (prostatic) urethra. The sperm, together with mucinous substances secreted by the two glands, the prostate and seminal vesicles, make up the bulk of the seminal fluid or ejaculate.

The prostate of man, a gland about four cm in diameter, is wrapped around the first part of the urethra at the neck of the bladder. Muscular contraction forces its secretion into the urethra through a series of ducts. In old age the prostate often enlarges, compresses the urethra, and causes a retention of urine within the bladder. The seminal vesicles are two sacculated organs at the base of the bladder each with a short duct that leads to the common ejaculatory duct. Marsupials, carnivores, and cetaceans have no true seminal vesicles, but they are particularly large in some insectivores.

These, then, are the male and female reproductive patterns in man. In general, we have shown that man has much in common with most other mammals, but one must remember that the details of the reproductive pattern in each mammal are unique.

11

Conception, Development, Pregnancy, Parturition, and Lactation

Every man is some months older than he thinks him; for we live, move and have our being . . . in that other World . . . the womb of our Mother.
(Sir Thomas Browne)

The precise timing of each successive event in mammalian reproduction is a sine qua non if each single process is to reach fruition and avoid disaster. Precise order at all levels of organization must characterize each process within each phase. Long before a pregnancy materializes, there unfolds a series of historical events that differentiate that pregnancy from any other, even in the same individual, and make of it a singular experience with its own unique environment, sequence of events, and final product. If the reader can grasp the drama of these events, the details that follow will not be abstruse to him. A knowledge of them makes the whole pageant rather more awe inspiring than less.

The Ovum

The ova of placental mammals are surprisingly similar. They are all small, with a diameter of 90 to 150 μ and, unlike those of birds and all other vertebrates, contain very little yolklike material. Except in the dog, a central nucleus undergoes the first maturation division while still within the follicle and the second at the time of fertilization. When ovulation occurs, the ovum is really a secondary oocyte, with a polar body underneath the encapsulating zona pellucida to which adhere granulosa cells from the follicle to form the corona radiata (Figure 11–1). In some rodents, a mucous coat is added to the ovum as it traverses the oviduct.

A human ovum has a diameter of about 140 μ, of which the zona accounts for some 10 μ. During surgical removal of the uterus or uterine tubes, living human ova, some of them fertilized, have been recovered and grown in tissue culture for a short time on synthetic media. However, needing the intratubal and intrauterine environment for further development, they have survived only a few days. Such techniques could, however, provide useful information leading to the successful transference of fertilized ova from one individual to another.

The life spans of ovulated mammalian ova are remarkably short. Ova of laboratory and domestic mammals remain viable for only 24 to 30 hours, when they begin to degenerate. They are fertilizable for an even shorter period. We are not certain how long nonfertilized human ova can survive, but some evidence suggests less than 24 hours. Only one ovum at a time is usually released from the human ovary, but occasionally two follicles mature and rupture together and release two ova, which if fertilized give rise to unlike, or fraternal (dizygotic), twins. Identical (monozygotic) twins arise from the fragmentation of a single egg. In

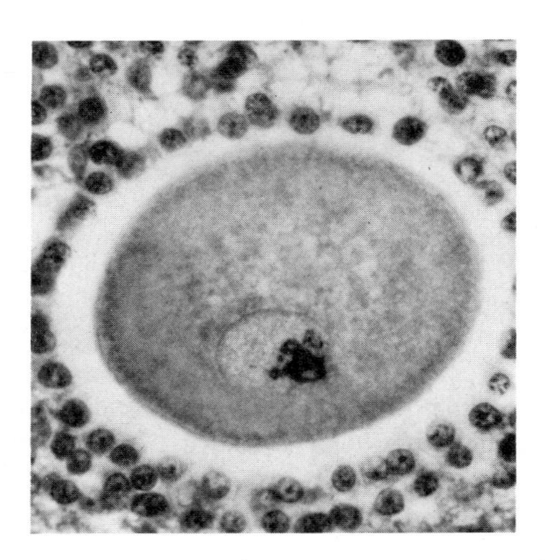

Figure 11–1. Fertilized ovum of a rhesus monkey showing male and female pronuclei. The male pronucleus is more darkly stained. Observe the transparent zona and corona radiata. *(Courtesy of Professor Richard J. Blandau, University of Washington School of Medicine.)*

the ovaries of some mammals, occasional follicles contain two oocytes (binovular follicles), which if released could also produce fraternal twins, though they usually degenerate. The administration of gonadotrophins at precisely the right time stimulates the ovary to liberate more than one ovum. This practice has proved successful in some infertile women, but they run the risk of obtaining multiple births; indeed, more than one family has been presented with quintuplets.

Spermatozoa

At puberty the seminiferous tubules of the testes contain both mature and developing germ cells and "nurse" or Sertoli cells (Figures 11–2, 11–3). The testes are the site of a continuing cycle of spermatogenesis, i.e., a continuous process of maturation from stem cells to sperm. Spermatogonia (stem cells), in the coiled seminiferous tubules, divide to give rise to primary spermatocytes, which divide again into secondary spermatocytes. These in turn split into two spermatids, which metamorphose into sper-

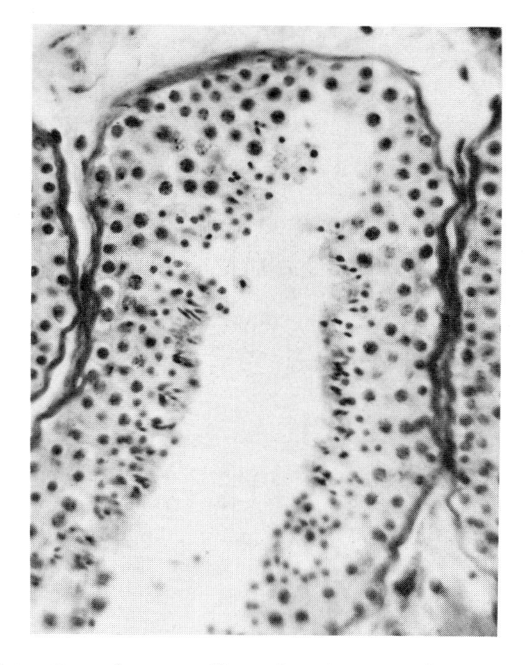

Figure 11–2. Maturation of germ cells within the seminiferous tubule of a human testis. *(Courtesy of Professor Edward C. Roosen-Runge, University of Washington School of Medicine.)*

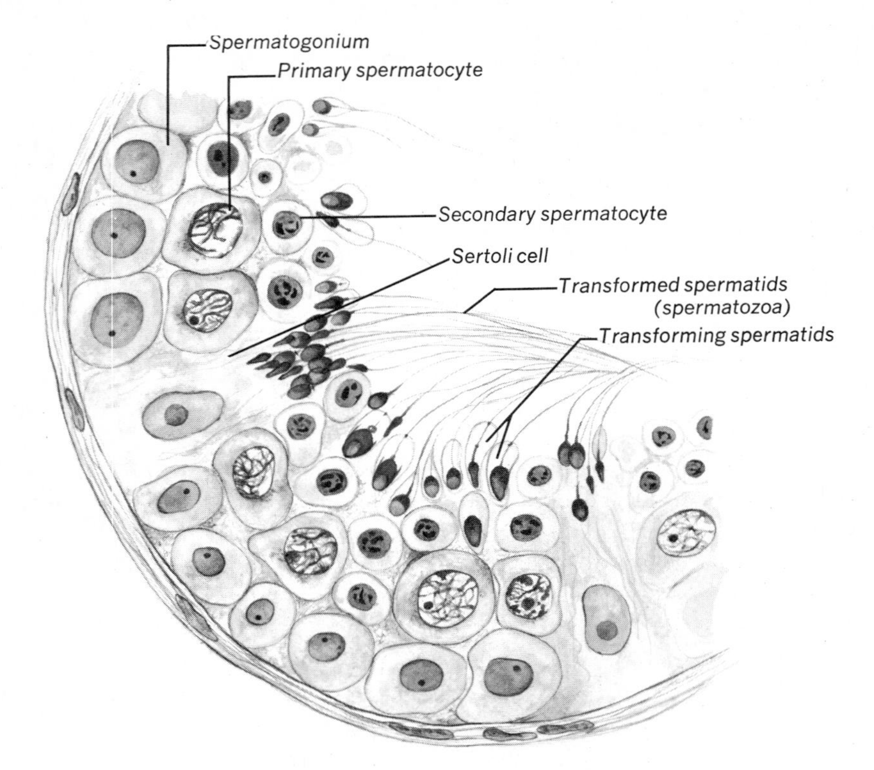

Figure 11–3. Diagram of the maturation of spermatozoa.

matozoa while attached to the Sertoli cells (Figure 11–3). Through the two subsequent (maturation) divisions in man, one primary spermatocyte gives rise to four spermatozoa, two of which contain 22 + X and two 22 + Y chromosomes. When first formed, spermatids resemble the parent secondary spermatocytes but are smaller. Later they undergo a metamorphosis (spermateliosis) and assume a shape resembling a tiny tadpole with a head, neck, and tail.

A mature human sperm is a very small cell about 54 μ long, with an ovoid, flattened head (4.6 to 5.0 μ long) consisting mostly of nucleus. The front is covered by a caplike structure, the acrosome, and the back by a postnuclear cap. The tail has a neck (0.5 μ), a middle (4.0 μ), a long main section bounded by a delicate proteinous sheath, and a terminal end-piece without it (Figure 11–4).

Although the spermatozoa of all mammals are constructed along this basic pattern, the head is different in each species (Figure 11–5). Human spermatozoa have a pear-shaped head. Those of other mammals are oval, hooked, and asymmetrical (some rodents), or spoon-shaped in

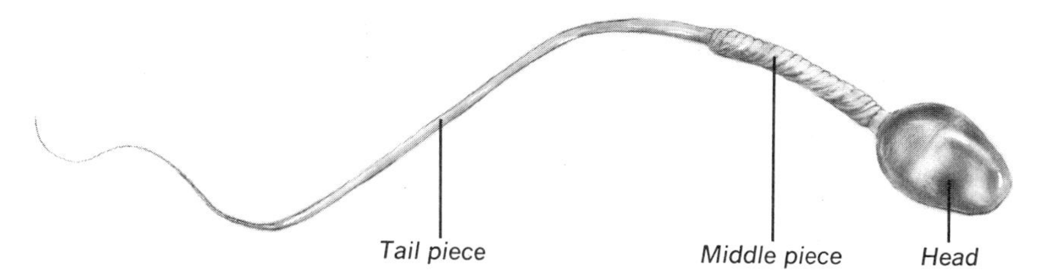

Figure 11–4. Diagram of a mature human spermatozoon.

profile (guinea pigs). Clearly the fundamental structure of spermatozoa in mammals has changed little, if at all.

When spermatozoa leave the testes, they are ineffective and must mature in the long coiled epididymis, through which they are slowly propelled by peristalsis. During this journey of about three weeks, they become increasingly mobile. Entering the ejaculatory duct from the ductus deferens, they are expelled at ejaculation together with the seminal fluid produced by the seminal vesicles and prostate. If not ejaculated, many lose their potency and degenerate. The first ejaculate after a period of abstinence contains many dead and degenerating sperm.

Man ejaculates an average of 3.5 ml of semen (range 0.5 to 11.0), each ml containing about 120 million sperm (normal range 20 to 200 million), or an average of about 400 million sperm per ejaculate. The ejaculated semen of rhesus monkeys contains about five times more spermatozoa than that of man. Although only one sperm fertilizes an egg, a man is likely to be infertile if his ejaculate yields less than 20 million sperm. All ejaculates commonly contain a number of abnormal sperm. When more than twenty-five percent of the total number are abnormal, the individual is infertile. This may indicate that the remaining, apparently normal, sperm or the chemical composition of the semen are also abnormal. The spermatozoa constitute one-tenth of the volume of each ejaculate. Both the amounts of total semen and the number of spermatozoa are subject to a great many internal and external factors. The volume of the ejaculate is small in mammals with small or no seminal vesicles, large in those whose glands are large (e.g., boars).

Seminal fluid provides a medium in which spermatozoa can survive and move. Together with secretions from the female genital tract, it also supplies a limited amount of nutriment. It contains enzymes and buffering substances that prevent it from becoming excessively acid inside the vagina. In most animals, and perhaps also in man, the final maturation of sperm takes place in the uterine tube by a process known as "capacitation."

Once released from the male reproductive tract, spermatozoa gain increased motility after the liquefaction of the coagulated seminal fluid

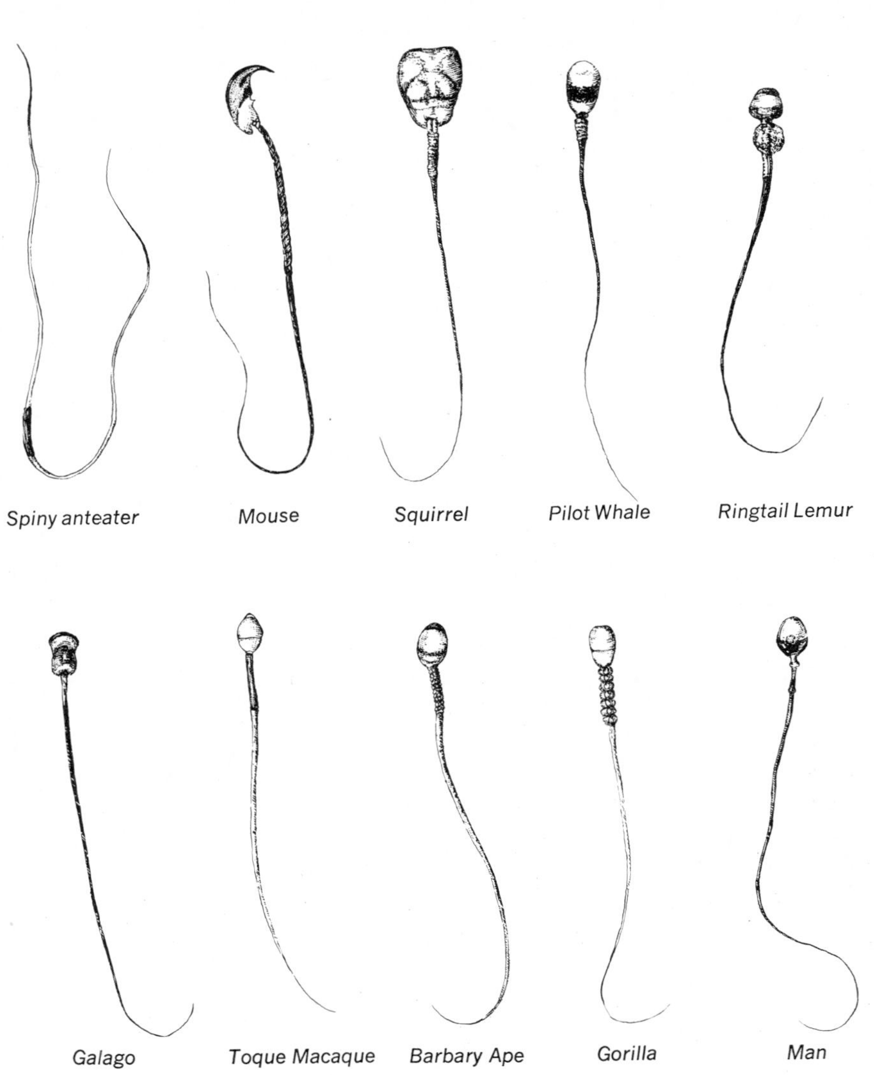

Figure 11–5. Spermatozoa from different animals.

around the cervix. Changes also occur in the characteristics of cervical mucus about the time of ovulation. Spermatozoa move the tail in an undulatory fashion, propelling and oscillating the head forward along a spiral path. Although it is difficult to determine accurately, human spermatozoa progress at a rate of about 1 to 3 mm per minute, yet they can travel from the cervix to the last segment of the uterine tubes in about 30 minutes (Figure 11–6). Particularly interesting is the fact that they move by preference against the prevailing current caused by the beating of cilia and to some extent against gravitational pull. Although motility is the basic characteristic of healthy spermatozoa, athletic prowess alone does not get them to their goal. Enormous numbers are eliminated at the cervix, more in the uterus and at the tubo-uterine junction; only a few reach the outer end of the uterine tube. Motility does not necessarily mean fertility, nor vice versa. Abnormal and sterile spermatozoa can be motile; even broken-off tails can move about. Normal spermatozoa under experimental conditions can lose motility before fertility. Seemingly, motility is essential only to enable them to conclude their journey; the contraction of smooth muscle in the uterus and in the tube is the most important factor in sperm transport within the female.

Once inside the female reproductive tract, human spermatozoa survive for only two to five days or less. Survival time varies in the

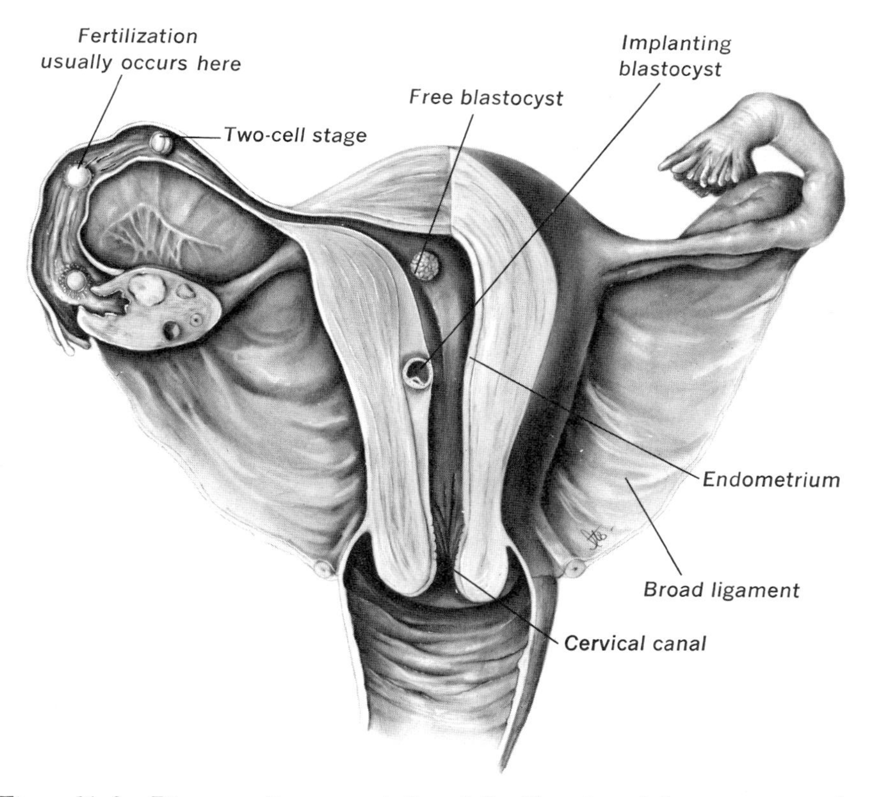

Figure 11–6. Diagrammatic representation of the liberation of the oocyte at ovulation, fertilization in the outer third of the uterine tube, and implantation in the endometrium. The size of the oocyte, cleavage stages, and the blastocyst have been deliberately enlarged.

different parts of the tract and appears to be shortest in the vagina. Durability may also depend on various factors related to the sexual cycle. Among domestic mammals, the spermatozoa of the stallion has the longest survival rate in the female tract (130 hours). Cornell University zoologist W. A. Wimsatt reports that in some bats spermatozoa are normally stored for as long as five to six months in the female reproductive tract and can fertilize newly released ova after the passing of winter and the period of hibernation. Provided that harmful elements are avoided, human spermatozoa can be stored at a temperature of +4°C when metabolism is just sufficient to maintain integrity. Bull semen can be preserved for years at temperatures of −79°C and below. Much investigation is in progress to determine optimal storage conditions for human spermatozoa. It is not preposterous to anticipate that a man will be able to beget children long after his death.

Fertilization

This occurs when a sperm unites with an ovum, becomes an integral part of it, and together with it initiates the development of a new individual (Figure 11–7). The ova of some invertebrates and occasional amphibians normally or experimentally (induced by heat, cold, electrical current, certain salts, and other agents) may develop without the participation of spermatozoa, a process known as parthenogenesis (Gk. *parthenos,* a virgin). Even ova of rabbits can divide parthenogenetically in a limited way. Parthenogenesis in its truest sense, however, has not yet been demonstrated in mammals.

Superficially at least, fertilization starts when a spermatozoon approaches an ovum. We are not certain whether capacitation occurs in man or whether human ova secrete substances that increase sperm motility to attract them. A singleness of purpose exists in germ cells: spermatozoa normally enter ova and no other cell, and healthy ova will permit the entrance only of spermatozoa and generally only of one. Although the latter usually penetrate eggs only of their own species, cross-fertilization between closely related species can occur, but the offspring of most such crosses are infertile or sterile.

When sperm arrive near an egg, the cells of the corona radiata become dispersed, probably through the activity of enzymes on the spermatozoa such as hyaluronidase, which breaks down the intercellular materials. Profound ignorance still prevails about events in human fertilization. Whether many sperm are needed to open up the path for the winner or whether one spermatozoon can make it alone remains unknown. Hundreds are known to reach the outer end of the uterine tubes where

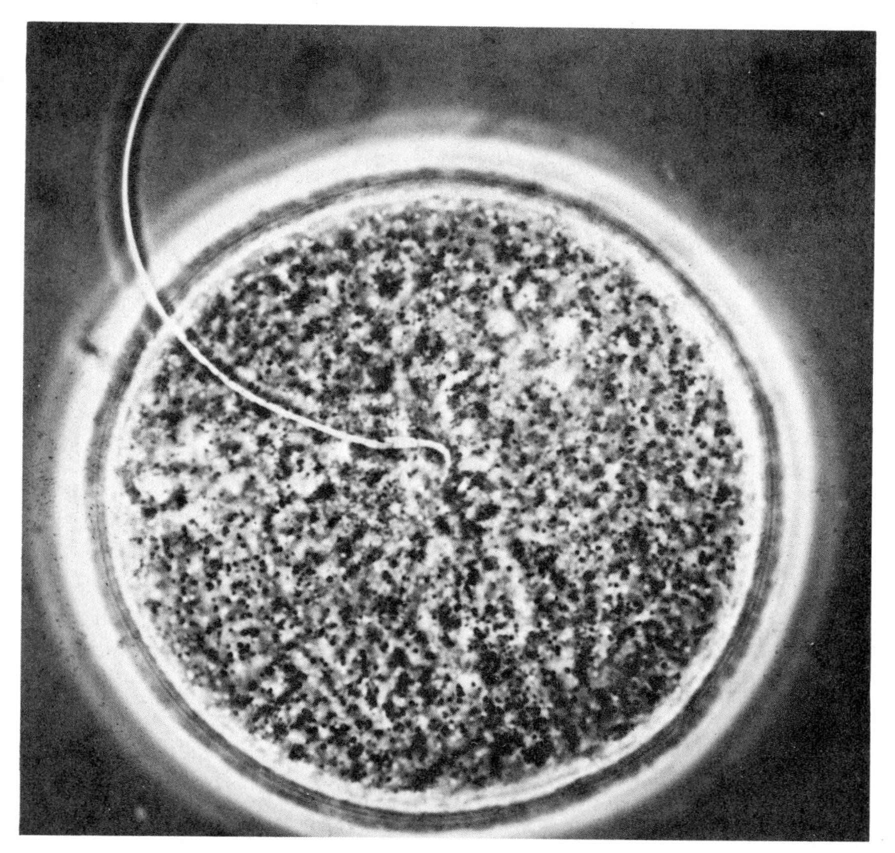

Figure 11–7. A living ovum of a rat after penetration by a sperm. *(Courtesy of Professor Richard J. Blandau, University of Washington School of Medicine.)*

fertilization takes place. Several may reach the egg simultaneously and even penetrate the zona or die in its substance. Usually, however, only one enters the ovum by piercing through the egg (vitelline) membrane. As the sperm penetrates it, the ovum, still a secondary oocyte, undergoes its second maturation division and expels a minuscular second polar body. The egg nucleus, with half the number of somatic chromosomes (haploid), is now called the female pronucleus. Once inside the egg, the spermatozoon loses its tail, which is absorbed, and the head swells to become the male pronucleus (Figure 11–1). The two pronuclei approach, fuse, and form the zygote—the genesis of a new individual. Up to this point, the processes of fertilization take about 24 hours. However, fertilized human eggs must face numerous hazards; many probably succumb before they become safely embedded in the uterine lining.

The diploid number of chromosomes (44 + XX, female or 44 + X

+ Y, male) is restored at fertilization, and the basic characteristics of the new individual are determined (genotype). Subsequent events, both inside the uterus and postnatally, also influence the final appearance of each individual (phenotype). Thus, all normal human embryos are programmed to become human beings, but the unfolding of the genetic instructions does not result in *exactly* identical products: all human embryos look alike, but adults are easily distinguished from one another.

The sex of the new individual is determined at fertilization. Normal human females have 2X, and males X + Y chromosomes. The presence and type of sex chromosomes can now be ascertained by relatively simple cytological tests on scrapings of buccal mucosa, blood, bone marrow, skin, and other tissues. Several types of aberration can occur in human sex chromosomes. Normally, a remarkable congruence exists between the continuation of the sex chromosomes and the structure of the gonads, accessory organs, external genitalia, and secondary sexual characteristics. Under certain conditions, however, the correct sexual phenotype does not develop. Sometimes the gonads fail to develop properly. Testes, for example, may develop in a female phenotype with an XY chromosome constitution. Male or female pseudohermaphrodites can also develop, where the male form with testes has various degrees of feminization of the reproductive organs and the female with ovaries has a masculinization of the reproductive equipment. Very rarely an ambisexual condition of true hermaphroditism occurs. Up to 1962, about eighty such anomalies had been reported. True hermaphroditism does not occur in man, but the origin of this word is interesting. Hermaphroditus, a minor Greek deity and the child of Hermes and Aphrodite, possessed female breasts and a male phallus and carried functional aspects of both sexes. Biologically, the term means the presence in one individual of both male and female anatomical and functional capacities (snails and slugs are true hermaphrodites). Human hermaphrodites so far examined have been reported to have either a testis and ovary on one or both sides of an "ovotestis," but these gonads were not functional.

Aberrations can also occur in the number and size of sex chromosomes. Individuals displaying various clinical syndromes are described as Turner-Klinefelter after the original accounts of H. H. Turner and H. F. Klinefelter. For example, 44A + XO, 44A + XXY, XXXY, or XXYY, and other arrangements may occur. A state of mosaicism with two or more cell lines with different numbers of chromosomes and different associations of sex chromosomes may also exist in an individual. One, for example, can have cells with XXY and others with XX. Most of these aberrations no doubt occurred during the earliest divisions of the zygote.

Although the ratio between the number of boys and girls born in any year might be expected to be equal, actually more boys are born. This is called the *secondary* sex ratio (at birth) as opposed to the *primary* sex ratio, which indicates the number of male and female zygotes formed at fertilization. The primary sex ratio is impossible to determine in man. In domestic animals such as pigs, it can be as high as 160 males:100 females. In a series of 907 human embryos and fetuses accumulated by one of us over 20 years, the ratio was 140 boys:100 girls. The secondary sex ratio in man varies slightly from year to year and in different countries (where such records are kept). Recently in the United States it was 105.3:100 and in Great Britain 105.6:100 live births. The preponderance of males continues through the second decade, and equality in the numbers of the two sexes (judged by phenotype) is reached by age 15 to 20 years. Thereafter, the number of women increases steadily until age 85, when the number of women is twice that of men.

Many hypotheses, scientific and otherwise, have been propounded to explain these sex ratios, most of which are unworthy of comment. Y-bearing spermatozoa, with a higher metabolic rate, may be more motile or longer-lived than the allegedly heavier X-bearing type and may, then, be swept first and more frequently up the female genital tract. Efforts to separate the two kinds of spermatozoa, and even to recognize them, have been mostly unsuccessful. Some have suggested that ova could exert differential attraction for X- or Y-bearing sperms, but this has not been proved. Regardless of the causes, "maleness" clearly has a lowered degree of survival. Sex-linked or sex-limited recessive lethal genes could cause an early death of male conceptuses; defective genes could accumulate more in males than in females because of less intense selection, but no one knows how much "maleness" is due solely to the Y chromosome or to a deficiency in the quantity of X. Much remains to be learned in this field.

Scattered observations suggest that the sex of the offspring depends on the time of coitus versus that of ovulation. If ovulation occurs before or within a day after coitus, the ovum is supposed to be more likely fertilized by a Y chromosome-bearing spermatozoon, which would produce a boy. If ovulation occurs three or more days after intercourse, the chances are that an X chromosome-bearing sperm would fertilize the ovum, to produce a girl. If this is true, there must be different biological potentials within the fertilization environment that bring about an excess of X or Y chromosomes at different times after insemination.

As long ago as 1899, both Landsteiner and Metchnikoff, working independently, suspected that sperm immobilization by antibody response influences fertility. Because of technical difficulties, investigations on female response to the antigenic properties of spermatozoa have not progressed much, but certain species are known to develop decreased fecundity in response to sensitization with sperm or seminal protein. In rhesus monkeys seminal protein is absorbed through the vaginal lining and produces a rise in the titer of circulating sperm antibody in animals previously sensitized to semen or seminal plasma. The uterus of some immunized female animals has been found to contain fewer spermatozoa after insemination. Immunization of females with spermatozoa, however, appears to have little or no effect on ovulation or the sex cycle. We do not know whether some differential effect on the performance of X- or Y-bearing spermatozoa occurs in such circumstances. Convincing evidence that infertility caused by immunization to spermatozoa can be alleviated by a period of continence or the use of a condom for some months also exists. Thus, if a brood of closely spaced children is of the same sex, a period of sexual abstinence improves the chances of producing a child of the opposite sex.

Other factors have been thought to affect the sex ratio of children. Urbanization, climate, social class, revolutions and wars, migration, illegitimacy, age of mother and father, length of marriage, number and sex of previous children have all been examined statistically, but the results show no correlation with sex ratio.

Cleavage of the Fertilized Ovum

Soon after fertilization the zygote undergoes a series of divisions that are collectively called cleavage. In man the whole zygote divides completely into two nearly equal parts. (In eggs with abundant yolk, cleavage is partial and incomplete.) The cells resulting from the division, the blastomeres, increase in number and decrease in size as division continues at fairly definite intervals. These divisions of the fertilized ovum continue within the zona pellucida as the ovum is transported along the oviduct. This passage of the zygote through the oviduct is a delaying action to allow the preparation of the endometrium of the uterus for its reception, but clearly there is a limit to the time spent in the tube. During this journey, the cleaving ovum receives nutrition from foodstuffs stored within it and from the secretions of the epithelial cells lining the plicated wall of the tube.

Muscular contractions of the ovarian tubes and mucous flow from the beating of the cilia of its lining guide the dividing ovum toward the uterus. As cleavage goes on, the blastomeres become progressively smaller until they have reached a size normal for somatic cells. As yet we do not know much about cleavage in human eggs. Possibly, as in the zygotes of some other animals, the first two blastomeres are totipotent and, if separated or appropriately stimulated, can each develop into a complete new individual (forming monozygotic, like or identical, twins). Should separation occur at a somewhat later stage in division and be incomplete, the result would be conjoined, or Siamese, twins. Such twinning is not always complete, as seen in Figures 11–8 and 11–9.

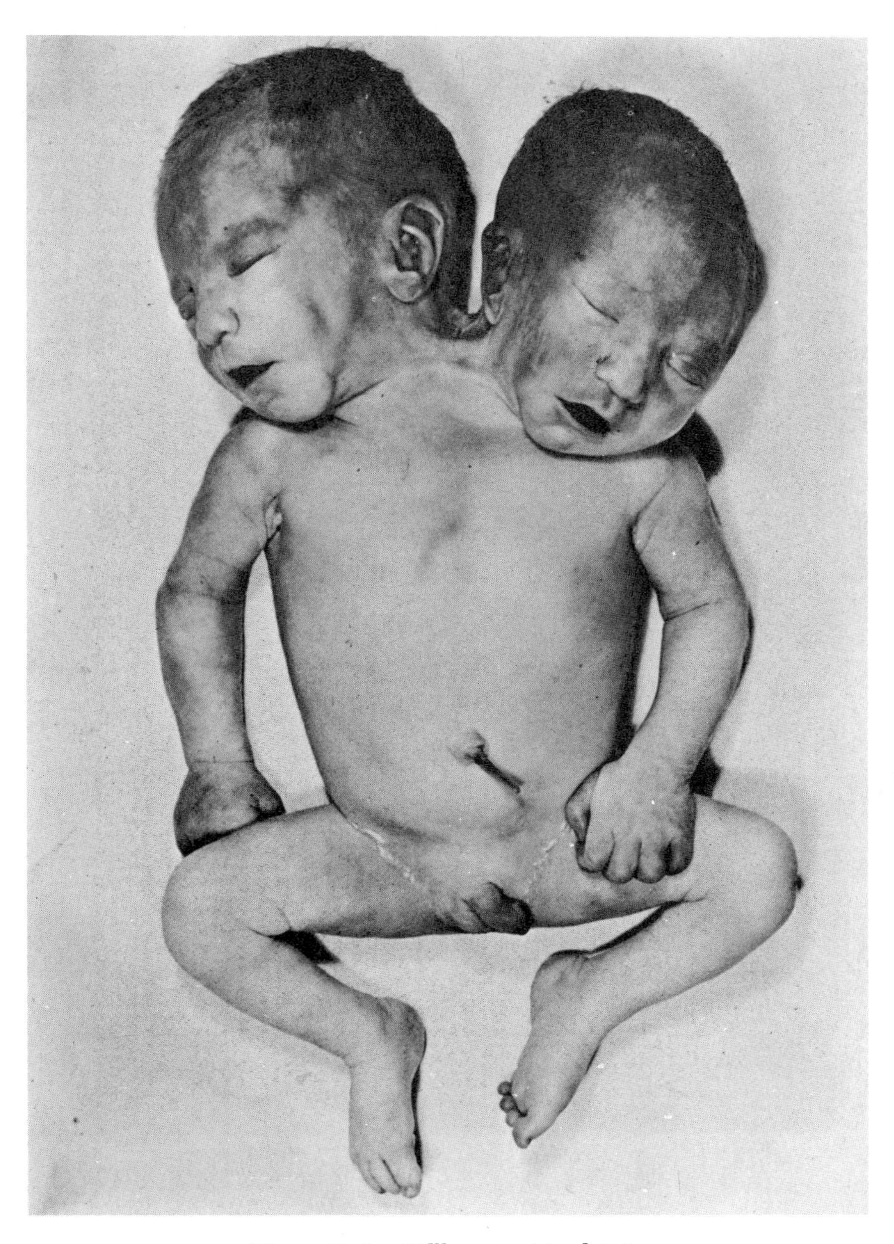

Figure 11–8. Stillborn conjoined twins.

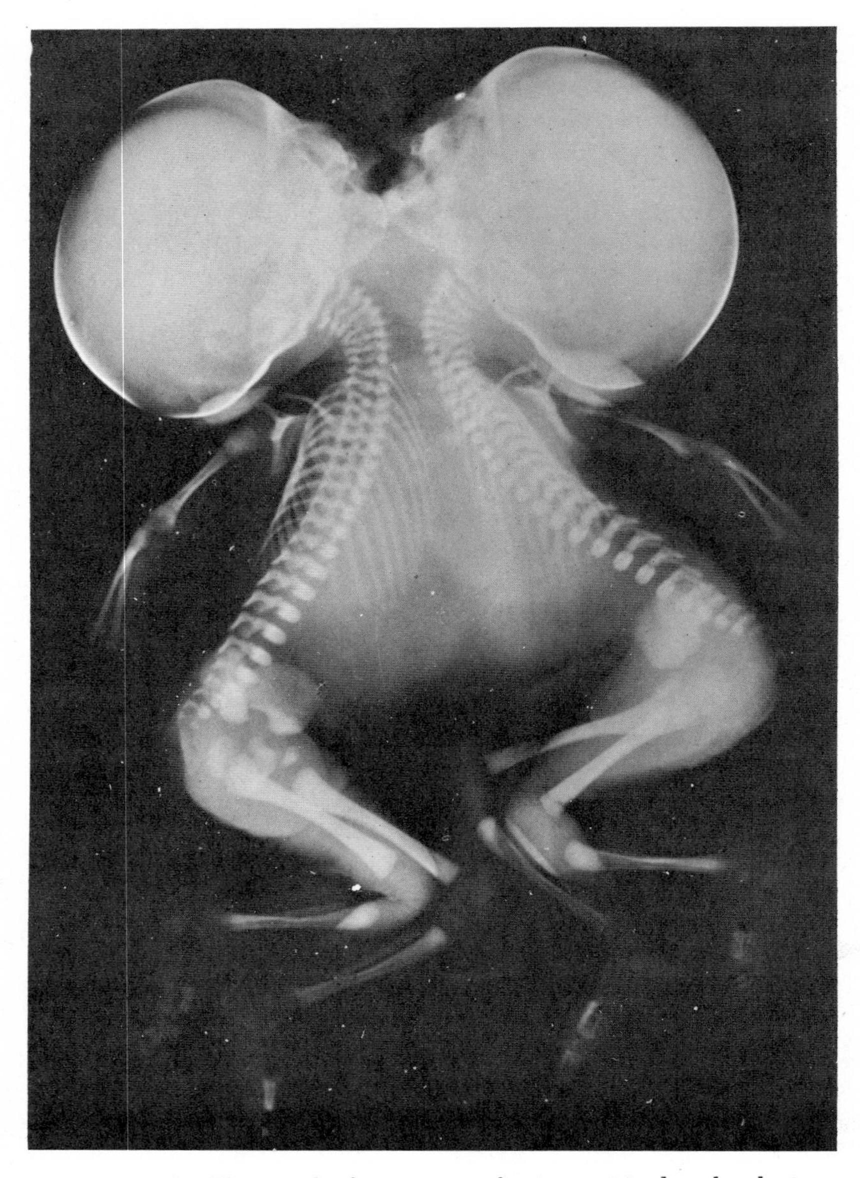

Figure 11–9. Photograph of a rare case of twins conjoined at the chest.

Cleavage is the visible manifestation of the initial formation of a complex organism and of the differentiation with each newly formed blastomere. It also indicates the power to manufacture new cell membranes and cell constituents. The movements of cells in relation to each other, which commence with cleavage, herald the increasingly more complex groupings and reorientations of cells that will occur as development advances. Cleavage results in the formation of 2, 4, 8, 16, and 32 cells within the zona until a solid cluster (morula) is formed. The zona restricts any increase in total volume and thus facilitates passage through the narrow tubo-uterine junction. Just before entry into the uterine cavity, the morula becomes transformed into a blastocyst (Figure 11–10).

Implantation

The free human blastocyst is a fluid-filled sphere of cells one layer thick (trophoblast) with a small group of cells, the inner cell mass, inside at one pole. Four to five days after ovulation it contains just over one hundred cells and has grown only to 0.09 to 0.1 mm in diameter. Exactly what happens during the next few hours in man is unknown, so information has had to be sought in nonhuman primates and other mammals.

A short time after its arrival in the uterus, the human blastocyst adheres to the surface of the endometrium. This most commonly occurs high up on the back wall of the uterus between the mouths of two uterine glands. At this time the zona pellucida is lost in some still unknown way. (Studies of rhesus monkeys have shown that progesterone may be involved in its removal.) The contact between the trophoblast and uterine epithelium may result in rapid adhesive chemical changes as well as in an interlocking of trophoblastic processes with tiny processes on the surface of uterine epithelium (Figure 11–10).

The next series of events results in the incorporation of the growing blastocyst within the endometrium. Exactly how this comes about in a human female is still unknown. When contact is made, the cells of the endometrium become engorged with glycogen and fat, and a marked increase in blood flow occurs. All these changes would seem to be advantageous to the blastocyst in that they provide nutriment for it, although they could be considered as efforts to contain its invasion. The trophoblast becomes two-layered, with a cellular inner layer (cytotrophoblast) and an outer one in which all the cells lack intervening membranes (syncytiotrophoblast).

The outer syncytial layer seems to insinuate itself between uterine cells and destroy them by enzyme action. As these cells die and break up, they form a material (histiotrophe) that is nutritious to the invading

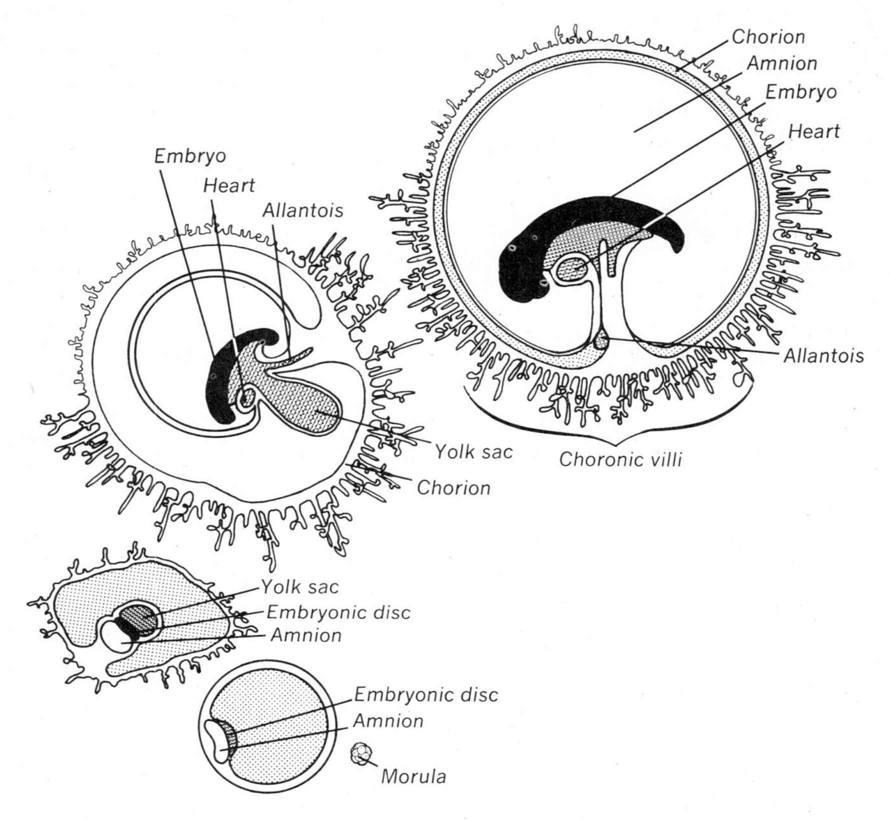

Figure 11–10. Stages in the earliest development of a human embryo.

tissue. Small projections develop on the outside of the invading syncytium to form rootlets (primitive villi) that reach even farther into the maternal layers. Endometrial cells seem to rearrange themselves around fluid-filled lakes (lacunae) that form in front of the invading blastocyst as if to assist the spreading of the syncytium. Shortly, the smallest maternal blood vessels either break or are broken down, and maternal blood comes into contact with the processes of the syncytial trophoblast. In some places the rupture of maternal vessels occurs into a lacuna, where the maternal blood may clot, becoming available for ingestion by the syncytium, or it may start to flow through the space, establishing a primitive circulation of maternal blood past the ever-increasing syncytial surface.

The syncytial processes (villi) project into the maternal blood spaces, and by about 16 days after conception embryonic blood vessels develop within them. When eventually the embryo's blood begins to flow through these vessels, the villi transfer gases and essential nutriments from the maternal blood around them to that of the growing embryo. As the conceptus becomes more firmly embedded, the formation of villi proceeds rapidly especially in the region deepest in the endometrium. Here the shaggy surface of the conceptus, known as the chorion frondosum, will become the placenta.

In about 0.3 percent of human pregnancies the blastocyst becomes implanted in the "wrong" place. Such ectopic implantations occur in the ovary, in the abdominal cavity, or, more commonly, in the uterine tubes. This tube is not constructed to contain a pregnancy, and the conceptus either dies or ruptures the tube with resulting hemorrhage that could kill the mother unless stopped surgically. Obstructions inside the tube or faulty transport of the cleaving ovum may cause tubal pregnancy. Possibly significant is the rarity of ectopic pregnancy in other mammals and virtual nonexistence in subhuman primates. Perhaps human reproductive processes are subject to more subtle influences than we realize.

A number of mammals exhibit a period of delay between the time the blastocyst enters the uterus and its final implantation. In some rodents delay may be related to lactation; if the number of young sucklings is over three, implantation of blastocysts formed after a postparturient ovulation is delayed by about one day for each additional young, up to a maximum of an extra nine. Another type of delay is seen in certain carnivores, probably in all pinnipeds (except the walrus), armadillos, and roe deer. In these mammals the period of delay may or may not begin during lactation and is continued for several months and even, in badgers, for up to two years. No evidence exists to date that delay in implantation occurs in man or in subhuman primates.

Gestation

The gestation period of mammals must be long enough to allow the newborn to survive at birth, take advantage of maternal care, and not damage the mother during birth. The less well developed at birth the neonate is, the more maternal attention and protection it will need. Young mammals vary greatly in their degree of dependence on one or both parents after birth. The gestation period of many rodents and carnivores is so short that the newborn are hairless with unopened eyelids and need succor in nests of some kind. By contrast, newborn guinea pigs, deer, and others can scurry about within a few hours of birth, and newborn dolphins can swim expertly within minutes. Newborn primates all require different degrees of maternal care. Young lemurs become practically attached crosswise to their mothers. Galagos and pottos keep their infants constantly on their breasts, carrying them safely clamped between their teeth when moving from place to place. Most monkeys and apes, however, nearly always hold their infants on their breasts, where the young cling to their mother's hair with their precociously strong hands. The amount of time an infant is totally or largely dependent upon the mother also varies in different species. Galagos remain with their mothers only a few months, monkeys and apes about a year or more until the mother has another young, when the yearling child and even young adult, though not dependent, still prefer the vicinity of their mother.

The human neonate is dependent longest on maternal care. He needs several months just to hold his head up without aid, six months or more to learn to sit up, and about a year to walk unsteadily. Except for their greater precocity, the infants of great apes are in a physical sense little different from those of man. Indeed, several attempts have been made to bring them up together. Unfortunately most of these "experiments" have been motivated more by emotion than by scientific zeal, and their observations have little value. The few reliable observations made, however, show that in some respects infant chimpanzees are a great deal more precocious than their human counterparts for the first two or three postnatal years and even at the end of this period have a much superior muscular and general locomotor coordination. However, young apes do not really learn very much according to human standards, and even the brightest of them seem pretty stupid. Human babies, though slow starters, accumulate, store, and coordinate information from the very beginning and by the age of four or five outdistance the brightest of great apes. At the end of the second year, when a human child begins to talk, it demonstrates that it is destined to develop mental abilities beyond those of any other living animal.

In any event, gestation must last long enough for a fetus to adapt immediately to the violent changes that accompany the "catastrophe" of birth. Within a few hours all mammalian fetuses are cast out of the warm, watery environment of the uterus into the harsh, outer environment. Oxygen, food, and all other essentials that were previously supplied through the umbilical cord from the mother must now be obtained by other means, and in the case of oxygen, without delay. As pregnancy advances many organs must complete their development and begin to function even before birth. The respiratory and cardiovascular systems, together with the urinary and alimentary organs, are called upon to function virtually to perfection at the time of birth.

The length and characteristic features of the gestation period should be such that the mother's strength and food reserves are not so taxed that each successive pregnancy reduces her efficiency and longevity. The length of gestation differs considerably among mammals (Table 11–1) and primates (Table 11–2). Human gestation varies more than that of most other mammals. During pregnancy many maternal organs alter greatly, especially the uterus. Excluding its contents, the human

Table 11–1. Gestation Periods in Some Mammals

	Days		Days
Virginia opossum (*Didelphis*)	12.5	Vampire bat (*Desmodus*)	90
Golden hamster (*Mesocricetus*)	16	Leopard (*Panthera*)	90
Common shrew (*Sorex*)	20	Lion (*Panthera*)	110
Mouse (*Mus*)	20 (first preg.)	Pig (*Sus*)	112
Dormouse (*Muscardinus*)	21	Tiger (*Panthera*)	113
Rat (*Rattus*)	22 (first preg.)	Sheep (*Ovis*)	150
Mole (*Talpa*)	30	Goat (*Capra*)	150
Rabbit (*Oryctolagus*)	31	Brown bear (*Ursus*)	210
Weasel (*Mustela*)	35	Hippo (*Hippopotamus*)	240
Hedgehog (*Erinaceus*)	35	Man (*Homo*)	267
Grey squirrel (*Sciurus*)	40	Common seal (*Phoca*)	275
Hare (*Lepus*)	40	Jersey cow (*Bos*)	278
Ferret (*Mustela*)	42	Porpoise (*Phocoena*)	320
Cat (*Felis*)	63	Dolphin (*Tursiops*)	330
Dog (*Canis*)	63	Horse (*Equus*)	340
Guinea pig (*Cavia*)	67	Sperm whale (*Physeter*)	365
		Giraffe (*Giraffa*)	450
Greater horseshoe bat (*Rhinolophus*)	70	African elephant (*Elephas*)	660

uterus has increased in weight sixteenfold at the end of pregnancy and its size has kept pace with the growing fetus. Small and pear-shaped before pregnancy, it becomes globular by three months, when it occupies virtually the entire pelvic cavity and displaces other viscera. By four months it extends upward out of the pelvis and can be palpated through the abdominal wall between the pubes and the navel. By six months it reaches the umbilicus, and at full term its top reaches as high as the lower margin of the thoracic cage (Figures 10–4, 10–5). A few weeks before birth the uterus falls a little as the fetus settles into the confines of the maternal pelvis.

As pregnancy advances, the number and size of the muscle fibers in the uterine wall (myometrium) increase, as does their ability to stretch. The uterine vessels become larger and coiled at first, but as the uterus distends they straighten out. Growth and distention of the uterus during pregnancy, as well as contractions of the uterine muscle and movements of the fetus, although not painful, are often felt by the mother. Early in pregnancy the stretching of the uterine tissues causes reflex disturbances such as nausea and vomiting in some women. Though unpleasant these

Table 11–2. Gestation Periods in Some Primates

	Days		Days
Tree shrew (*Tupaia*)	46–50	Rhesus monkey (*M. mulatta*)	150–180
Lemur (*L. catta*)	120–140	Barbary ape (*M. sylvana*)	200–210
Loris (*L. tardigradus*)	160–170	Chacma baboon (*P. ursinus*)	180–190
Bush baby (*G. senegalensis*)	120–146	Sacred baboon (*P. hamadryas*)	154–183
Tarsier (*T. syrichta*)	180	Langur (*P. entellus*)	170–190
Howler monkey (*A. seniculus*)	139	Gibbon (*H. lar*)	210
Spider monkey (*A. paniscus*)	139	Orangutan (*P. pygmaeus*)	220–270
Woolly monkey (*L. lagothricha*)	139 (also 225)	Chimpanzee (*P. satyrus*)	216–260
Marmoset (*C. jacchus*)	140–150	Gorilla (*G. gorilla*)	250–290
Japanese macaque (*M. fuscata*)	140–156	Man (*H. sapiens*)	267 Average
Crab-eating macaque (*M. irus*)	160–170		

disturbances are rarely dangerous. Later in pregnancy the pressure of the uterus against neighboring structures may be somewhat distressing and, depending on the physique of the mother, incapacitating. After childbirth, when a women first gets up, she may experience a sensation of falling forward since she has become accustomed to carrying some twenty extra pounds in front of her center of gravity.

The cessation of the menstrual period is usually the first indication of pregnancy, but soon after other signs appear. The breasts become swollen and tender and their veins prominent. The nipples enlarge and together with the areolae become more deeply pigmented. Changes also occur in the mother's cervix, uterus, and skin. Increased excretions of gonadotrophin can be detected during the early weeks by various "pregnancy tests." The urine or blood serum of pregnant women, injected into female mice, rats, rabbits, and frogs or toads, induces ovulation in these animals. Reliable immunological techniques, using antihuman chorionic gonadotrophic serum, have been recently developed.

Pregnancy is not so easily detected in other mammals. In Old World primates fur conceals changes in the skin, and a thick fatty layer prevents palpation of the uterus. Because hormonal changes are either absent, different, or not so marked, the biological tests for human pregnancy are not applicable. Naturally, we do not know whether other female mammals become "aware" that they are pregnant. Later in pregnancy certain behavioral changes occur: alterations in response to males, increased apprehension and even aggression, and nesting activity. But these are probably responses to the hormonal climate of pregnancy rather than to an appreciation of the developing conceptuses in the uterus. Knowing so little about their general behavior patterns, we cannot tell what specific changes occur in expectant mothers of other primates.

The uterus is more than a protective purse for the conceptus; very early in gestation it becomes a complex and fascinating laboratory where activities center around the microscopic life that has just begun there. Intermittent, involuntary, and scarcely appreciated muscular contractions very likely aid blood to circulate through the placenta. The general tone of the uterine muscle is an important factor in keeping the placenta in position. As the uterus enlarges, the muscle fibers rearrange themselves so that those in the thicker middle layer tend to align themselves spirally about the organ. Such an arrangement enhances the expulsive power needed to eject the conceptus at parturition. As gestation advances, the cervix becomes "softer" and easier to dilate. The important uterine vessels function especially to supply the fetus with the necessities for successful growth and to remove its waste products.

As pregnancy advances, the mother undergoes numerous changes, many of which reflect the altering hormonal climate; others are hard to explain. Even the mother's attitudes change. Contrary to popular belief, a woman does not necessarily have a feeling of well-being, a more lively disposition, and an enhanced sensitivity, even during a normal pregnancy. Acute observers may notice such subtle changes as increased authority, better powers of choice and decision, and occasionally an imperial smugness. Pregnant women frequently experience longings for particular foods or sudden compulsions to perform self-imposed, often irrelevant tasks. Like so many other human reactions, these and other unpredictable behavior patterns depend greatly on the experiences and the temperament of the woman.

Pregnancy has a noticeable effect on the pigmentation of the skin, the degree of darkening depending on the natural complexion. Pigmented areas often develop locally: a darkened linea nigra usually appears on the center of the abdominal wall below the navel, and the facial skin becomes somewhat browner. In some cases an almost masklike change occurs in facial coloration and is known as the chloasma (greenness) of pregnancy. Most of the increases in generalized pigmentation fade after parturition. The skin over the body of the breasts sometimes assumes a webbed or mottled pigmentation. The sebaceous glands at the periphery of the areola enlarge and look like small tubercles (Figure 11–11). In the later months,

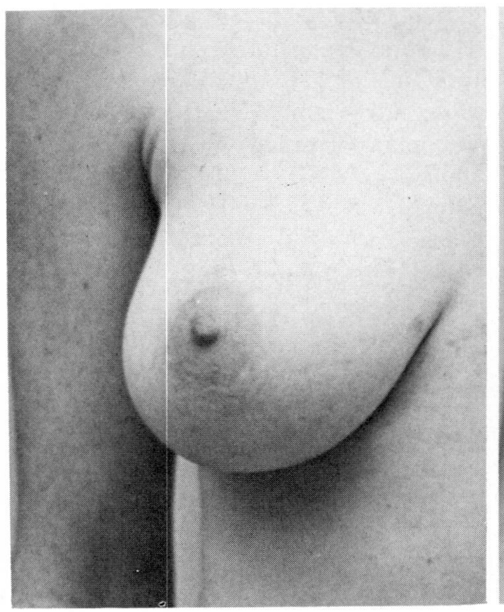
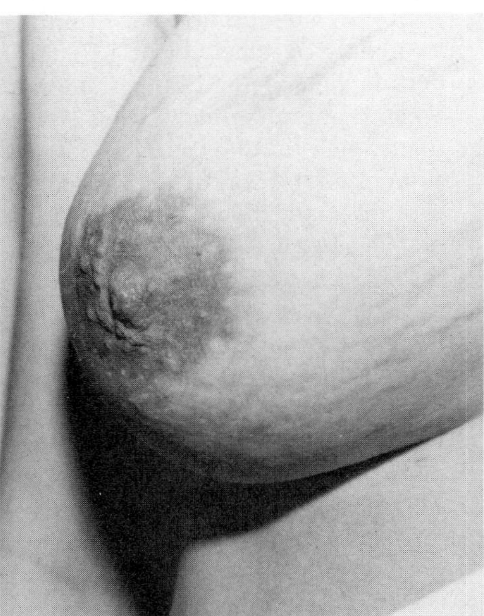

Figure 11–11. Comparison between the breasts of a nonpregnant woman and of one who has just given birth. The areola and nipple of the lactating breasts are larger and more heavily pigmented than those of the quiescent breast. Note prominent tubercles of Montgomery.

a watery secretion is sometimes exuded from the nipple. Some scalp hair may be lost, though more often this occurs after parturition, and the nails may become brittle.

Changes in the circulatory system reflect the increased demands not only of the uterus but of other organs as well. Cardiac output, total blood volume, and the volume of blood in the veins at any one moment all increase. Engorged and dilated veins become visible in the skin, particularly in the neck, abdominal wall, and legs. Varicosities and hemorrhoids often accompany or follow pregnancy. Sluggish venous flow, especially in the legs, leads to increased tissue fluid and swelling. The weight of the pregnant uterus pressing on pelvic and abdominal veins often interferes with venous return and may reduce the movements of the diaphragm. The heart, usually capable of dealing with additional demands, can use up its reserves and may have difficulty in meeting extra stresses. The upright human posture poses problems that are not experienced by four-footed mammals. Although one does not advocate that pregnant women progress on all fours, the temporary adoption of such a position can sometimes prove therapeutic, especially during the late months of pregnancy.

During pregnancy almost every maternal endocrine organ undergoes some alteration. The pituitary nearly doubles in size, the thyroid and adrenal glands enlarge, and their increased and altered activities are probably largely responsible for many of the changes mentioned here. The parathyroid glands become much larger to meet the growing need of the fetus for calcium. Provided that the mother's diet is adequate, no net loss of calcium from her skeleton and teeth occurs because of transfer to the fetus. The turnover of salts and other substances in the mother's bones and tissues is probably increased. The whole question of the hormonal interaction between mother and fetus is a matter of some concern but about which we are still uncertain. Studies on experimental animals have established in these animals, and even in some women, the susceptibility of certain developmental processes to specific hormones. However, we do not yet have enough information on the extent to which normal hormonal changes in the mother affect the growing fetus and even influence the intelligence of the child after birth. Clearly these hormonal changes are potentially more dangerous at specific periods during fetal development. On the other hand, some endocrine disorders in the mother appear to be alleviated by pregnancy, but usually recur after parturition. This amelioration may be due to the combined influence of fetal endocrine organs and of the placenta.

Embryonic and Fetal Growth

Embryonic development progresses continuously, uninterrupted by pauses or static periods. An embryo is a living organism, which embryologist G. L. Streeter said, "in its time takes on many guises, always progressing from the smaller and simpler to the larger and more complex." The age of an embryo, gauged by the probable time of ovulation, cannot be known precisely to within hours. The youngest human embryos can be obtained only rarely, most often by chance. When such precious specimens are found, each is carefully preserved, sectioned serially, and minutely described. (Particularly early specimens are frequently named after the investigator who described them.) Quite apart from a natural curiosity about the development of human embryos, other reasons exist for studying the early fetal stages. Many of the specimens recovered may be abnormal, and information on what has happened and why is useful. Many of the congenital defects of children are due to the cessation of one or more embryonic processes before their completion.

Embryos are particularly sensitive to certain adverse factors (called teratogens if they cause fetal abnormality, from Gk. *teras,* monster) at specific times in their development; knowledge of which periods are critical has enabled us to offer better care to mothers. The devastating effects of thalidomide on human embryos have served to alert us to the necessity of testing drugs thoroughly before they are administered during pregnancy. An even greater problem is posed by those drugs that do not cause gross malformations but can have subtle or adverse effects on the central nervous system. Species differences in embryonic reactions to teratogens are known, and the presence or absence of various factors can ameliorate or aggravate these reactions. No two embryos develop at precisely similar rates and no two newborn children are identical, not even identical twins. Just as no two human embryos have precisely identical genetic constitutions, so the conditions to which they are exposed in the uterus are never exactly the same.

Since early human embryos are available for experimental investigation only rarely and fortuitously, we have to depend on the embryos of other animals to provide the information needed to interpret the events that occur in man. Investigations of the developing embryos of laboratory animals have become increasingly important as we seek to determine how teratogenic substances exert their deleterious action. Even so, we unfortunately still know far too little about the embryology of most nonhuman primates, except for scattered, unrelated articles and the excellent series on the rhesus monkey published in the *Carnegie Contributions to Embryology.*

The most dramatic changes in the embryo occur during the early phases of development. For the sake of convenience the successive stages of development have been classified and are referred to as horizons, meaning epochs or periods of time that cover certain events. G. L. Streeter, availing himself of the unequalled collection of human embryos housed in the Carnegie Institute of Embryology in Baltimore, first applied the concept of horizons. Today, detailed descriptions of twenty-three horizons, covering the first seven weeks of the human embryo, are available. Here the development of the human embryo will be briefly illustrated first by weeks and then by months; the horizons and then the main changes in growth will be indicated.

First Week

The events during the first week of a human pregnancy have already been described.

Horizon	I	Single-cell fertilized egg.
	II	Segmenting egg in uterine tube.
	III	Free blastocyst in uterus.
	IV	Blastocyst implanting.

Second Week

Implantation begins 5 to 6 days after fertilization and continues during the second week, the blastocyst gradually becoming embedded in the endometrium. An amniotic cavity, composed of primitive ectodermal cells above and endodermal cells beneath, develops within the inner cell mass above a pear-shaped embryonic disc. Growth of endodermal cells by the 11th day establishes a primary yolk sac which by 13 days forms a definitive yolk sac (Figure 11–10). A primitive streak appears on the ectodermal surface of the narrower caudal part of the embryonic disc by 14 to 15 days. Mesodermal cells now form from the sides of the primitive streak and spread out between the ectoderm and the endoderm of the embryo. This mesoderm will give rise to the connective tissue, bones, muscle, and so on of the embryo.

Horizon	V	Blastocyst implanted but without villi.
	VI	Distinct yolk sac. Villi appear on the chorionic sac.
	VII	Embryonic disc defined. Chorionic villi branching.

Third Week

Now that the early human embryo is a three-layered disc, events succeed one another rapidly on its surface, inside, and beneath. On the surface of the ectoderm in the broader, anterior region, a head process extends ahead of the primitive streak. From a point close to a surface node described by a German embryologist, Viktor Hensen, exactly in the midline, the notochordal canal has tunneled forward into the head process. The notochordal canal forms a cord, or process, which is the most primitive skeleton of the embryo, the precursor of the vertebral column. All mammals, and for that matter all chordates, sometime or other in their lives possess a notochord. By 19 days, just ahead of the primitive streak and above the notochord, the thickened midline strip of the neural plate develops (Figure 11–12). From its grooved center, the edges rise up on each side to form a furrow, the neural groove, with a neural fold on each side.

Horizon	VIII	Hensen's node present. Primitive streak region active.
	IX	Neural plate defined; neural folds forming; notochord present and elongating.

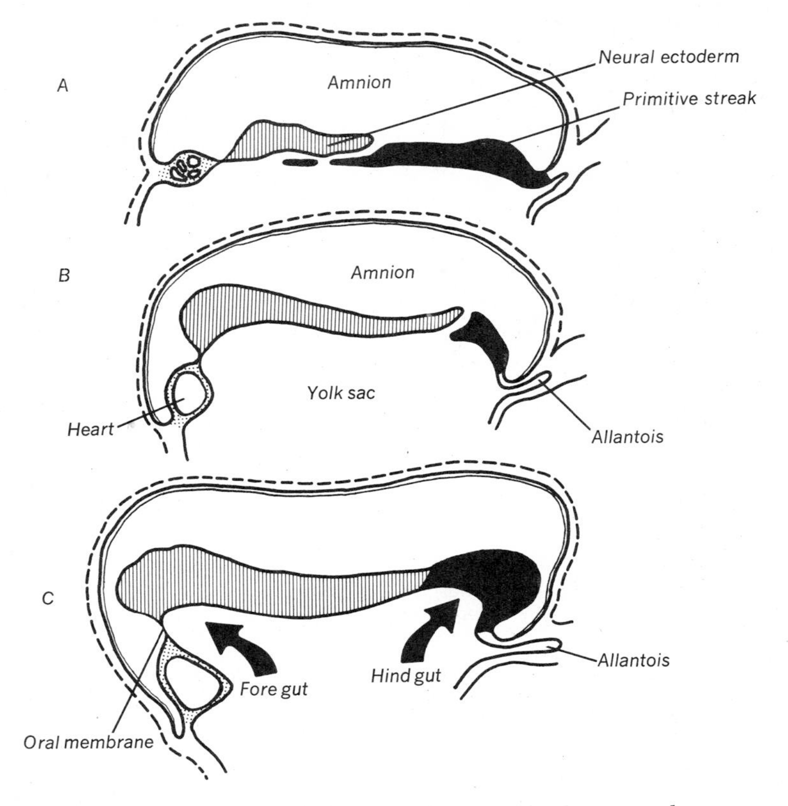

Figure 11–12. Early developmental stages of a human embryo.

Fourth Week

From the 20th to the 28th day several important developments take place synchronously. The neural folds rise, meet on the midline, and close over to form a neural tube, which later gives rise to the brain and spinal cord. As the mesoderm proliferates from the primitive streak, it moves up to the sides of the notochord where it divides into blocklike masses, somites, arranged in successive pairs and giving the embryo a segmented look (Figure 11–13). Pairs of somites start to form near the head end and more are gradually added behind them. At 21 days the embryo has 3 pairs of somites and is about 1.5 mm long; at 22 days it has 7 somites and is just over 2.0 mm in length. Eventually about 44 pairs of somites are formed. These, supplied by spinal nerves, give rise to muscle masses and contribute to the formation of vertebrae. The somites represent the serial repetition of homologous parts (like that seen in the segmental arrangement of earthworms) of which primitive animals are constructed. Although much modification of the early segmental pattern occurs in human anatomy, one can still recognize its arrangement in the ribs, vertebrae, spinal nerves, certain parts of the vascular system, and, though much modified, in the limbs.

Horizon X Somites forming.

Early in the fourth week (Figure 11–13) the yolk sac is shrinking. Because of active growth elsewhere, the embryo appears to be folding over on itself, with both the front and hind ends "tucked under." What precisely ordains these movements is not yet clear, but they result in changing a flattened embryonic disc into an enclosed cylinder. The yolk sac continues to shrink as it helps to form a narrow, almost tubular foregut, a larger midgut region, and a curved, backward-directed hindgut from which projects a hollow cul de sac, the allantois. The whole embryo has now grown away from the encapsulating chorion and is attached to it by a stunted body stalk. Figures 11–10 and 11–12 show these movements diagrammatically. Figure 11–10 shows how, at a later stage, the umbilical cord is eventually formed from the connecting stalk. All placental mammals have an umbilical cord.

As the head-fold rises up, the developing heart, which was ahead of it, moves to a ventral position beneath the elongating foregut (Figure 11–10). First closed anteriorly by a buccopharyngeal membrane, later the foregut is open as the membrane breaks down. This gives rise to the oral cavity—the pharynx, esophagus, and stomach. Later the lungs and trachea also grow from it, and many important structures are formed from its lining. From the midgut the intestines differentiate as far as the distal third of the transverse colon. The remainder of the alimentary canal grows from the hindgut. At first, much of the simple, uncoiled small and large intestines protrudes from the belly of the embryo as an embryonic hernia since as yet there is no proper abdominal wall. The allantois gives rise to a part of the urinary bladder.

The mesoderm lateral to the somites becomes divided like a sandwich into an inner layer that attaches itself to the shrinking yolk sac and an outer part that fuses to the under surface of the ectoderm. The layer of ectoderm and mesoderm folds under the body of the embryo to form the abdominal wall and encases the now coiled intestines into a newly formed abdominal cavity. The split in the mesoderm remains as the coelom, another characteristic feature of all vertebrates. Occasionally the anterior abdominal wall fails to form properly about the umbilicus, leaving the newborn child with what is called a congenital umbilical hernia.

Horizon XI 13 to 20 somites present. Folding of head and tail region.

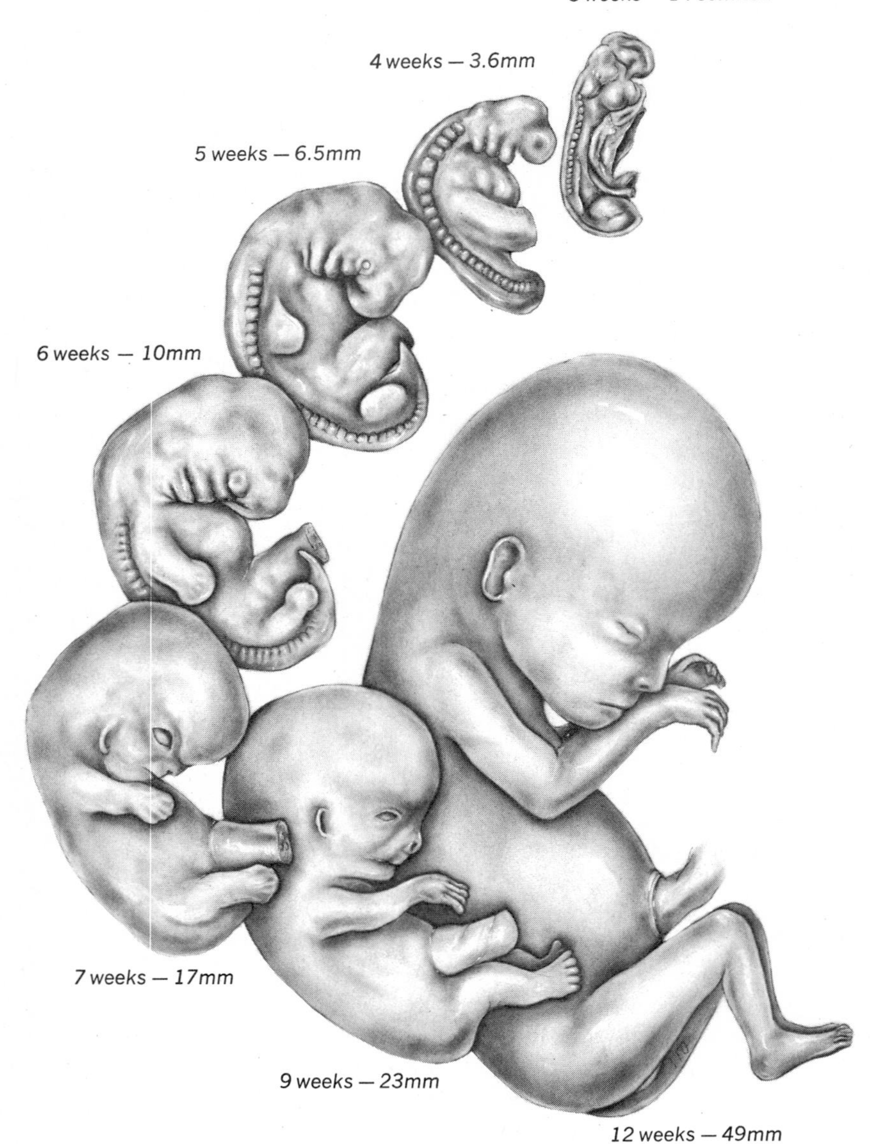

Figure 11–13. Human developmental sequences.

The embryo is now about 24 days old. Since it is curled up, somewhat in the form of a C, within the amniotic cavity, it has to be measured from its crown (the most anterior part, or head end) to its rump (the most posterior part of the curved tail end). The measurement, known as the C.R. length, is now 2.5 to 3.0 mm (Figure 11–14). A series of pharyngeal or branchial (gill) arches appears on each side of the developing primitive mouth. (In adult fish, these gill bars have slits to permit the passage of water from which oxygen is removed by diffusion into the gill capillaries. Pharyngeal arches do not have a respiratory function in reptiles, birds, or mammals, either adult or embryonic, and no slits form between the arches.) In the mammalian embryo, each pharyngeal arch has an external cover of ectoderm; a middle layer of cartilage, muscle, nerves, and blood vessels; and an inner lining of endoderm. The formation of particular structures in the head and neck is due to these elements. Human embryos have five such arches separated by four grooves; the fate of the posterior two arches is difficult to follow in human embryos.

By the sixth week, the second arch overlaps the last three and obliterates them. The resulting fusion corresponds to the operculum of bony fishes. After about 14 days the remaining arches are absorbed or molded into other structures and the embryo loses all resemblance to primitive ancestral forms (Figure 11–13). The arches "inherit their pattern of growth through some one common protovertebrate ancestor, of fishlike general character though not exactly like any present-day fish," wrote embryologist G. W. Corner. "The descendants, as they evolved into various classes, necessarily inherited such a pattern, but they worked it over into new forms suitable for their needs."

Horizon XII Age 26 ± 1 day. Length 3.2 to 3.8 mm. Three pharyngeal arches and 20 or more somites visible. Obvious tail present.

Fifth Week

At the start of the fifth week the human embryo has a C.R. length of about 5 mm, is still very small, and has no resemblance to even a miniature human being (Figure 11–13). As yet the head is not perceptively larger than that of other mammalian embryos but will be so during the next two weeks. With the enlargement of the heart the embryo attains a progressively rotund appearance. The heart is now being called on to supply blood to the embryo and to the ever-increasing bulk of the placenta, which is relatively larger in late embryonic life than at any other time. The critical developments of the heart occur

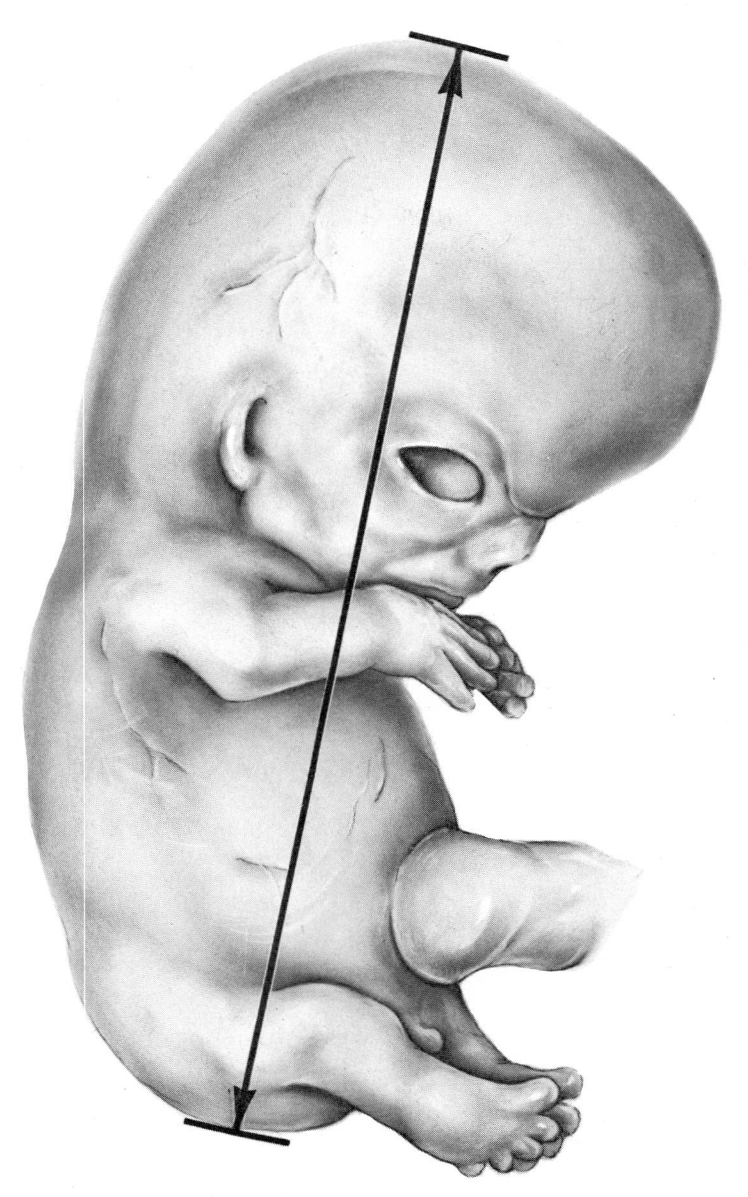

Figure 11–14. Diagram showing how crown-rump measurements are taken.

during the next 14 days: two closely applied partitions subdivide the atrium internally into two chambers, and a third separates the two ventricles. The complete separation of the chambers is essential in the evolution of terrestrial animals because it separates the pulmonary circulation and allows it to function independently of that supplying the body (systemic). In the fifth week, the number of somites increases to about 30. Limb buds appear laterally from the body wall; in man and most primates the upper limb bud develops first and more rapidly. Having reached its greatest relative degree of development, the tail now retrogresses. Optic vesicles appear on the sides of the head, and a series of little hillocks marks the early external ear. The brain enlarges, and the outlines of its vesicles can be seen through the surface coverings. The nostrils, too, are discernible as two widely separated pits.

Horizons XIII and XIV	Age 28 to 30 days. Length 4.0 to 7.0 mm. Arm and leg buds present; ear and eye regions defined.
XV and XVI	Age 31 to 34 days. Length 7.0 to 11.0 mm. Limbs enlarging and nostrils appearing; main parts of brain forming.

Sixth Week

The head is noticeably larger and facial processes develop to form the nose and upper lip. The head is still bent forward and there is no neck. The anterior limbs continue to develop, changing from blunt, finlike configurations to arms having forearms and platelike hands with furrows delineating the fingers. A few days later a similar differentiation takes place in the hind limb buds. The body, now a more compact mass, loses its C-form and attains a marked dorsal convexity. The tail becomes inconspicuous and buttocks appear.

Horizons XVII and XVIII	Age 35 to 37 days. Length 11.0 to 16.0 mm.
XIX and XX	Age 39 to 40 days. Length 17.0 to 23.0 mm.

Seventh and Eighth Weeks

At the outset of this period, a human embryo is about one inch long. Despite its small size, the major organ systems and body parts are distinct or will be so in the next two weeks. The large head lifts slightly and the body straightens. The face has formed; eyelids begin to grow over the developing ball of the eye; the hillocks around the outer ear canal are fusing into a minute pinna; a chin begins to jut forward; and a constriction marks the neck. Fingers and toes appear (Figure 11–13). The body still has a protuberant appearance, partly because of the large heart and developing liver, a relatively huge organ in all mammalian embryos whose early function is to form red blood corpuscles in lieu of the still unformed marrow in the developing bones.

Horizons XXI to XXIII Age 43 to 47 days. Length 24.0 to 30.0 mm.

The embryonic period ends on the 56th day, when the embryo is 35 to 40 mm long. Thenceforth it is called a fetus, and subsequent changes will be largely of growth. The head and brain are relatively massive, and the body is waxlike, shiny, hairless, and slightly transparent (Figure 11–15). At this stage, external appearances give no clue to its sex.

Third Month

From the 8th to 12th week a fetus grows to about 56.0 mm in length (some two inches). The body straightens out still more but the head is bent forward, the arms are folded across the chest, and the knees drawn up against the abdominal wall. This characteristic fetal position is often adopted by adults when asleep, an interesting phenomenon that provokes some speculation but which is in reality one of the best positions in which to conserve heat. The abdominal wall is now completely formed. With the growth and elevation of the nose, the face is less flat; the eyes are directed forward instead of laterally; and the eyelids fuse over the eyeballs. Ossification centers appear in the developing bones, and nails can be seen first on fingers and then on toes. Genital swellings develop into genital organs, the first external indications of the sex of the fetus.

Fourth Month

The fetus is now about 110 to 112 mm long (4½–5 inches). Individual characteristic features that are unique and foretell those of the adult begin to appear (Figure 11–16). The umbilical cord is attached relatively higher on the abdominal wall as the lower part of the abdomen and pelvis develop. Hair follicles and the precursors of cutaneous glands are found in the skin. General sensory organs appear, and the ears and eyes differentiate further.

Figure 11–15. A fetus about two months old attached to the placenta above by the coiled umbilical cord. Note the large umbilical vessels.

Fifth and Sixth Months

By the end of the fifth month, a fetus measures about 160 mm (6½ inches), and one month later the C.R. length is some 205 mm (8 inches). Head hairs already present for some time become conspicuous, and the body is covered by a fine down (lanugo). Eyelashes and eyebrows grow somewhat later. The skin is still lean and taut, but some wrinkles, folds, and creases are just discernible. As cornification commences, the surface of the body becomes more opaque. Some of the superficial cells, mixed

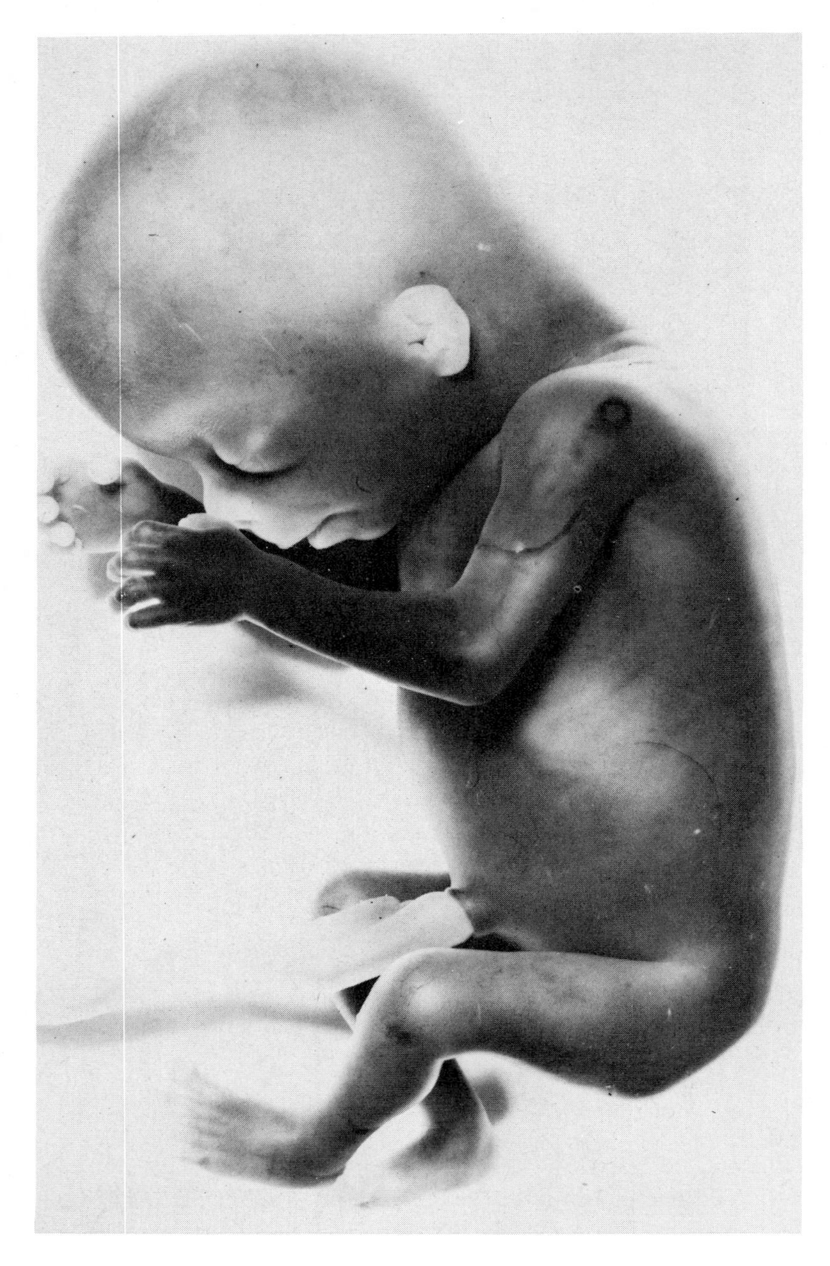

Figure 11–16. Human fetus about four months old.

with early secretions of sebaceous glands, form a cheesy material (vernix caseosa) over the entire surface of the skin. Washed off with the first postpartum bath, the vernix nevertheless protects the skin and perhaps facilitates the passage of the fetus through the birth canal. Early in the fifth month, spontaneous muscle movements are felt by the expectant mother, described as a "quickening," a sensation somewhat akin to the fluttering of a butterfly inside the lower abdomen. Later these movements become stronger (there is no truth in the old wives' tale that male fetuses move more vigorously). Ossification centers become more widespread, and the framework of the skeleton is laid down. The first elements of developing teeth begin to be calcified.

Seventh to Ninth Months

At seven months the fetus measures about 240 mm (9½ inches), at eight months, 280 mm (11 inches), and at nine months, 310 mm (12–13 inches). The fetus is considered "viable" at seven months (28 weeks), i.e., its chances of surviving a premature birth are good. A progressive accumulation of subcutaneous fat and a general thickening of the skin gradually transform the wizened fetus from a raw-red, dried-up miniature of a human child into a reasonable facsimile of the chubby babe-in-arms so fondly depicted by Rubens and Botticelli. At eight months the eyelids have reopened; the testes, on their way to the scrotum, have reached the inguinal canal although they are occasionally found in the scrotum. At this stage the fetus can perceive light and taste sweet substances. It is still "deaf," for though the auditory apparatus is fully formed, the ossicles of the middle ear are surrounded by fluid and cell debris that prevent vibration until air enters the pharyngotympanic tube and middle ear cavity. By nine months the baby is chubbier, with plumper and more rounded body and limbs. Head hair is now prominent, and the fingernails have almost reached the tips of the fingers.

The average weight of a human neonate (full term, about 38 weeks and a mean of 267 days) is 3.5 kg (ranging from 6 to as much as 12 pounds). Its length is 20 to 21 inches (50–53 cm) from the top of its head to its heels, and its C.R. length (or sitting height) is about 12 to 13 inches (30–33 cm). The circumference of the large head is about 33 cm (13 inches), with a relatively small face and jaws. At birth the head may be somewhat temporarily distorted or molded into a "sugar-loaf" form from having had to be squeezed through the birth canal. The nose is broad, and the cheeks are puffed out by a pad of buccal fat designed to facilitate sucking. The jaws have unerupted teeth, though occasionally the tips of the lower central incisors will have just emerged from the gums. The neck is often indistinct, and the shoulders, set high on the thorax, emphasize a bullnecked appearance, giving the look of a fixed shrug—or resignation to the "catastrophe of birth"? Since muscular control and strength are still minimal, the head tends to loll forward or sideways. The thorax, with its relatively huge heart, is barrel-shaped. At the lower center of the plump abdomen is attached the cord, later to be cut. Neonates lack a "small of the back," but their hips are slender with rounded buttocks; their arms and legs are almost equally long. The feet are turned inward, a fact that gives early walking movements a bow-legged, waddling effect. When the sole of the baby's foot is stroked, he elevates the big toe (Babinski reflex). After he begins to walk, he develops plantar flexion. Nails have reached the tips of fingers and toes, or project beyond them. Most of the lanugo hairs have been shed but head hair is abundant.

At birth dramatic changes occur within the lungs and the cardiovascular system. When the baby takes the first breath of air, the respiratory tract and the air sacs of the lungs begin to open. This "pneumatization," which is helped by vigorous crying, continues steadily for several days until it is completed. When the cord is tied and cut, the venous shunt through the liver (ductus venosus) closes down since no blood is now returning via the umbilical vein. The foramen (ovale) in the interatrial partition of the heart, as well as the arterial shunt between the left pulmonary artery and the aorta (ductus arteriosus), closes. Thus, the pulmonary circulation is completely separated from the systemic. Should these shunts fail to close at birth, venous blood is mixed with oxygenated blood and the infant is known as a "blue baby." If the condition is severe, the child dies.

The Placenta

Eutherian mammals have a placenta (Gk. *plakous,* a flat cake) so named by Realdus Columbus of Padua, a sixteenth-century disciple of Vesalius, whose work provided much of the groundwork for William Harvey's experiments. Anatomist H. W. Mossman of the University of Wisconsin, who has spent a lifetime studying fetal membranes, defined the placenta as "any intimate apposition or fusion of the fetal organs to the maternal (or paternal) tissues for physiological exchange." The placenta is therefore an "outgrowth" of the embryo that attaches to special portions of the parent organism that are adapted for the purpose. Obviously, the placenta is an essential feature of viviparity (giving birth to living young), yet in several species of animals, the two have evolved independently. Like other mammalian organs and structures, the placenta appears to have evolved from an originally simple pattern in lower forms, each pattern adapting and modifying to sustain the steadily increasing complexity and intimacy with the maternal organs. The placenta is the most diverse of all mammalian organs in both external form and internal arrangement: it effects exchange between mother and child in both directions, prevents undesirable substances from being transported to the fetus, and it may even produce hormones.

The human full-term placenta is a flat disc molded against the inner, usually upper and posterior, surface of the distended uterus. It is about nine inches (23 cm) in diameter, one inch (2.5 cm) thick in the center and weighs about a pound (45 gm). A number of clefts subdivide the outer, or uterine, surface into 16 to 20 subsidiary lobes (cotyledons), which fit into depressions in the uterine wall with wedges of uterine tissue filling the spaces between the lobes. From the edges of the placental disc, the fetal membranes continue outward to enclose the fetus within an ovoid chorionic sac composed of the chorion (laeve) on the outside and the amnion on the inside. This membrane continues on to the inner or fetal side of the placenta proper and then covers the umbilical cord, which is usually attached near the center of the placenta (eccentric and marginal attachments also occur). The umbilical vessels fan out over the fetal surface of the placenta in characteristic patterns.

During early development of the human placenta, a large number of fingerlike processes appear on that part of the chorion that first penetrates the endometrium and give it a shaggy appearance (chorion frondosum). Some of the villi extend to attach to the uterine tissues while others join into a latticework almost resembling a sponge. The uterine arteries open directly into the interstices between the villi, filling the intervillous spaces with maternal blood, which percolates through the spongework of the labyrinthine placenta and eventually returns to the uterine veins. Gases, nutriment, and other substances are transferred across the covering tissue of villi (the trophoblast) into the placental capillaries, and excretory products pass in the opposite direction. Fetal blood passes from the single umbilical vein to the fetus and is returned to the placenta by the two umbilical arteries. Normally the blood within the maternal circulation never mixes with that of the fetus, and vice versa. However, in certain toxic conditions that damage the villi, some leakage can occur.

A mammalian placenta has several component parts. This discoidal human placenta, briefly described above, is really the chorioallantoic placenta. Birds have a well-developed allantois, the rich vascular beds of which serve for respiratory, absorptive, and excretory purposes. It is variously developed in groups of mammals and may be very large, sausage-shaped, and full of fluid, as in ungulates. In primates, including man, the allantois is a small, short, hollow process projecting into the cord at the umbilicus. Its lining cells within the embryo help to form the urinary bladder, and the rest degenerates before birth to form a slender strand (urachus) extending from the apex of the bladder to the umbilicus. Although very small, the connective tissue of primate allantois gives rise to the placental vessels. Thus the human discoid placenta is still considered to be chorio-allantoic even though there is no large allantoic sac.

The fate of the early yolk sac already described varies in different groups of mammals. In some it grows out and eventually makes contact with the chorion, forming a nonvascular, bilaminar structure, as in many bats, rodents, and the armadillo. Material from the maternal tissues is absorbed through it by diffusion. Should vascularization of this structure occur, a chorio-vitelline or yolk sac placenta is formed with its own circulation passing up the cord through vitelline vessels. Among primates, marmosets have the largest yolk sac, but a chorio-vitelline placenta does not develop. The human yolk sac, which is small, is prevented from reaching the chorion by the large extraembryonic coelom (Figure 11–10). Even so, it is an important structure because the first blood is formed there and it stores germ cells before they migrate to the embryonic gonads.

The form and construction of the chorio-allantoic placenta also vary in different mammalian groups. Unfortunately, these variations are of slight value in phylogenetic studies and are difficult to correlate with functional characteristics. Pigs, whales, and horses have a diffuse placenta in which the chorion is spread over the whole uterine area and the entire sac adheres to the uterine wall. In deer, goats, sheep, and cows, only restricted regions of the chorion show villous processes, known as cotyledons. The villi have various shapes, long and slender in deer, and digitiform in sheep. They fit into depressions in special aglandular regions (caruncles) of the uterine wall, which eventually become deep and complicated crypts. There are, then, series or bits of placenta (placentomes), each supplied by branches from the umbilical cord splitting to send branches to opposite sides of the uterus. The placenta of tarsiers, the great apes, and man is a single discoid mass. The human placenta is occasionally bidiscoidal (as it is in monkeys), but usually one lobe is very much smaller than the other. This is explained as being a cotyledon that has become separated from the main mass and is an accessory placenta. Only rarely does an annular human placenta develop or one that resembles a diffuse type.

Inside a human placenta the maternal blood literally washes against the outer covering of the villi, and the relationship is hence termed hemochorial. Tarsiers and all anthropoid primates including man have hemochorial placentas. The type of relationship and the intimacy that develops between the chorionic cells and the maternal elements are essentially related to the method of implantation. In lemurs and lorises the placenta is diffuse and attached to the uterine wall on which the maternal uterine epithelium has persisted. Such a relationship of the cellular elements is referred to as epitheliochorial. In tree shrews and among carnivores generally, another type of relationship, endotheliochorial, is found in which the chorionic elements come into contact with the endothelium of maternal vessels.

The human placenta is caducous, i.e., shed at birth and deciduate in that some maternal tissue is also torn away at the separation. In the contradeciduate type found in cattle, the placenta is not shed (apart from the falling away of the umbilical cord) and is absorbed within the uterus. In the nondeciduate type, found in pigs, horses, whales and dolphins, and lemurs, the placenta is shed but without loss of maternal tissue.

In a human context there are certain advantages in the shedding of the placenta. It reduces the risk of tearing the cord and losing fetal blood, some of which is, in fact, returned to the fetus as a result of uterine contractions. It also reduces the volume of the intrauterine contents, allowing greater uterine contraction, retraction and sealing of uterine vessels, and reduction in the size of the scar that marks the placental site.

The structure of the covering layers of placental villi is designed to allow diffusion of certain vital substances. The outer layer, in contact with the maternal bloodstream, is a continuous sheet, devoid of cell boundaries. This syncytiotrophoblast has numerous microvilli on its outer surface. It is actively concerned in numerous transfer mechanisms and synthesis of substances. It manufactures steroid hormones, and its output of estrogens increases steadily as pregnancy advances. The next layer, the cytotrophoblast, is thin and cellular but becomes attenuated with the progress of pregnancy. It is concerned with the production of chorionic gonadotrophin during the first part of pregnancy.

The elements of the trophoblast, together with wisps of subjacent connective tissue and the endothelium of the fetal capillary, comprise the placental "barrier." Through, or across it, must pass all the gases, water, salts, and foodstuffs needed by the growing fetus, as well as all the products of fetal metabolism that are returned to the maternal circulation. Substances of low molecular weight pass or are transferred more easily. Many of these pass unaltered by diffusion under pressure along a gradient. Others are modified or assisted during transfer and come under the influence of various enzyme systems and of lysosomes. Certain substances, such as amino acids, appear to be transferred against a gradient when the concentration is higher on the fetal side of the barrier than on the maternal. Certain drugs, anesthetics, alcohol, and several hormones also pass across the placental barrier. Depending on their concentration and the time of gestation, these may or may not affect the growing embryo or fetus. Under certain circumstances, some substances can effect drastic alterations in development and even kill the conceptus. Thalidomide, for example, can, at a particular time in development, affect the growth of limbs. Just how these teratogens exert their effects is not yet known.

The intact placenta normally provides an effective barrier against the passage of bacteria to the fetus. Some viruses, however, or their toxic products, do get across. German measles (the virus rubella) contracted by the mother early in pregnancy can affect the development of eyes, ears, teeth, and others organs. Proteins in immunological quantities can pass across the barrier as unaltered molecules. The Rh agglutinogen passes from the fetal to the maternal side, and the antibody, or isoagglutinin, can return to the fetus, occasionally with severe results. Some maternal hormones are able to cross the placenta. Cortisone in small concentrations is suspected of causing cleft palate, and steroid hormones can influence the developing gonads, accessory reproductive organs, and also the developing nervous system. Normally the placenta seems able to protect the sexual development of the embryo and probably does so by breaking down excessive quantities of maternal hormones. However, how a male fetus can survive and prosper within a completely female environment remains a mystery.

Little is known about the fine structure of the placenta in nonhuman primates mainly because of the need to foster carefully any pregnant animal. Elizabeth Ramsay of Baltimore, who has made careful studies of blood flow at the utero-placental junction and in the placenta of rhesus monkeys, estimates that some forty veins and twenty arteries communicate with the intervillous space.

Parturition

Birth, or labor, is usually divided into three stages: during the first, the lower segment of the uterus and the cervix become dilated; during the second, the fetus is expelled; and in the third, the placenta is voided. Then follows the period of puerperium, when the uterus involutes, the placental scar heals, and lactation is established. Altered blood, debris, and fluid, known as lochia, drain from the uterus.

The first stage lasts about 18 to 24 hours in primigravid women (first pregnancy) and 6 to 12 hours in women who have already had at least one other child. As uterine contractions become increasingly powerful, the cervix gradually dilates and is drawn up into the lower uterine segment. Actually, the uterus begins to contract long before labor, but the contractions are slight and painless. Increases in estrogen concentration and a decrease of progesterone activate the uterine muscle, which becomes sensitive to oxytocin secreted by the posterior lobe of the pituitary. These factors and others operate to bring about contractions. The dilation of the cervix, the rupture of the chorionic sac and loss of amniotic fluid, and the sinking of the fetus toward the cervix increase the expulsive force exerted by the uterine muscle. An afferent reflex nervous excitation of the hypothalamus probably occurs through pathways from the uterus, neighboring viscera, and pelvic regions, with a direct stimulation of the posterior pituitary, which liberates oxytocin ("swift birth"). This peptide stimulates the sensitized uterine muscle to contract, though the responses vary greatly, even from hour to hour.

Labor identical with spontaneous labor is also initiated with prostaglandins. These unusual substances, discovered by Goldblatt (1930), and subsequently by Euler (1934 and 1935) and others, are secreted by the seminal vesicles. Similar substances have also been found in the seminal plasma of some other mammals including nonhuman primates. Prostaglandins have striking pharmacological properties, and it is surprising that serious research on them did not begin until about 1963. It has since been found that these are a family of lipids with a wide variety of pharmacological actions and are present in many, if not all, human tissues. To mention some of their properties: they stimulate smooth muscle, depress peripheral vasodilatation, and inhibit lipolysis, platelet aggregation, and gastric secretion. Since they are potent smooth muscle stimulators, they are used clinically as effective abortifacients and as inducers of labor.

The human fetus can be positioned in several different ways within the uterus (the lie). The celebrated eighteenth-century collector, obstetrician, and anatomist William Hunter, after gathering information for twenty-five years, wrote the first detailed treatise on the gravid uterus. The fetus may lie head down with its vertex presented to the exterior (and its occiput directed anteriorly to right or left, or posteriorly in like manner). It can also be the other way up and as a breech present its buttocks toward the cervix, again directed to either the mother's front or back. More rarely, the fetus may lie transversely across the mother's abdomen. The commonest position (about 55 percent) is that in which the fetus lies with its vertex down and occiput directed to the mother's left. The position is ordained by several anatomical factors involving the "packing" of the fetus, the amount of amniotic fluid, the shape of the uterus, and the pressures exerted by the maternal viscera.

A fetus can be expelled most easily when it is born as a vertex presentation with the occiput to the mother's front. When some other lie is discovered before labor begins, various maneuvers can be performed to turn the fetus into a better position. Abnormal positions of the limbs or the cord can complicate labor, but obstetricians have devised methods of dealing with most of such exigencies. The reason for preferring the human fetal head to be born first is that the antero-posterior diameter of the head (usually about 5½ inches) is the largest part of the body. If, therefore, the head dilates the birth canal first, the rest of the body passes through more easily and there is less likelihood of distress to the fetus as a result of delay or constriction of the cord, which might happen with a breech presentation.

Four-legged, fetal mammals, contained in a bicornuate uterus, are "packed" in a more elongated fashion with the limbs tucked up at front and back. The head relatively is not so large as that of a human fetus, and there is usually less danger associated with fetal arrangements at birth. Less likelihood of trouble with multiple births also exists since the several fetuses are arranged linearly along the horns of the uterus and do not share the same chamber.

The second stage of labor results from increasingly powerful and frequent uterine contractions and from the reflex abdominal contractions that the mother exerts. This involuntary "bearing down" is stimulated by a distention of the vagina and vulva by the advancing fetal head. It is characteristic also of labor in the majority of female mammals. Once the second stage has been reached, most mammals are incapable of inhibiting the process. Some evidence has been found, however, that certain marine mammals, like seals, can postpone labor should bad weather interfere with safe delivery. In some mammals the ligaments connecting the pelvic bones, especially at the symphysis pubis, become markedly relaxed in response to relaxin, a nonsteroid hormone probably produced by the ovary and the placenta. Not much is known about the production of relaxin in women, although this hormone from other sources has been used to soften the human cervix and pelvic tissues.

As a human fetus passes through the birth canal, some degree of molding occurs. This is particularly marked on the head during a vertex birth and is proportional to the pressure exerted. When the head is pressed down against the dilating maternal tissues, a girdle of contact is formed. The central region of the scalp swells (caput succedaneum) from the accumulation of fluid. These effects usually subside a day or two after birth. The fetus is adapted to this pressure because its skull bones are pliable and separated by a membrane that allows some degree of overlap without danger to the underlying brain. However, there is a limit to the degree of molding and compression it can withstand.

The fetus also undergoes some degree of rotation as it passes through the birth canal. The shape of the maternal pelvis determines this rotation; the inlet is widest transversely and the outlet widest antero-posteriorly. The fetal head is at first markedly flexed, but after rotation within the pelvis it becomes extended as it passes to the outside. This movement is made possible by the occipital bone at the base of the skull, which intervening cartilage and membrane still separate into component parts allowing molding and even hingelike movements.

Uterine contractions usually cease immediately after the expulsion of the fetus. Shearing forces at its base cause the placenta to become separated, and this is accompanied by various degrees of hemorrhage from the maternal vessels. Continued contraction causes a coiling of the uterine vessels that reduces the blood flow and checks the loss of blood. Even so, women are liable to postpartum hemorrhage. As the result of renewed strong uterine contractions and an effort by the mother, the separated placenta is expelled, usually unaided. During the intervening weeks, lochia continues to flow while the placental scar is repairing. In the subsequent six months, the uterus slowly shrinks but seldom returns to its original size.

Parturition in nonhuman primates is usually performed with great expediency and lack of complication. Although it is difficult to be sure of this, the animals do not seem to be in great discomfort during the usually short labor. Most nonhuman primate infants are born head first, and, as soon as the head has emerged, the mother often pulls the neonate out with her hands. When the placenta is expelled shortly afterward, the mother eats it. On rare occasions, a breech birth may be troublesome even in lemurs, when the young may asphyxiate if it becomes lodged for too long in the birth canal.

Multiple pregnancies complicate human parturition. Twinning occurs in many primates with different degrees of frequency, e.g., in lemurs and galagos it is frequent, in marmosets, usual. Among the macaques, twinning is about as frequent as in man, and it is known to occur in chimpanzees, gorillas, and orangs. Human twins are either monozygotic or dizygotic. The arrangements of the fetal membranes in monozygotic twinning may show double or single amniotic cavities. These features give some clue as to exactly when the single zygote divided into separate entities. Should a division be incomplete, one of the many types of conjoined (Siamese) twins develops. Dizygotic (fraternal) twins are three times as common as monozygotic. The incidence of twinning in man *at birth* is one in eighty-five, although an unknown number may die during the early stages of pregnancy. Since premature birth is almost the rule in twins, they have a more hazardous start in life. Dangers at parturition include hemorrhage from too early a separation of the placenta after the birth of the first twin, knotting of the two cords in the monoamniotic condition, and interlocking of the fetal heads during parturition. Occasionally the second fetus may be born days after the first, but there is always the risk of hemorrhage. If human twins survive these and other perils, they are as normal as the rest of the population.

Lactation

A girl's breasts begin to develop at about ten years or earlier, enlarge conspicuously at puberty, and continue to grow until late adolescence (Figure 11–17), their development being induced primarily by estrogenic hormones from the ovary. The actual size of the breasts has very little relation to the amount of mammary tissue, which is about the same in all individuals. This points out once again that large breasts are a characteristic feature of women; they are a secondary sex character and part of the ornamental features of their body contours. Even in a large-breasted virgin girl, the total amount of mammary tissue may not be more than a spoonful, the rest being mostly fat and connective tissue.

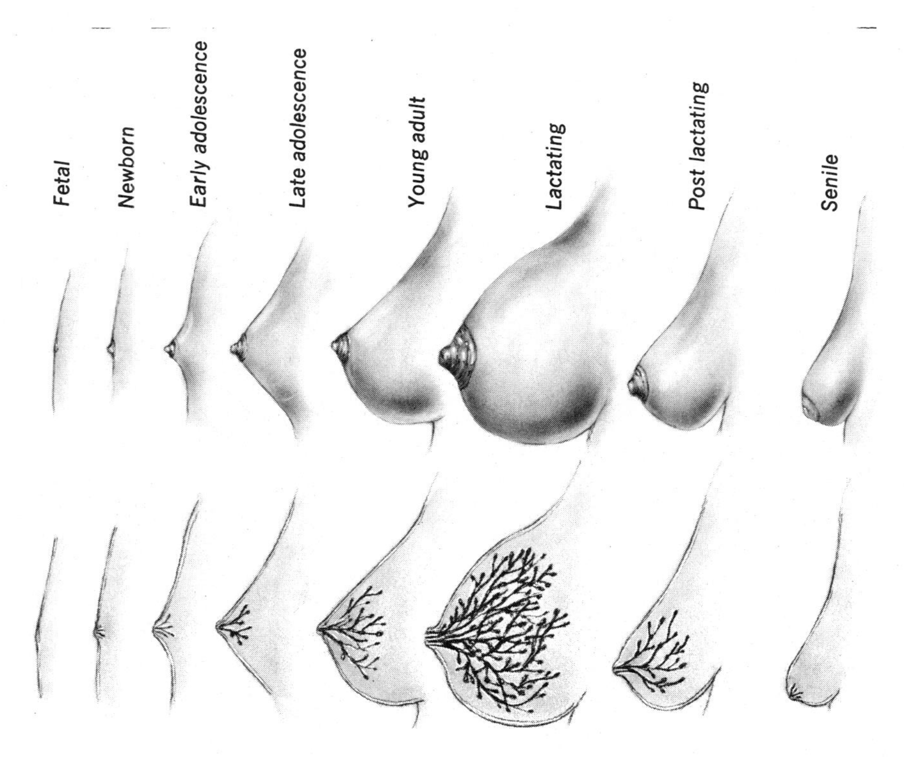

Figure 11–17. The life history of breasts.

Thus, one cannot predict from the size of a woman's breasts whether or not she will produce an adequate amount of milk for her infant.

Mammary tissue begins to grow relatively early in pregnancy and steadily increases until term. This growth is maintained by the large amounts of estrogens secreted by the placenta, together with some growth hormone from the anterior lobe of the pituitary. Growth hormone stimulates the functional mammary tissue and induces the accumulation of additional fat and connective tissue around it. The combined action of placental estrogen and growth hormone can bring about the growth of breast tissue only to a certain level, beyond which the simultaneous action of progesterone is also required. A mother's breasts are fully developed near the end of pregnancy, when they begin to secrete small amount of fluids (colostrum). This has the same composition as milk except that it contains little or no fat. Immediately after the baby is born, lactogenic hormone from the pituitary, up to now somewhat suppressed by the large quantities of estrogens and progesterone from the placenta, stimulates the production of milk to more than one hundred times that of colostrum.

Although continuously secreted, milk remains stored within the glands. Unless breasts are stimulated by suckling, milk will not flow or be "let down." When the baby suckles the nipple, relays of nerves wound around the lactiferous ducts transmit sensory impulses to the hypothalamus of the brain, which releases the hormone oxytocin and some vasopressin. Under the influence of these two hormones, the small musclelike cells around the engorged mammary glands contract and eject the milk into the lactiferous sinuses, and the negative pressure caused by the suckling brings the milk out through the duct orifices at the tip of the nipple. Once the baby has begun to suck, it requires less than a minute for these phenomena to take place.

Lactation is a delicately controlled system; even psychogenic disturbances can inhibit the secretion of oxytocin, which automatically stops the formation of milk. For this reason the mother is wise to remain serene during the period of lactation. Should the mother not nurse her child, her mammary glands stop forming milk within one or two weeks, and what milk had been stored within the glands and not expressed is eventually resorbed. Thus, a continued secretion of milk is maintained only by continued milking. Normally, lactation ceases in seven to nine months when the child is weaned, but many exceptions are common, including cases in which mothers have successfully nursed their children for several years. Women can produce nearly 1 liter of milk a day when the flow is fully established: higher yields are not uncommon, and a record output of 3 liters in 24 hours has been reported.

A striking relation exists between lactation and the behavior of the uterus. The uteri of women who nurse their infants involute much sooner than those of women who do not. In fact, the uterus of a lactating mother is sometimes reduced to an even smaller size than it was before pregnancy. In contrast, that of the nonnursing mother sometimes remains flaccid and relatively large.

A woman's menstrual cycles usually stop altogether during the early months of lactation, but this point is unpredictable. Some lactating women can become pregnant only a few months or weeks after giving birth and continue to nurse their infant until the next one is born. This phenomenon often occurs in nonhuman primates. Galagos and lemurs, for example, go into what is called postpartum estrus only a few days after the young is born. At this time they are receptive to the males and can become gravid. One galago rewarded us with two and a half successful pregnancies in one year (the gestation period is about 130 days).

The phenomenon of lactation is far from being completely understood. No doubt, the vagaries of its function, as well as the bother and accompanying discomfort, have discouraged many modern women from nursing their infants.

12

His sexual behavior

At the present time, when seven out of every eight scientists who ever lived are now living, expertise is on the upgrade.
(Alex Comfort)

Among the unique biological attributes of the human male animal are his almost perennial concern with sex and his never-ending search for sexual experiences. Man has shrouded sex in dogmas, taboos, rituals, symbols, religion, fantasies. He has both sanctified it and reduced it to vulgarity. He writes poems about love, erects giant monuments, paints pictures, carves stone, wood, and metal, subjects himself to torments and even death in the name of his beloved. His attitudes toward sex span the gamut from total denial to total indulgence, from the infernal to the paradisiacal. Man's fervor for sexual experiences, real and fancied, is reflected in his literature, most of which recounts his joys and sorrows in love. How he narrates these experiences determines whether his literature is good or bad, a work of art or a piece of vulgarity, a sensitive analysis of emotional storms and experiences or pornography. Sex, then, may well be the most complex aspect of man's nature.

Reduced to its basic biological function, sexual behavior is the mechanism that makes possible the coming together of germ cells at fertilization. The struggle for survival of all species is geared to ensure the succession of generations. The young maintain the race sturdy and indestructible. When old and no longer reproductively competent, individuals eventually outgrow their usefulness to the species and either die naturally or are eliminated. The continuation of the species is the essential, and the individual remains important only as a link in genetic succession. The numerous courtship antics of man and other animals are preliminary steps directly aimed at reproduction. Male birds display to females only during the breeding season. Most male and female mammals are largely indifferent to each other except during the period of estrus, when the female becomes receptive to the male, in some cases aggressively so. Correspondingly, male interest (called the rut in mammals such as deer) in the female is usually coincident with the receptivity of the female. In most animals, then, only when fertilization is possible is overt sexual behavior likely to occur. In man, however, the situation is different.

Sexual Attraction

The important first step in the consummation of the sexual act is the selection of a mate. Most of the specific details that determine this selection are amorphous and do not lend themselves to analysis. The human male animal is ever alert, concupiscent, and ready to spring into action regardless of laws and social mores. Whether or not he follows his impulses is immaterial, since the particular society to which he belongs has conditioned him and largely determines to what degree he is free to act. Why this man is attractive to that woman is extremely difficult to fathom, as, indeed, is male beauty in the abstract. Standards of acceptability seem to vary. The tall, square-jawed, even-featured, muscular aggressive man—ever the symbol of masculinity—is often rejected without a second look for a short or tall, lean or fat man. Tastes vary even in one society and with each individual.

Masculinity, the quality that women ultimately seek, may be embodied in a violent or meek temperament, an aggressive or retiring behavior, a loud voice or gentle speech, an athlete or a scholar. With women, more so than with men, the mental qualities of the future mate play a greater role in the selection than do his physical traits. In modern societies a woman's choice of mate is strongly conditioned by her own upbringing and her reactions to behavior patterns in her own and other parents and, we suspect, to the images of life presented through mass media as entertainment and as example. Expediency, money, status, and escape are not as influential today as in the past.

Undoubtedly easier to identify are those traits in a woman that arouse interest, desire, and lust in men. Anatomical standards vary in different societies; obesity, for example, unattractive in most societies, is agreeable in a few. In all, however, most men, regardless of race or society, are attracted to the youthful rather than the aged woman. The woman who lacks the more overt physical masculine traits is most likely to attract men. Hairless and smooth body skin, good contours, flawless childlike complexion, round breasts and buttocks, slender waist, well-rounded limbs, and most important, an attractive, even-featured face are all acceptable standards of beauty. Carriage and bearing also play an important role in attracting a sex mate.

Of all these general external features, the breasts and genitalia of women are certainly among the most important releasers of desire. But here too many cultural and individual differences exist. In our European and American cultures, large, rounded, firm, and uptilted breasts excite desire in men (Figure 12–1). Yet, some men prefer them pendulous or small. The perineum, and particularly the vulva, are almost universal sex stimulants for men. Among some aborigines who wear no clothing, the women wear a loincloth just large enough to conceal the vulva. In some primitive societies, the deliberate exposure of the vulva by a woman usually excites male eroticism. It does not seem to be particularly important for men to cover their genitalia.

This is not universally true and is subject to customs and culture. The early white settlers in Australia, for example, and particularly the missionaries, strongly objected to the naked aborigines and made them wear clothes. However, aborigines still shed their clothes in the privacy of their own homes or when they go to the bush, out of sight of white men. Still, since clothing has now become a status symbol, successful aborigines cling to rags long after they have outlived their usefulness and are a menace to health. Aborigines of either sex, and members of certain cultural groups, are sometimes reluctant to undress even for medical examination.

Man's concepts of sexual beauty and attractiveness in women are in general nonobjective and often exceedingly personal, excluding the obvious erotic anatomy considered above.

Given the proper impetus, normal adult men can perform successful coitus daily and sexual athletes have been known to indulge several times daily without harm. Although the modern European and American man lives in a largely monogamous society, his sex drive still seems to be guided by an adventurous zeal. Most men, whether consciously or not, seek new sexual experiences. With his habitual sex

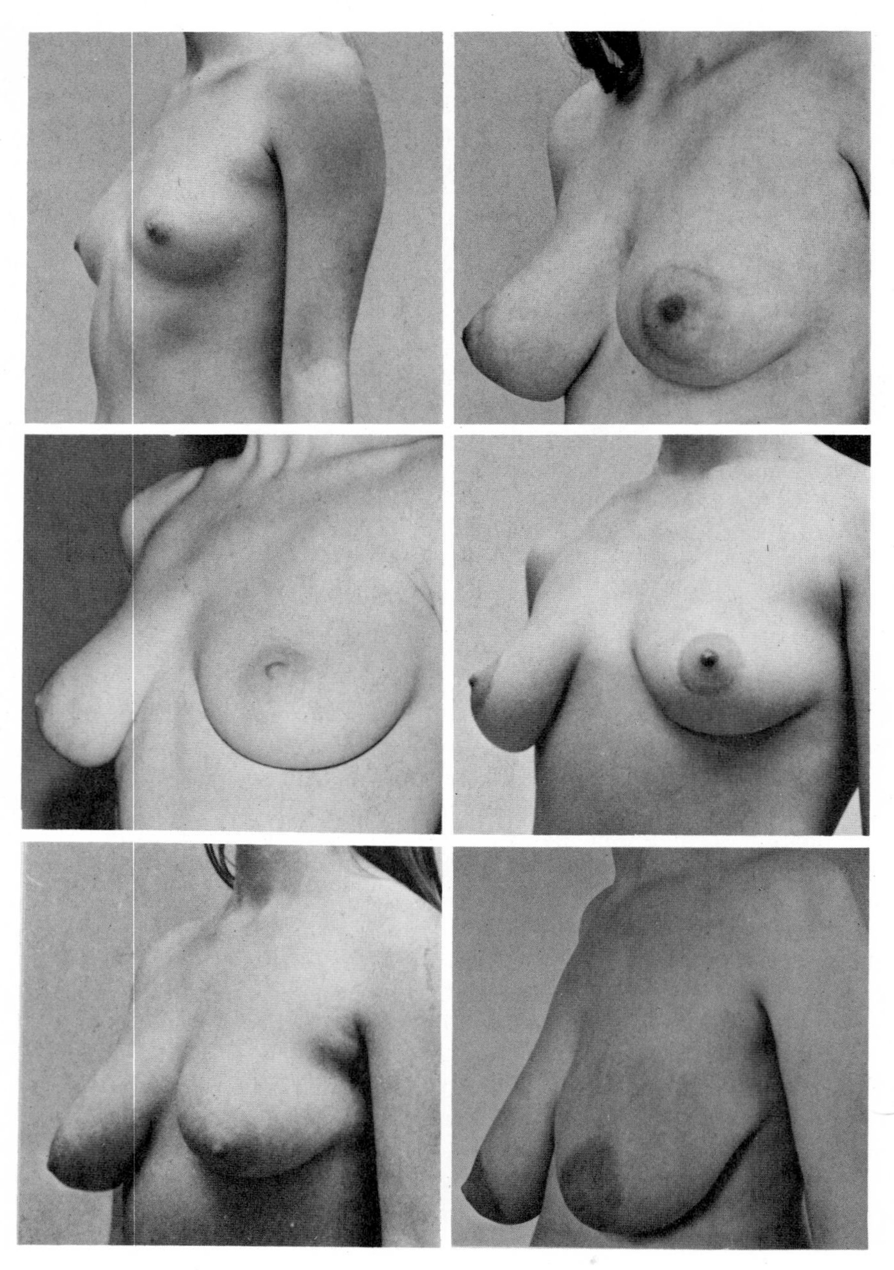

Figure 12–1. Different sizes and contours of breasts in young women of about the same size.

partner he may have daily coitus whereas with new partners his capacities may greatly increase. Such prowess is not unique in man; it has been commonly observed in laboratory rats, and in the northern fur seal G. A. Bartholomew and P. G. Hoel have recorded mating to be maintained at a rate of once an hour for periods lasting up to three weeks. In contrast to men, women who are capable of orgasm can often experience a second, third, fourth, and fifth orgasm. Man's sexual drive undergoes a gradual decrement with age and presumably ceases outright in senility, the last phase of life when, feeble and incompetent, he is "sans eyes, sans teeth, sans everything." Concupiscence, however, may remain active, even when the male is no longer capable of consummating the sexual act. Individual differences vary greatly; some exceptional men are capable of siring offspring in their seventies, but others may pass through a relatively abrupt "climacteric" before their fifties.

The sex drive and behavior of women is somewhat different from that of men. As female animals they are, under normal circumstances, in the unique position of making the final decision of accepting or rejecting the advances of the male partner. The use of antifertility drugs and devices has given women more independence in the pleasures of the sexual act and have divorced them from the fear of pregnancy. Women seldom need to be overtly aggressive: their power lies in their ability to release sexual aggression in men through provocative gestures, facial expressions, and tone of voice. Regardless of the amount of ovarian hormones released into their blood, which should make them receptive only near the time of ovulation, or of the menstrual flow, which makes coitus untidy, or of cultural mores, which often dictate when sexual relations are permissible, women are variously receptive to male advances throughout the cycle. This receptivity is almost peculiar to the human species, as is also acceptance of the male during pregnancy.

The menopause (p. 252) is followed by a period of physiological readjustment that is almost always accompanied by numerous and varied psychogenic disturbances; some women become frigid and unresponsive and largely indifferent to sexual advances. On the other hand, others seem to experience a reawakening of interest and may even become sexually aggressive. There are even those who seem to enjoy orgasmic pleasures unknown to them in their younger years. These changes in behavior have sometimes been attributed to elimination of the fear of becoming pregnant, but this may be an oversimplification of a very complex aggregate of psychological states. Like men, some women continue to participate in the sexual act in advanced age, but as a rule the participation is passive or willful rather than a part of their total involvement.

The patterns of human sexual behavior point clearly to the many undefinable qualities of sexual attraction. In man, then, the selection of a sexual partner is guided at least as much by the intellect and by mores and culture as it is by physical attraction. This fact suggests that the possible means of attracting a sexual partner are certainly not limited to the mere visual and physical. Both men and women possess an equally rich repertory of wiles.

Modulations and inflections of the voice and speech—song, poetry, and impassioned declaration of "love"—are universal sex lures. Feats of strength, demonstrations of courage, and leadership are usually effective. The bearing of gifts and other fond blandishments are all parts of the armamentarium of lures that men employ to gain acceptance by the women of their choice. Women, on the other hand, except when they are actively seeking sexual gratification, are much subtler than men. Prominent among their behavior patterns is their apparent assumption of the total feminine repertoire: soft speech, shyness, apparent disinterest, averting the eyes of the wooer, and turning the back to him—all actions designed to inflame the ardor of the suitor. In our modern society feminine dress has assumed an extremely important role in this respect; styles, whether designed to reveal or conceal, always emphasize just enough of those anatomical features that evoke eroticism and arouse desire. Perfumes, once used to mask stale and excessive body odors compounded by long periods without washing, are no doubt erogenous in their own right, as are the body odors themselves, which may even act as pheromones (p. 374). The wearing of ornaments, the use of innumerable cosmetic agents to accent or change color or form, personal cleanliness, bearing, and walk are important items of female attraction.

Unlike human beings, most other mammals mate only during a restricted season. Sexual behavior by the pair is restricted to the mating season, after which the gonads are largely dormant and the sexes lapse into mutual indifference. During their sexually active periods, other mammals also show certain preferences in choosing sexual partners, even if the situation is in some forms made more complex by the establishment of harems and "pecking" orders. Although the final choice generally rests with the male, in some species the female decides who among the many suitors are the preferred ones. To analyze the factors influencing man's choice of a mate is difficult enough, but to analyze the behavior of other animals is harder still.

When subhuman primates are bred in captivity, the first important consideration is the selection of a compatible pair. In the breeding of rhesus monkeys, the males are kept in separate cages. Since experience has taught that to be effective the male must not be stressed or disturbed, the females are brought to him. However, the male must first be given a chance to accept or reject, because, if he is antipathetic toward the female, he may kill her. The female is therefore offered to him in a small holding cage. If the male responds amicably to her, he will register his approval by lip-smacking, clicking sounds, and other signs; if he does not, his behavior will be hostile.

In some species of subhuman primates, the sex skin and female genitalia become swollen near the time of ovulation. The buttocks of pig-tailed macaques *(M. nemestrina)* and Celebes black apes *(C. niger)*, for example, attain disproportionate sizes (Figure 12–2) and brilliant red hues in response to endogenous or exogenous estrogenic stimulation. The vulva of most primates studied is anointed with substances that attract the attention of the males, who in turn diligently inspect it, sniff and lick it as a prelude to coitus. Vulval secretion is probably a combination of substances secreted by the uterine and cervical glands, by the greater vestibular glands and those of the labia, since the vagina itself is devoid of glands. We have noted repeatedly that female lemurs accept males only during or just before ovulation and, similarly, males are largely indifferent or even hostile to females except during that brief period. Vaginal lavages of ovulating female lemurs smeared on the perineum of nonovulating ones render males at once sexually aggressive. The vulval excretions of rhesus monkeys at the time of ovulation never fail to evoke sexual responses in males. Castrated females have no such excretions, and males are utterly indifferent toward them; these same females become sex objects after they are injected with estrogens, or

Figure 12–2. The greatly swollen buttocks (sex skin) of a female pig-tailed macaque *(M. nemestrina)* about the time of ovulation.

when vaginal lavages from normal ovulating females are smeared over their sex skin or perineum. Finally, proof that these are real pheromones is that males rendered experimentally anosmatic have no interest in ovulating or estrogenized females, but become libidinous when their olfaction is restored.

Even women experience a systematic fluctuation of olfactory acuity during their menstrual cycle, with a peak of sensitivity at the time of ovulation. If they are exposed to musky odors during ovulation, they experience a lowering of their osmatic threshold. (It may be of some significance that men normally excrete androstan in their urine, which has a strong musky odor.) This osmatic sensitivity at ovulation is abolished by ovariectomy but restored by estrogenic therapy. Another phenomenon that calls attention to the possible subtle or even subconscious action of pheromones is the synchronization of the menstrual cycles of women who are close friends or who room together and even of girls who live in the same dormitory. What mechanism accounts for these events is not certain, but pheromones are prime candidates.

The release of sexually stimulating pheromones is concomitant with the physiological readiness of the female, and the two are the most potent factors in evoking sexual interest in males. Male chimpanzees, for example, rarely make sexual advances until the females are midway through the estrous period, when they actively seek the attention of males. This, however, is not always the case; some particularly tempestuous male chimpanzees will, when they please, force a female into coital activity whether she is ready or not. Among most subhuman primates the male is so much larger and more aggressive that a female may be rendered submissive even when she is disinterested.

There are doubtless other factors, such as general appearance, facial expressions, vocal utterances, gestural signs, stance, and dominance in a troop, besides other unknown ones, which function in the selection of compatible sexual mates. Where the females have a choice, e.g., among some baboons and chimpanzees, they seem to prefer males who are endowed with large genitalia and whose copulatory performances have lasted a long time.

Other animals share with man certain tendencies in preliminary sexual exploration or courtship. The major general difference between most mammals and man is that sexual activities in the former occur only during certain seasons. But whether seasonal, monthly, or daily, male interest in the female focuses upon the genitalia. Even bull elephants have been observed inspecting the vulva of several cows before mounting one. Secretions from specialized cutaneous glands or from glands in and around the perineum of either sex give off certain odors (pheromones) that are sexual attractants, as discussed above. Vocal utterances, patterns of behavior, and perhaps even physical appearance are all important means for bringing animals together.

Sex Play

Courtship leads directly to the more active preludes to coitus. In most human societies this is preceded and accompanied by an assortment of sensory stimulations that can be collectively referred to as sex play. Like the search for and selection of a sex partner, many of the differences in the expression of sex play are dictated by the mores and taboos of the specific society and by the zeal of the participants. In spite of obviously different expressions, however, several basic behavioral patterns are common to most human societies as well as to other mammals. These activities all have basic significance and cannot be dismissed as meaningless. Lovemaking, or sex play, serves largely to arouse desire to such a level that each partner is ready for coition. Largely because of cultural restrictions, women usually become aroused more slowly than men, and most of the preliminary activities are geared to them, although they are pleasurable to both partners.

The more general expressions of sex play involve many uses of the mouth. Among Europeans and Americans the apposition of the mouths of the two partners, kissing, is a universal prelude to other sexual activities. The degree of intimacy in kissing mostly depends upon previous experiences, family and social backgrounds, adventurousness, and the level of excitement of the participants, as well as upon accepted social practices. As in the entire repertory of sex play, there is close correlation between the complexity of involvement of oral play and the degree of education or enlightenment of the participants. Dull-witted or uneducated men come abruptly to the point without preliminaries, their only interest being copulation. Some anatomists consider that the facial musculature of Europiforms has developed further than in other perhaps more enigmatic-looking groups. These increased differentiations with enhanced neuromuscular control, together with an abundant sensory innervation of the facial regions, especially the lips, are undoubtedly significant if this type of play is to be pleasurable as well as demonstrative. Oral exploration is not limited to mouth-kissing, but often extends to mouthing the genitalia of the partners, sucking, licking, nibbling, and light to vicious biting. Even in those societies that do not practice kissing (e.g., those who rub noses), the partners use the mouth and tongue in many ways, either before or during coitus.

Tactile exploration of the entire body, particularly of the erogenous areas, is practiced by nearly all societies. Such explorations range from gentle stroking with the fingers and palms to rough manhandling. A good deal of general sensory pleasure is derived from the contact of the body skin. Women's breasts are not only symbols of femininity but also objects of visual stimulation (Figure 12–3); and licking, mouthing, and sucking

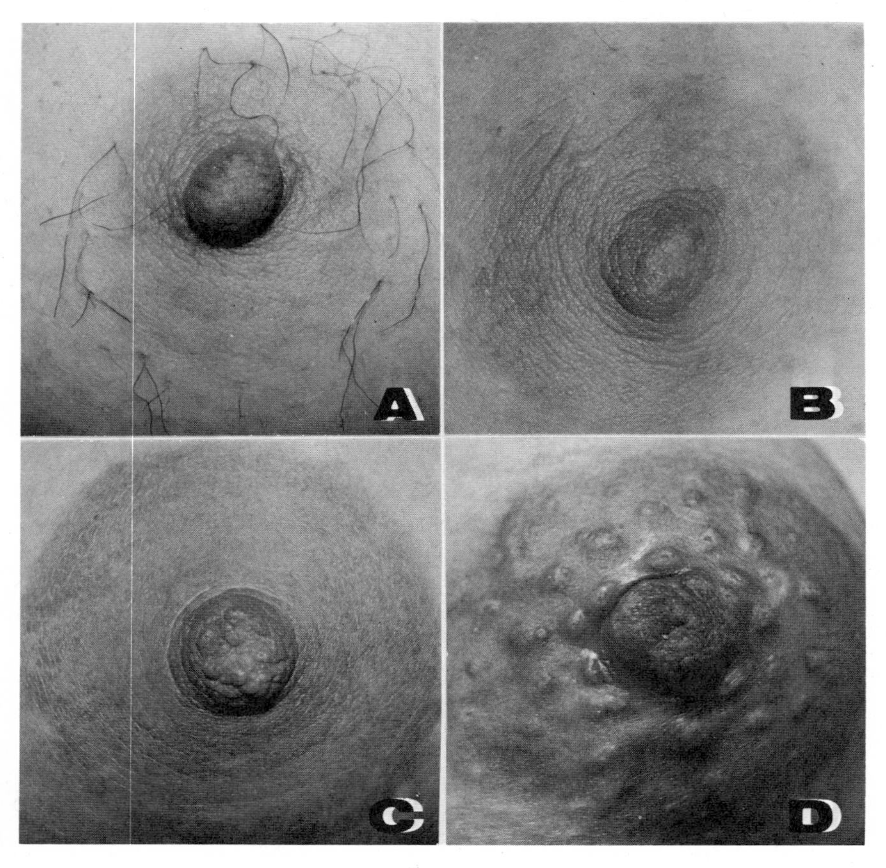

Figure 12–3. Differences in the size, shape, and pigmentation of nipples in young women of about the same age. The tubercles of Montgomery are either prominent (D) or practically indistinguishable (A, B, and C).

of the nipples and stroking them with the fingers are also universal practices. There is an apparent paradox in the sensitivity of breasts. Whereas they are generally considered to be erogenous areas, those who have carefully tested their tactile sensitivity and discrimination have found them singularly insensitive. No one, however, has studied the specific sensitivity of the tip of the nipple, where the orifices of the lactiferous ducts are encircled by numerous sensory end-organs. The sensibility evoked by suckling or rough manipulation is recorded by the many sensory nerves around the lactiferous ducts and sinuses (Figure 12–4). Except at the tubercles of Montgomery—which mark the opening of the ducts and which, like the lactiferous ducts at the tip of the nipple, are rich in sensory nerves—the areolae have few nerves and are relatively insensitive. One cannot minimize the fact that psychic factors also stimulate the breasts. Interest in the breasts, even more than kissing, is peculiar to human beings, and the males of other species are apathetic toward them. Evidence has revealed that stimulation of the nipples does more than evoke a pleasurable sensation. Reflex nervous pathways link the nipples to the central nervous system and thus can affect the uterus and possibly other reproductive organs in a variety of ways. For example, the suckling of the newborn baby brings about a reflex contraction of the uterine muscles that is identical with that after an orgasm.

Probably the most significant precoital activity of men is the exploration of the vulva with the fingers and often with the mouth and tongue. Eventually these activities are directed toward and concentrated on the clitoris, which, when stimulated, becomes turgid and even erect. The size and turgidity of the clitoris and labia minora are variable because the extent of the cavernous or erectile tissue is different in each individual. In some women, the entire hood formed by the labia minora and most of the labia themselves are richly supplied with erectile tissue. Limited studies point to a possible racial difference, Europeans being the least endowed. It should be mentioned, parenthetically, that the clitoris and the ventral part of the labia minora have the largest concentration of sensory nerve endings of any part of the body, male or female.

Visual sensations are integral parts of sex play; hence sexual activity is likely to have its roots during the daylight hours. However, because of convenience and social activities, restriction and responsibilities, sex play and coition in most clothed human societies take place at

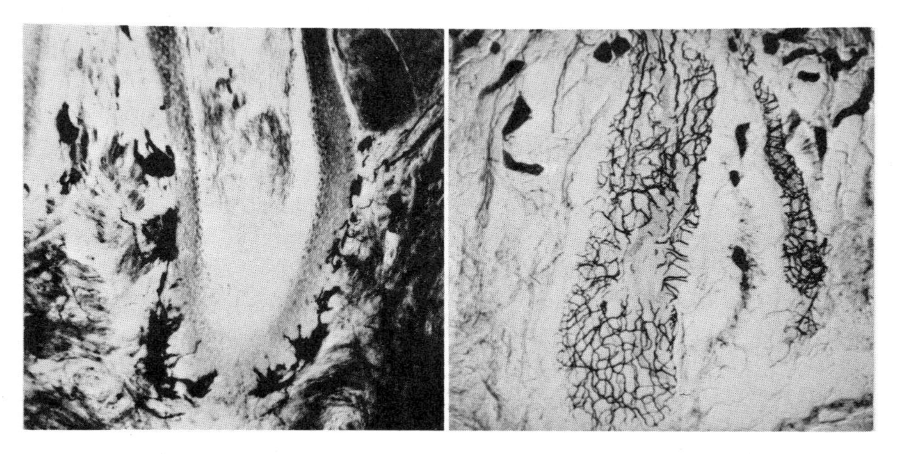

Figure 12–4. Right: sensory end-organs around the orifice of a lactiferous duct at the tip of a nipple; left: skeins of sensory nerves wrapped around two lactiferous ducts.

night. The practice of putting the light out during consummation of sexual relations is probably a gesture to the mores learned by a people so conditioned that they are embarrassed at the spectacle of themselves, educated intellectuals, behaving like animals.

Mentioned elsewhere is the fact that the sight of the female genitalia and the nipples and areolae are powerful sex releasers and that man, though microsmatic, produces and responds to pheromones. Olfaction, then, is an important accompaniment of sex play and intercourse. The natural sebaceous and sudoriparous odors from clean skin, the axillary odor, in addition to the peculiar odor from the fresh vulva all set the stage for the final sexual act. Man often uses perfumes, which act as pheromone reinforcers, for their alleged aphrodisiacal properties. In all of these varied and unnamed performances, the female partner has in most societies been a passive participant. The extent of her active involvement has depended upon the state of her arousal and her conditioning by attitudes of home and society. Western sexual culture, however, has changed so much during the last few years that there are practically no restrictions and men and women are often equally sexually aggressive.

Sex play, whose function is to raise the level of readiness of the partners for successful coition, has a psychological and physical basis. Women, who usually become aroused more slowly than men, require these stimulatory activities so that gradually their final readiness for intercourse becomes marked by intense desire and by the secretion of proper amounts of vulval fluids that make the vagina physiologically ready for intercourse. Precopulatory activities, then, whether at symbolic levels through language or through intimate physical contacts, allow the sex partners to attain an ideal degree of emotional excitement to synchronize their readiness and actions.

Sex play develops slowly in children. Except for curiosity, boys, until about the age of ten or eleven, seem to be disinterested in girls. Or perhaps their specific sex play is more subtle and less recognizable to adults since children are rarely concupiscent. Young boys often show a greater interest in the bodies of mature girls and women than in those of girls of their own age. The intensity of sex play blossoms at puberty and becomes progressively more elaborate and meaningful in fully adult individuals.

Subhuman primates also experience a comparable precoital behavior of variable degrees of complexity. Although too little is known about the physiological significance of these activities, they appear to be important preludes readying the sex partners for coitus. Monkeys and apes, aside from sniffing and nibbling the ears and lips of the sex partner, have nothing that resembles kissing. Among the prosimians, male galagos often thoroughly lick the face, mouth, and ears of females, and lemurs often lick each other's tongue. These forms of "kissing" occur between mother and child as well as between sex mates. Other male mammals commonly sniff and lick the face and mouth of the female, but this is probably not comparable to kissing.

The use of the lips and tongue in the exploration of the female genitalia and perineal area is common practice in all subhuman primates and infraprimate mammals. Male orangs and chimpanzees mouth, suck, and lick the vulva of the partner, sometimes for hours. The females in their turn may sniff and lick the male genitalia or manipulate the penis when they are actively soliciting copulation.

Physical contact and tactile stimulation are commonly shared precoital activities of nearly all mammals. Monkeys and apes groom each other as a gesture of friendliness, and this activity probably attains greater significance in sex partners. Remembering that hairs are to a greater or less extent equipped with sensory end-organs, grooming may arouse a vivid rather than passive sensation. Prosimians do not groom with the fingers but use their long tongues and special, comblike incisor teeth.

Coitus

The aim of all courtship and sex play is intromission and ejaculation, which, in a strictly biological sense, is the only action that matters. Man, however, probably because of his highly developed means of communicating his emotions and because of his rich fantasies, often regards the final sexual act as a detail of the entire elaborate pattern of heterosexual relationship. Moreover, since intercourse can occur at almost any time and not only at ovulation, it is a pleasurable and even a necessary experience, independent of its purely reproductive function. This means that the repertory of sex play and coitus is important in shaping the family and society. In spite of the increasing vagueness of such existential experiences as parental love and the inexplicable bonds between lovers and spouses, the parents' dependence upon their mutual sexual gratification is the most important binding element of the family. This is to some extent true even today when man has generally emerged from Victorianism, and marriage partners have both the license and the opportunity to practice adultery. In some social sets in the United States, married couples belong to groups that practice promiscuous sexual behavior openly. An interesting note is that when such couples leave children at home with baby-sitters, the sitters are given a telephone number where the vagrant parents can be reached in case of emergency. What such promiscuity has done to the stability of the family can only be guessed, but the prognosis seems doubtful.

Being bipedal, upright animals with genitalia directed forward, men usually perform and consummate coitus face-to-face, a unique pattern of action dictated by the exigencies of anatomy. This is the standard practice, but since human beings are explorers, they have discovered and practice innumerable other positions. Some of these, however, derive from need other than mere adventure. During intercourse the rhythmic in-and-out action of the penis causes friction against the labia of the vulva and particularly against the clitoris, which, without this friction, remains unstimulated and renders coitus unsatisfactory for the woman. Since the location of the clitoris and the slant of the vagina vary widely in individuals, sex partners seeking gratification explore positions in which the woman can also get maximal satisfaction. Thus, many of the coital positions that depart from the standard face-to-face are guided by a practical approach, rather than by the need for new experiences. Here, too, the enlightenment or education of the partners has much to do with the varieties of performing intercourse, its mutual success, and not a little with fertility. Infertile couples can often be given good advice on the best position to adopt to insure maximal chance of impregnation.

Notwithstanding its basic biological nature, the art of coital performance is mostly learned. The inexperienced male lover often fumbles, and his performance is of short duration and therefore unsatisfactory to his partner. The refinements and variations of the basic pattern are mostly learned. Although all aspects of the coital act should be pleasurable to the participants, the pleasure is actually anticipatory to the climactic moment when, after a burst of overwhelming emotion, each participant ceases, temporarily exhausted but satisfied. In the male this climax is marked by the ejaculation of semen. In this sense all the details of courtship, precoital play, and coitus itself are preparatory to the climax that terminates in satiety.

The nature of the human orgasm has been discussed and analyzed by ancient and modern sexologists. Their explanation, however, still leaves much to be desired. Its sensory intensity is accomplished by powerful muscular contraction, then sudden release of tension followed by an acceleration of heart beat and often profuse sweating. In males the orgasm undoubtedly assists expulsion of the ejaculate and reflex emptying of the accessory organs that supply the seminal fluid. Much debated is whether an orgasm in women promotes transport of spermatozoa in the female genital tract. Orgasm is not essential for fertility. On the other hand, whether it improves fertility is not known. Its effects must only be temporary and immediately could only affect sperm transport and viability. It would hardly have much influence on implantation six days later, although subsequent orgasms might. Some have argued that what preliminary orgasms fail to do, pregnancy does later, that is, procures the proper development and functioning of the entire female reproductive apparatus with all its associated organs.

The major basic difference between male and female partners is that man's orgasm automatically accompanies ejaculation, but women have nothing comparable to it, and it is more difficult to achieve. Many women have to learn to have an orgasm and some never experience it, in spite of repeated performances. Moreover, many factors, past and present, profoundly affect it. Hence, precoital repertory becomes particularly significant in preparing a woman for her orgasmic experience.

For both partners to attain the greatest success, the entire delicate performance must be synchronized on both the physical and the emotional level. Given the wide range of individual differences, some women are easily aroused and participate freely and with abandon, achieving the desired climax easily and repeatedly. Others, perhaps because of unpleasant early experiences, phobias, and social taboos, are inhibited and cannot be aroused regardless of their partner's or their own attempts to do so. Many women achieve an orgasm only after prolonged sessions of love play and intercourse, while others enjoy it several times in a single experience. This makes the staying power of the male partner all the more desirable. The length of each coital activity varies from seconds to ten or fifteen minutes. Sex partners who are so well attuned to each other that they can achieve repeated successful sexual relations are likely to be the best-fitted permanent companions. However, this aspect of man's nature is so replete with vagaries and exceptions that generalization is dangerous.

Unless the man is highly aroused, ejaculation and orgasm are accompanied by a relaxation of the penis that precludes a second intromission for some time. The human penis differs from that of many other mammals, including the great apes, in that it has no skeletal element. Prosimians, monkeys, apes, and notably carnivores, including seals, have a cartilaginous or bony element that gives the penis a permanent stiffness. All cetaceans lack such an attribute. Even though orgasm comes automatically in man, he can learn to control it, so that the experienced man can delay it sufficiently to synchronize it with that of his sex partner. This skill, together with some of the other considerations mentioned above, differentiates the sexual behavior of man from that of all other animals.

Children, particularly those who have witnessed coital activities in adults, given an opportunity, will attempt to perform them. However, in our Western societies they are closely watched and seldom given such opportunities. Their attempts, then, are usually of an experimental nature and are motivated more by curiosity than by libidinous desire. Seemingly, children's coital attempts are a part of their learning mechanism. In some aboriginal societies, however, girls are routinely deflowered and taught coitus at a comparatively early age. Whether encouraged or delayed, the action, both physical and emotional, is largely a learned experience.

Coitus in all other animals, even in primates, is essentially similar to, but less elaborate than, that in man. What distinguishes the primates is that the act is performed frequently. Among other mammals only lions copulate frequently and sometimes several times a day. Those who have watched social groups of baboons and macaques have noted the universality of the action, particularly in the early morning hours and at dusk when the animals are quiet. This is in contradistinction to other social animals such as ungulates, which, except during the breeding season or rut (p. 258), are rarely seen mounting. Most primates have no clearly defined breeding season. Regardless of preparatory activities, which sometimes seem to be endless, the coital act in nonhuman primates is very brief, rarely lasting more than fifteen seconds. Male macaques have been observed to copulate successfully with three consecutive females within a period of forty-five seconds. Nothing in nonhuman primates is comparable to the orgasm that accompanies the fulfillment of the coital act of man. Male macaques occasionally shiver lightly after ejaculation, but the females show no sign that the experience has been a pleasurable one. During coitus they may reach back with one hand and gently touch the male genitalia, but that is all. Thus, the comparatively lengthy duration of the act, the emotional involvement, and the orgasm appear to be distinctive attributes of human intercourse.

Among other animals the coital habits of birds are particularly interesting. After weeks of sex play, displays, fighting off rivals, guarding their territory, and finally nest building, the male mounts the female on the ground or on a branch or in flight, and the two cloacae come into contact for an instant during which the sperm is delivered to the female. The duration of coital activity is brief in most mammals, notably the ungulates in which there is only a very brief intromission. Copulation, however, is more elaborate in dogs and cats. Dogs remain conjoined for some time after, largely because of the peculiar anatomy of the penis and the construction of the sphincter around the vulva. Cats have complicated courtship displays marked by much clamor and fighting among rival males. During copulation the male holds onto the nape of the female with its teeth inflicting pain and sometimes wounds, a behavior pattern particularly seen in mink, a mustelid. Because of the peculiar structure of the male cat—the penis mostly directed posteriorly, the short front legs, and the general loose-jointedness of the body—these animals fumble a great deal before intromission. The biting of the nape by the male may help to steady him. The penis of cats, like that of many prosimians, has a cartilaginous or bony skeleton and is covered with backward-directed horny barbs. During the in-and-out coital motion, these barbs no doubt stimulate or even damage the vulva and vagina. At the conclusion of the act, the female writhes for some time on the ground, an action that has sometimes been interpreted as comparable to orgasm.

Male monkeys begin to mount one another when still infants,

but they become more discriminating as they grow older. Adult female macaques sometimes spend hours grooming young males who often show their gratitude by embracing them in play-biting and by mounting them repeatedly. These females usually react nonchalantly but encourage rather than shun these overtures. Among many of the social nonhuman primates, when two males are fighting, the victor usually mounts the vanquished at the end of the conflict. Most males of social nonhuman primates live in a society where they have to develop the proper attitude to perform coitus successfully. Animals raised in confinement are usually inept physically and emotionally. The coital act, then, must be learned through social interaction, and the attempts of the young are no doubt a part of their sex education and experience. Young infraprimate mammals often attempt to mount each other indiscriminately. This is mostly observed in male dogs, but even bullocks and heifers make such attempts in their ponderous ways. Most of the repertory of sexual behavior, while rooted in what might be called instinctive behavior, has to be learned. This learning process probably lends significance to the early sex play of children and later to the endless explorations and experimentations of adolescents and young adults.

Birth Control

Because man is exceedingly fertile and the world today faces the specter of overpopulation and because social and economic restrictions demand that the birth of children be limited or prevented, man has had to devise methods of contraception.

There are two principal methods of preventing the birth of young. One is to prevent spermatozoa from meeting the oocyte; this is prevention of fertilization. The second is to arrest the development of the ovum at some stage after fertilization; this is a form of abortion. It need scarcely be said that despite its obvious disadvantages abstinence from sexual intercourse is the most reliable prevention. Even though the sexual urge seldom brooks long delay, many of the methods in use today demand preparation or foresight before intercourse. A perfect contraceptive would be one that could be used or taken orally after the act, without risk of side-effects, when the participants could perhaps take a clearer view of events. A by-product of increasingly efficient contraceptive methods has already liberalized social attitudes toward sexual behavior by removing the fear of pregnancy, reducing the number of unwanted children, and avoiding the septic "back-street" abortions. These methods and attitudes have also influenced the altered place of women in modern society. However, as an eminent authoress has declared, until men are able to bear children, methods of birth control will not fundamentally alter the real difference between the sexes. Contracepitve methods are essential for the many millions of couples who are not yet ready for parenthood; they help families space out the births of their children and are a salvation to those mothers who desire no more children. Whether as an outcome of contraceptive methods or not, the place and position of women in society has altered, and one may well ask why a modern woman, married or not, should not enjoy safe sexual relationships. Males have done so with impunity from time immemorial. Any description of sexual behavior between men and women must consider practical matters and must be blunt about methods of contraception. All earlier remarks in this chapter about preliminary love play, eroticism, and orgasm are relevant only to ideal situations and to romantic lovers on celluloid; real life is more prosaic. Modern times demand that young people be equipped with appropriate sexual information and that when a relationship with one of the opposite sex becomes deeper than an exchange of glances, notes, and letters, there should be an understanding of what can happen, how to control affairs with sensitive restraint, and what might be the result of haste and ignorance. The human sex drive and the search for sexual satiety, though powerful and immensely satisfying, are potentially dangerous.

Prevention of Fertilization

A method that depends on knowledge of the female reproductive cycle (chapter 10) prohibits intercourse except during what has become known as the "safe period." It assumes that ovulation will occur on day 13 to 15 of a regular 27- to 29-day cycle, that an oocyte can be fertilized only up to 36 hours after ovulation, and that spermatozoa are capable of fertilization only up to 2 days after ejaculation in the vagina. The days of the cycle are counted from the *first* day of one menstrual flow to the *first* day of the next. On these assumptions it would be unsafe to have intercourse during days 11 to 17, or even days 9 to 19. This leaves the period from the end of menstrual flow (usually day 5) until day 9 and the period from day 19 until day 28 when intercourse can occur "safely." However, no one can predict with certainty that a cycle will be exactly like the previous one. Ovulation has been recorded as early as day 8 and as late as day 18 (in known but probably rare instances, from a second follicle). Furthermore the length of the cycle often changes. This method is clearly unreliable when cycles are irregular. Other "natural" methods include withdrawal of the penis from the vagina before ejaculation or a vaginal douche immediately after intercourse. Both methods are unreliable and frustrating and are always followed by the well-founded anxiety that some spermatozoa escaped before the withdrawal or survived the douche.

Several types of devices have been provided commercially to catch, trap, and contain spermatozoa within the vagina; these are often used together with a spermicidal cream (usually containing phenyl mercuric acetate) or with a foaming pessary inserted into the vagina. One type consists of a thin diaphragm within a circular spring, known as a Dutch cap, which can be compressed to allow insertion into the upper part of the vagina where, if it fits properly, it should prevent spermatozoa from reaching the cervix. A smaller type of diaphragm can be fitted directly over the cervix as a cervical cap, but it is more difficult to introduce correctly and has to be the right size and fitted by a doctor. It is inserted, together with the spermicide, and must be left in position for several hours after intercourse for the spermicide to be effective. Despite its cumbersomeness and the necessity of anticipating the imminence of intercourse, the method can be mastered and is effective. Spermicides alone, as creams or foaming pessaries, are mostly unreliable. One of the most ancient of all contraceptives is the condom, or French letter, a sheath of thin rubber or plastic slipped over the penis before intercourse. Some of these have a small dilatation or teat at the end to safely receive the ejaculate within the vagina; the sheath must fit securely over the penis so that no leakage occurs. The appropriate moment during preliminary love play for putting on the condom correctly is just before coitus; a lubricant, perhaps a spermicidal cream, is often needed. This commonly used method is dulling to both partners and destroys the spontaneity that is an integral part of the sex act. Moreover, there is the worry that the device has slipped or even burst; however, with some skill, correct timing, and care, the method is effective.

In recent decades, research for effective methods has been directed toward preventing the production of germ cells at their source. In men, spermatogenesis can be suppressed by taking such substances as diamines, but these are slow acting and can inflict severe damage on the testes. Hormonal preparations that need to be injected are sometimes effective, but they too can have damaging effects on other organs. To be effective, these methods require abstinence from alcohol and temporarily but virtually emasculate the man; understandably, therefore, they have not enjoyed wide acceptance. In women, however, research has been dramatically successful in regulating the control of ovulation by the anterior lobe of the pituitary. Because of the feedback effect of ovarian hormones, ovulation can now be suppressed by oral pills that contain a mixture of estrogens and progestagens. The pills are taken well before ovulation, starting from day 5 to the cycle and continuing to day 25. These hormones suppress the production of FSH and LH (p. 243), the follicles do not mature, and ovulation does not occur. After day 25 the effect of the pill is released, and menstruation occurs as in a normal cycle but the quantity of loss is usually reduced. Not only does the pill suppress ovulation, but it also has inhibitory effects on the uterine tube and on the endometrium, which is not prepared adequately for implantation even if an oocyte were fertilized. The pill also causes changes in the cervical mucus that make the passage of spermatozoa difficult. There are many proprietary brands of pill, most of which differ in the proportions and in the nature of the estrogenic and progestational substances. There has been much concern about both the short-term and long-term effects of these pills. They should definitely not be taken by women who have diabetes or liver disease; moreover there is the very real possibility that they induce thrombosis and clotting in blood vessels. A woman who has a personal history of blood clotting in vessels at any site should consult a doctor before taking oral contraceptive pills. Women sometimes experience troublesome side-effects from a particular type of pill such as fullness or tenderness of the breasts, nausea, headache, skin eruption, chloasma (p. 281), gain in weight from water retention, and loss of libido. Medical advice on the type of pill can rectify such troubles, but it must be remembered that relief from the anxiety of becoming pregnant, and perhaps indulgence, can increase appetite. On the other hand, the "safety" afforded by the pill removes the woman's classical excuses for avoiding intercourse with her husband or lover. Contraceptive pills have proved effective, and when the risk of their use "is set against that of death due to pregnancy and delivery the use of oral contraceptives proves safer than nearly all the other possibilities that are open to a married woman" (Peel and Potts, 1969).

Another method of birth control is to prevent the blastocyst from implanting in the uterus by the insertion of some form of intrauterine device (IUD), nowadays plastic loops, coils, or spirals. These are generally suitable only for women who have had children and whose cervical canal is thus open enough for their insertion. The device expands within the cavity of the uterus and causes changes in the endometrium hostile to the processes of implantation; it can also interfere with normal tubal function and thus with the normal timing of events. Intrauterine devices have not been universally accepted because they can easily be expelled from the uterus and can cause a low-grade chronic irritation and even infection that can lead to permanent sterility.

Surgical methods are permanent or nearly so. In men, vasectomy involves the cutting and removal of a portion of each vas deferens (p. 260) at points close to the base of the penis. Vasectomized men have no spermatozoa in their ejaculate. Their testes, however, are still active and produce spermatozoa, which, having no outlet, are resorbed. This apparently simple and effective method has some negative aspects because of immunological side-effects. However, vasectomy is practiced widely, especially in India. It is a practically irreversible measure, although it is said that reanastomosis has occasionally been successful. It must be remembered that after years of resorbing spermatozoa, the body treats them as foreign bodies and forms antibodies against them; in this case, reanastomosis would fail. Vasectomy is a form of functional sterilization but not equivalent to castration. There are conflicting reports of both increased and decreased libido after vasectomy, but this may be psychogenic. Apparently this method is enjoying increased popularity among some sailors. Our informant advanced the explanation that their wives, having ceased taking the pill and thus risking an unwanted pregnancy, forsook infidelity during the sailor-husband's absence at sea!

The removal of a segment or the whole of each uterine tube in women is a sterilizing method comparable to that of vasectomy. After being ovulated, the oocytes disintegrate and are resorbed. This procedure is usually performed on women for whom pregnancy would be dangerous because of illness or on women who already have a large family: it is almost always irreversible. Menstruation continues normally unless under certain circumstances it proves advisable to remove the uterus and leave the ovaries.

Abortion

Any procedure that terminates life in an embryo or a fetus up to the age of seven months is technically an abortion; after that age a fetus is legally viable. Abortions can occur spontaneously or they can be induced. (It is of more than casual interest that a significant percentage of embryos aborted spontaneously are abnormal.) Depending on one's country, legal abortions are accepted, forbidden, or not mentioned. They have been, and are, performed for social, medical and psychiatric, and eugenic reasons; according to the circumstances any reason that produces the desired result could be a good one. However, one aspect of abortion is seldom discussed or remembered: it wipes the slate clean for all persons but one.

In the early stages of pregnancy, dilatation of the cervix and curettage (D and C) of the uterus is a common procedure. Uterine aspiration with a suction pump is an added refinement of D and C; it saves time and is more gentle to uterine tissues. Hysterotomy (opening of the uterus) can be performed through the abdominal wall, or through the vagina, between the thirteenth and fourteenth week: such treatment inevitably weakens the uterus, and all future births should be in a hospital. Hypertonic saline (20 percent) or glucose (50 percent) injected into the amniotic cavity through the vagina or, after sixteen weeks, through the abdominal wall, induces labor and expels the fetus. Care must be exercised in this procedure and in the use of pastes and medicated soaps that can be introduced through the cervix into the uterus to cause expulsion of the fetus about a day later. The possible dangers are the penetration of the placenta with ensuing hemorrhage or infection, or only partial success, which must be followed by more radical evacuation of the uterus—and with more drastic effects on the mother's nervous system. Drugs that cause abortion have thus far proved unreliable and for obvious reasons have scarcely been investigated. Intravenous injection of prostaglandins are generally effective. The complications that can follow abortion are numerous and vary with the circumstances.

We have pontificated about numerous aspects of man's imperfections in this book. On abortion we declare that (1) if it *is* to be done, the earlier the better; (2) if it *has* to be done it is a reproach to someone—to the parents, to the participants, to teachers, doctors, state lawyers, to contraceptive services, to the weakness, carelessness, and thoughtlessness of men and women; (3) it is ironical that an increasing number of medical students are being trained to be terminators of life, when medical practice is dedicated to sustaining it. Some gynecologists and obstetricians are uneasy about the present attitudes toward abortion, and perhaps they should be. Except for therapeutic reasons, abortion must be considered a failure; but having been performed, it does give another chance for sensible sexual behavior, which nowadays is virtually its only excuse. We also realize that to many abortion is sinful. This attitude is understandable from many points of view—moral, religious, and social. The alternative to abortion, however, is not irresponsible procreation, but a responsible accommodation to the need for zero population growth advocated by concerned sociologists and to one's own existential need to procreate.

Aberrations in Sexual Behavior

Since the desire for sexual experiences is one of the most compelling of man's drives and occupies much of his conscious or unconscious time, deviations from normal sexual behavior are not surprising. Yet, the exact nature of some of these deviations is extremely difficult to define, mainly because behavior patterns in sexual gratification are highly personal. Some generalities, however, can be made. Biologically speaking, human sexual behavior is motivated by pleasure and reproduction, involves partners of opposite sexes, and culminates in penile-vaginal intercourse. Those individuals who prefer partners of their own sex and those who substitute other means for sexual intercourse are considered deviants.

However, before such a judgment is made, one should ascertain whether these substitutions are practiced to the exclusion of "normal" behavior. For example, whereas most healthy boys practice masturbation during puberty and adolescence as a normal part of their development, a man who prefers masturbation to heterosexual intercourse is not always behaving normally.

Psychologists and psychiatrists do not always agree on the causes of sexual deviation or perversion. Some believe that unnatural early experiences and abnormal relations with their parents can have a strong effect on the development of children's behavior. Others have postulated that too much love or too much hate for either parent can also influence adversely the development of sexual behavior. A boy who is either excessively hostile or excessively affectionate to his mother or father can become a homosexual. Actually, the point here is that an overidentification of a child with the parent of the opposite sex can condition his own sexual behavior. The real causes of the many behavior patterns that are considered abnormal are still unknown. There are a great many more disorders than one might imagine, but we will mention briefly only homosexuality.

A homosexual individual is one who is mainly or excessively attracted to members of his own sex. Many homosexuals (of either sex) appear to be in all ways physically fit; the disturbance is in their psyche. All suggestions that this disorder is due to an insufficiency of sex hormones have been ruled out. Homosexuality in men is not necessarily accompanied by feminine behavior; it is found among athletes and otherwise aggressively masculine individuals. Intelligence, sensitivity, and creativeness are as high in homosexuals as in any group of human beings. Cynics have declared, from Roman times, that a male is right in admiring and loving only another male because of the inferiority in all worldly virtues of the opposite sex. This is, of course, foolishness. We do know from experiments in rats, guinea pigs, and rhesus monkeys that the basis of sexual behavior is established early in embryonic life; it is stamped on the central nervous system (mainly the hypothalamus), usually by the embryo's own hormones. Substances that can have lasting effect on the future individual, however, do go through the placenta. Many homosexuals, aware of their condition, are unhappy but can, at best, only suppress it. Attempts to cure it with hormonal therapy have largely failed. Psychiatric treatments are successful only in some cases. Whereas until recently society was often brutally cruel to homosexuals, it has now come to accept them. Indeed, most homosexuals no longer feel crippled by their condition and some flaunt their activities openly and defiantly. This may be an expression of newly acquired freedom from social sanction. However, homosexuality is still an unresolved social problem.

Man's Sexual Needs

The preceding discussions may have left the reader with the feeling that the principal, if not the only, purpose of sexual behavior in man is the discovery of compatible sex partners. These, once united, would share sex life together without excursions into sexual relations outside the established "family" bond. Indeed, the laws in the United States, England, and Europe are not charitable toward extramarital affairs. This is perhaps understandable. Human infants and children are totally dependent upon parental care for a very long period, in fact, for as much as one-sixth or one-seventh of their lives if one assumes the average life span to be sixty to seventy years. In primitive tribal life, consisting of relatively small groups, a woman could care for infants and children regardless of whether or not they were her own, but modern society is not yet prepared to return to this indiscriminate mothering. This means that for most modern men and women of modest affluence, monogamy is not only a limitation of the law and church but a practical expedient, if only because of the severities of income tax.

From the preceding discussion one might also come to the conclusion that man is naturally monogamous and that sex and love are on equal terms. Yet some scholars of sex believe that the nature of man is such that monogamy is not always wanted, not easily attainable, and often unsuccessful. Experts such as H. Benjamin also believe that,

> [the] man who does not care for variety in sex, or does not include it in his pursuit of happiness, is in the minority. He has either been rarely fortunate in finding a perfect mate, or nature has provided him with a "low level" of psychosexual constitution.

Society has made hypocrites of us through its laws and mores, and it is unwilling to admit, or cannot admit, that most human beings desire and seek novelty and variety in sex. If this were not a natural urge, it would not have survived taboos, persecution, and hostile laws. None of the efforts to suppress it, nor pietistic sermons and threats of a fiery afterlife by ascetics and zealots have succeeded in eradicating it. Since normal man is bound to seek some gratification or outlet for his sex drives, premarital and extramarital intercourse or other physical intimacy by men and women is common enough to be a rule rather than an exception.

The fact that such activities are not overt or are denied seems to be largely based on fear of social sanction and of the law. If the urges discussed here were not, indeed, particular to man, prostitution as an institution would not and could not exist. Some form of prostitution has existed throughout recorded history. The word, generally used in a derogatory sense, denotes indiscriminate sexual relations, without affection, usually anonymous, and for pay. A prostitute is a woman (or a man) who performs these services professionally. Some prostitutes work full time at it, others have other employment and devote part time to it to supplement their income. Hired for temporary service, with detachment, and without moralistic considerations, a prostitute provides relief from sexual demands in a normal physiologic way, without the involvement of "love." Prostitution existed among the Egyptians, Romans, Jews, and Arabs. It has survived in the Orient, India, Europe, Africa, England, the United States, and in every remote place where human society has developed. Ridicule, discrimination, abuses, and harassments have not succeeded in abolishing it. Prostitution flourishes particularly in societies where the sexual mores are restrictive and punitive. It fares poorly where promiscuous pre- and extramarital relations are condoned; the effects of improved contraceptive methods on prostitution are difficult to assess. With the attainment of women's freedom of expression in sexual activities, prostitutes would seem to be forced to seek another profession. But man has also become more affluent than at any other time in history, and he has abundant free time on his hands. Both conditions favor libidinous and licentious behavior.

Aside from moral considerations, the "profession" has apparently existed only because man is a superlatively sexual animal, perpetually thwarted in his attempts to satisfy his basic need for variety in sexual expression. Moralists have too easily resorted to facile labeling: self-indulgence, sexual promiscuity, gluttony, adultery. Seldom have they elected to assess man's nature realistically and to regard sexual outlets as integral parts of his nature. If restless man cannot satisfy his basic need for new sexual experience through marital relations, he will do so through extramarital ones or prostitution.

We conclude with the reminder that sexual behavior is a matter of concern only when it pertains to human beings. Endowed with only a memory of sorts and perhaps no foresight, other animals have relatively fixed patterns of sexual behavior. For survival, these animals must obey the dictates of circumstance and will act according to their successive and diverse desires.

Only man seems to need substitutes for his sexual needs. It is surprising that he has survived so long a time confronted with continued frustrations. But perhaps, who knows! he may have turned these to his advantage by devoting his energies to other activities.

13

MEANS OF COMMUNICATION

Words are not merely the vehicles in which thought is delivered; they are part of thinking.
(Peter Medawar)

Among the many particularities of man, none is more distinctive than articulate speech. Language, the very stuff of speech, is a system of abstract logic that enables man to express his rationality. Since all human beings have evolved a language, it would be interesting to discover when articulate speech as we know it was born. Studies of this moot question are to some degree conjectural, but Philip Lieberman and Edmund S. Crelin (1971), who reconstructed the skeletal features of fossil Neanderthal man, showed that his supralaryngeal vocal apparatus was similar to that of a newborn modern human being. Since the authors had earlier shown by means of acoustic analysis that newborn human infants, like nonhuman primates, do not have the anatomical mechanisms necessary for articulate speech, Neanderthal man could not have been adequately equipped for speech. Yet, the level of Neanderthal culture was such that his limited phonetic ability was, no doubt, more advanced than that of modern nonhuman primates. Furthermore, since his brain was sufficiently well developed, he probably had some sort of symbolic phonetic means of communication.

In contrast, Cro-Magnon man, who succeeded Neanderthal man in the Dordogne, a department in southern France rich in the remains of fossil man, had laryngeal skeletal structures similar to those of modern man and no doubt a formalized language. One could speculate that Neanderthal man gave way to Cro-Magnon man because the latter was brighter and had a more advanced culture.

Thus, one of the most distinctive achievements of man over the other animals is his superlative development of methods of imparting information through speech. E. L. DuBrul has even defined human societies as "biotic entities whose integrative modality is speech." And Lieberman and Crelin have summed up by saying, "Man is human because he can say so." Man's speech can be so modulated and emphasized and the syllables so accentuated that he can express even the subtlest shades of meaning and interpretation. Language, or the means of communicating feelings, disposition, and information by articulate speech, is a relatively recent phenomenon, although in its broadest sense it is not exclusively human. The distinguished Swiss anthropologist A. H. Schultz regards human speech as a quantitative perfection of the highly specialized development of man's central nervous system, controlling the anatomical speech apparatus in the larynx, the tongue, the jaws, and the lips. Schultz, moreover, believes that if monkeys and apes had lacked the ability to make sounds, they would never have become the intensely social animals they are. Primates produce a surprising variety of sounds whereby they maintain contact between the members of a group, between mother and child, and between individuals and the entire group to alert about possible danger, to frighten enemies, to warn rival groups away from territories, to mate, and to announce food. The amount and variety of information exchanged by the vocal sounds emitted by social Old World monkeys probably surpass those of any other mammals. Although monkey sounds may seem identical, they actually consist of conscious, deliberate, discrete communications. Schultz states that "The orgies of noise, indulged in specially by howlers, guerezas, gibbons, siamangs, and chimpanzees, seemingly so repetitious and meaningless, are probably at least as informative to the respective species as most after-dinner speaking is to *Homo sapiens*." This is a deliberate oversimplification since even soporific human speech can convey abstract ideas.

The many attempts to teach chimpanzees to enunciate human words have been singularly unsuccessful. Much too great a fuss is made when one of these animals manages to articulate some human "word," which must be as meaningless to the animal as it is to parrots and mynas. Furthermore, in spite of hopeful claims, attempts to train chimpanzees to produce simple sentences with objects have proved largely useless. Still, vocal communication among mammals is often clear and even subtle. John B. Theberge (1971) in a beautiful study on "wolf music," found that these interesting animals transfer information on a universal and an individual level. Universal information is limited to a certain species and requires symbols that are the same throughout that species. Individual communication transpires among individuals that learn to recognize special qualities in the animals with whom they associate. The author shows that by howling, wolves can communicate to other wolves their locations, identity, and even something about their emotional state. Porpoises and dolphins, whose intelligence is reputed to be next to that of man, or at least equivalent to that of a chimpanzee, allegedly are able to communicate, but research has not yet demonstrated what precisely is being communicated. Some form of biological communication is probably imperative for the survival of all higher animal forms. However, speech per se is not the real issue here. Man has an organized language made up of repeatable symbols, which when recorded convey his thoughts, even in his absence or long after his death.

Speech, however, is not the only vehicle for expressing emotions and imparting information. Postural attitudes—crouching, gesturing, grimacing with eyes and brows—and physical contacts can also transmit information. In each instance, the observer receives patterns of signals, which he interprets in a known context or within a framework of accepted associations. All animals have behavior patterns whereby they somehow communicate at least the basic signals for survival such as greeting, mating, threat, danger, and flight. Nonhuman primates, which share many of these patterns both with the other mammals and with man, use various facial expressions: knitting the brow, opening the mouth and baring the teeth partially or fully in threat or attack, grimacing and smacking and protruding the lips in greeting, grinning with or without frowning in threat, greeting, or excitement. These patterns are accompanied by growls, clicks, bleats, whines, and other vocalizations, and by certain postural attitudes like dancing, pounding the chest, shaking branches, and walking backward. Each of these actions has its own specific meaning. Like dogs, cats, and other mammals, when angered, prosimians with well-developed ear muscles flatten the ears against the head and prepare to attack.

The atrophy of the human external ear and the muscles that move it eliminates these appendages in social communication. Animals also signal with their tail, its position or movement indicating the current disposition of the animal. As man lost these means of communication, he replaced them by facial and other gestures and by vocal sounds. Apes and monkeys also have certain social habits, such as grooming, to indicate social acceptance or contentedness and presenting the buttocks to an aggressor to convey submission and willingness to cease hostilities. This attitude is also struck by females to desirable males.

According to S. L. Washburn and I. De Vore, certain primates, like baboons, depend mostly on gestures to mediate their social relations; since vocal utterances are subsidiary to gestures, they are relatively rare. Among members of baboon troops, whose social groups are so compact that the animals are usually in sight of each other, most communication is gesticulatory and without vocal accompaniment. Certain vocal sounds, like a warning bark to indicate danger, do convey a meaning independent of gestures, but generally they are used primarily to draw attention to an animal and to emphasize its gestures and bodily postures. Mimetic and postural gestures usually accompany vocal signals of defiance, anger, fear, frustration, and contentment. The importance of such gestures can be judged from an interesting experiment. When the vocalizations emitted by some primates during previous stressful situations were recorded and played back to the same troop under quiet conditions, they did not always elicit the earlier response. A compilation of the different postures, gestures, and facial expressions of monkeys and apes would be useful, for they represent, as Schultz says, "an intricate and voluminous *silent vocabulary* of great aid in social intercourse."

Much research has been done on the mathematical and practical engineering aspects of communication. The stimulus of war and the incentives of commerce and the academic world have played a significant role in developing what is known as "communication theory." Besides the means for communication among organisms, each animal is equipped with mechanisms for chemical communication within itself. Numerous biological feedback systems include those between the various endocrine organs and the pituitary gland, between adjacent cells, and between the component parts of any one cell. Messages, signals, and information are communicated continuously and in some form or other in all biological systems at every level of complexity. When a cell performs its proper function, we must assume the existence of chemical signaling systems between its membrane, cytoplasm, and nucleus and between the individual component parts of the cytoplasm and nucleus. The evidence is incontestable that such communication exists in neurons and that the cells of endocrine organs signal to specific cells in specific organs while at the same time receiving orders from other cells. Fragmentary evidence suggests that embryonic growth is subject to the control of various systems, although too little is known of the mechanisms.

Structurally, a communication system consists of a transmitter, a medium to conduct the signals transmitted, and a receiver. The message must of necessity be so prepared as to have content and meaning and, equally, when received must lend itself to interpretation and analysis. This theoretical arrangement can be diagrammatically represented:

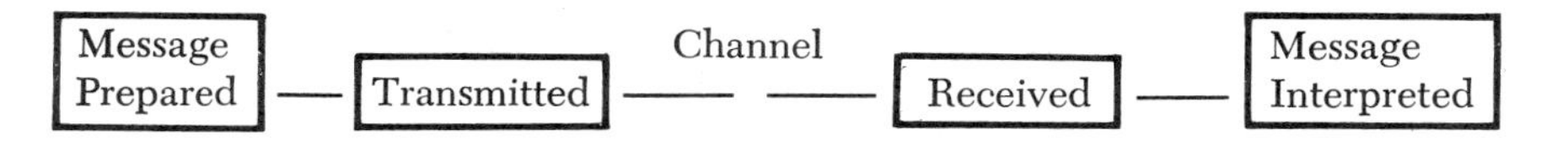

When a man speaks, the message is prepared in the cerebral cortex after the raw materials necessary to formulate it have been acquired from an inherited component, from learning, or from the reception of some immediate signals. That part of the cerebral cortex believed to be the site of origin of the neural impulses that cause speech is called the motor speech area. Situated on the lower and posterior part of the frontal lobe, it includes the lower part of the inferior frontal and precentral gyrus. Curiously, and so far inexplicably, right-handed people have their motor speech center on the left side of the brain. A baby, but only a baby, can develop a speech area on the other side if injury occurs to one side. This area was first defined by a French surgical anatomist and anthropologist, Pierre Paul Broca. In 1864 he realized its importance and noted that damage to it resulted in *aphemia,* or difficulty in speaking. (*Aphonia* means total loss of voice; *aphasia* is the word now commonly used for inability to speak, and *dysarthria* describes a difficulty in articulation.) Broca was a firm supporter and even proponent of the theory that particular functions are fairly closely localized in particular areas of the cortex. He even drew a topographic map of the positions of the various areas, but neurologists nowadays are not so explicit in localizing function in the cerebral cortex.

The initiation of the complex mechanisms that result in spoken words depends upon an intact Broca speech center, which overlaps into the precentral or motor area for muscles of the head and neck. Articulate language, possessed universally by all extant races of man, could not have been highly formulated by primitive human groups since even today lower forms of language are inseparable from lower forms of thought. Analyzing the possible evolution of human speech, Schultz concludes that in the

> . . . perfectly adapted arboreal life of monkeys and apes, the limited variety of sounds, together with the great variety of meaningful gestures and facial expressions, is fully adequate for all social life within such close contact as permits seeing and hearing these detailed means of communication. As soon as the early hominids had ventured into open spaces, had begun to use and even make tools, and had cooperated in hunting, the total variety of all means of expression needed additions, which could come only from an increase in sounds, since the compara-

tively little changed anatomy had already been fully used for all possible gestures. Gestures have always persisted in human evolution, but they have become overshadowed by an infinitely greater variety of sounds in increasing numbers of combinations.

Phonetics

Notwithstanding the complexity of his speech, man is related to an apparently inarticulate group of animals. Since the great apes are man's closest biological kin, it is somewhat surprising that their speech is even more limited than that of monkeys and some lemurs. Some prosimians, such as lorises and pottos, that have been kept in captivity for years, have not been heard to utter a single sound other than a soft growl when they are disturbed, but ring-tailed lemurs have a relatively extensive vocabulary.

Articulate speech is produced only when an individual makes a number of different noises, each unit of which can be placed in various juxtapositions to other units and thus into symbols and words. Emotional content, value, urgency, and other meanings can be given to the sequence of noises by spacing the symbols, by alterations in speed, pitch, and volume, by changes in pronunciation, variation of emphasis, and so on. In English and in many other languages, the noises of articulate speech consist of consonants and vowels. Some languages make more use of gutturals, of changes in tone (e.g., the several ways of enunciating "chang" in Chinese), and of clicks of the tongue (e.g., the Zulu language). But regardless of differences, all languages make use of the same sets of anatomical structures. Phonetics is that branch of language study that is concerned with the way speech sounds are formed by the organs of speech and put in order or sequence, the way they can be stressed or accentuated by the alteration in pitch, and the way the speech sounds affect the ear.

Quite apart from its application to the study of the anatomical production of speech, phonetics has practical applications. Among others, it can assist in curing speech defects, in aiding the deaf to speak, and in helping to construct alphabets for the blind. It helps to teach one to read and to understand a language and is a boon to students of foreign languages, dialects, and philology. Moreover, it is used in systems of shorthand, telegraphic communications, and other methods of signaling words.

Speech, then, consists of sequences of sounds, or particular types of air-vibrations produced by the organs of speech and heard by the ear. The speech organs are the larynx and vocal cords, the pharynx, tongue and palate, the jaws, cheeks, teeth, and lips. Other structures, such as the paranasal air sinuses, act as resonators and give passive assistance to speech. A gradual emptying of the lungs during exhalation, which causes a current of air to flow past the moving speech organs, generates speech sounds. In addition to this usual method of producing speech sounds, other noises can be made by drawing in or expelling air between lips and teeth or through the nose. As the air is exhaled, the shape of the passages enclosed by the speech organs is altered, and sounds are produced. If the air passage remains open, only "breath" sounds are produced, but if the speech organs move in one of several ways, a speech sound is produced.

The acoustic effect of any one speech sound on the hearer is technically known as a phone. If a phone carries meaning by itself, it is a phoneme. When, as is more common, two or more phones are put together, they produce the more complex morphemes. The production of a sequence of speech sounds in the form of a speech-chain is therefore heard as a sequence of phones. Two main classes of phones exist: consonants, produced by complete or partial closure of a part or parts of the air-passage; and vowels, formed by the continuous and free passage of air through the air channels. Different kinds of consonants and vowels are therefore the results of the various ways in which the speech organs modify or halt the flow of exhaled air.

The Larynx

The most important of the speech organs, the larynx, or "voice box" (Gk. *laryngos*, the gullet), is primarily responsible for the voice. If this structure is surgically removed, one can learn to speak by expelling air from the esophagus. Although all land-living tetrapods with lungs have evolved a larynx of sorts, not all of them have a "voice." The respiratory system develops as a hollow outgrowth from the floor of the primitive pharynx behind the tongue region. The opening to the respiratory tube, the glottis, is guarded by muscles in its walls that are capable of closing it. In mammals, the epiglottis, a flap containing cartilage, forms an effective lid at the back of the tongue that can fold over the glottis. During swallowing the muscular glottis and epiglottis automatically close the opening to the respiratory system to prevent food or fluid from entering it.

The larynx, a complex structure in the form of an enlarged vestibule surrounded by cartilage and muscles, is situated around the glottis and leads into the trachea. In many amphibia and most reptiles, the simple larynx and glottis are unsuitable for complex voice production. Frogs and toads, and some reptiles, can produce a croak or hiss when they expel air violently through an almost closed glottis. Some species of frogs use inflatable buccal or gular pouches as resonators to modify their croaks. Although birds have a simple larynx, the variety of sounds they make is the result of a structure comparable to another larynx, the syrinx, at the base of the trachea where the bronchi join.

Vocalization in birds serves the same function as that of stereotyped postures and movements, and both function in communication. Many, if not all, birds have evolved complex repertoires of subtle vocal displays, each vocalization communicating some information. In addition to a number of calls, most birds have a complex song. Every male acquires a unique set of song patterns, so that no two are identical. Once established through trial and error, a bird's song patterns remain the same for life. Bird song specialists claim that an individual bird can be distinguished by its song pattern as reliably as by its color markings. Though frogs, toads, and nearly all mammals can use the larynx to produce sounds, only man has two paired vocal folds—"false" and "true"—developed below the glottis, which, with appropriate muscular control, enable him to refine phonation and articulate speech. In cetaceans, the larynx is shaped like a whistle and is permanently held by a muscular sphincter in the posterior nares. Since these mammals have no vocal cords or folds, it remains a mystery how dolphins and porpoises produce the strange array of sounds recorded underwater.

That the construction of the complex larynx has enabled it easily to evolve into a voice-producing apparatus seems apparent. Yet the larynx has other functions than that of producing sounds. Primarily, it provides rigidity in the opening of the air passage and prevents solids or liquids from entering it via the glottis. Extrinsic muscles are attached to it and to the closely related hyoid bone from origins on the mandible, base of the skull, and thoracic region. Thus, the larynx can be moved; for example it is brought upward and forward during swallowing, out of the pathway of swallowed foodstuffs. The internal and external laryngeal muscles can control the flow of air entering or leaving the lungs and thus play a vital role in the production of voice. In some animals the muscular connections of the larynx are strong enough to help fix the thoracic cage during tree-climbing. However, the sounds the larynx is capable of producing are feeble and hardly recognizable as the human voice until the pharynx, mouth, nose, sinuses, and other structures amplify and modulate them.

The skeleton of the human larynx is composed of cartilages, the largest of which are the shield-shaped thyroid cartilage and ring-shaped cricoid cartilage (Figure 13–1). The thyroid cartilage, attached to the hyoid bone above by muscles and the thryro-hyoid membrane, articulates by its lower horn to the cricoid cartilage, which is attached to the trachea. The thyroid cartilage rocks backward and forward on the cricoid because of the way in which the two cartilages are articulated. Perched on the

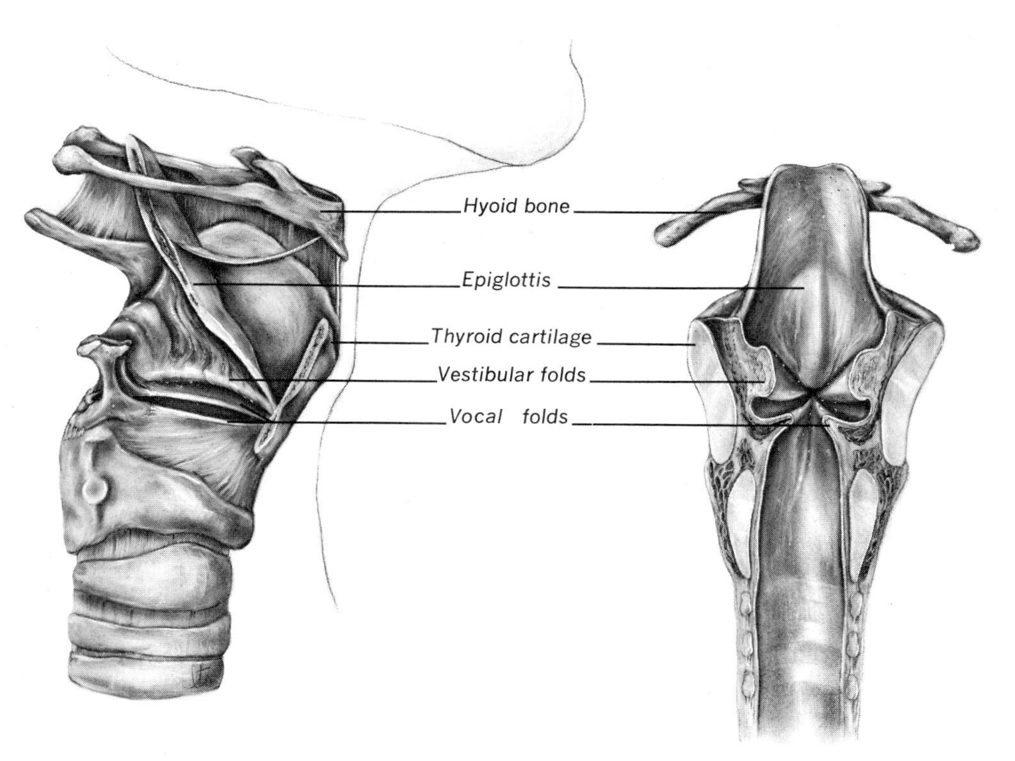

Figure 13–1. The human larynx with the thyroid cartilage and glottis removed to show its various parts. Left: lateral aspect; right: seen from above.

upper surface of the back of the cricoid cartilage are two small three-cornered arytenoid cartilages with which a number of intricate little muscles are associated. "Arytenoid" means pitcher-shaped and is derived from the fact that the apex of the arytenoid cartilage in man is directed slightly backward, resembling the spout of a pitcher. The true vocal folds stretch on each side from one of the processes of the arytenoid to the thyroid cartilage in front; together with the adjacent sides of the processes of the arytenoid cartilages, they form the boundaries of the glottis. The space between is called rima glottidis and is 22 to 24 mm long in men, 14 to 16 mm in women.

Both the vocal folds and the sides of the arytenoid processes are covered by a mucous membrane, tightly adhering to the vocal folds and giving them a pearly white color in life. A number of small muscles are attached to the arytenoid cartilage and to the thyroid and cricoid cartilages in such a way that the arytenoids can be pulled toward each other, away from each other, and rotated. These movements will open or close the rima glottidis in various ways. At the same time, the tension of the vocal folds and of the vocal ligament that lies below it can be altered by the tilting movement of the thyroid cartilage on the cricoid.

Although the basic structure of the larynx is essentially similar, many modifications in each order and often even in each genus are evident. Only the human voice can be modified by muscular control of the upper air passages in such a way as to permit articulate speech. The progressive differentiation of the larynges of mammals has culminated in that of the anthropoids, but in man some simplification has also occurred. G. Kelemen emphasizes that the evolution of the larynx is frequently not in step with the rest of the organism. The primitive kangaroos have small vocal folds, whereas some rodents, which are phylogenetically higher, have none. Sphincteric devices appear in various sites of the respiratory tract of frogs, reptiles, and birds in different phylogenetic sequences, though no clear ascending scale of complexity is found. Only man has double paired vocal folds.

Among the primates, the anatomical and functional performance of the larynx varies considerably (Figure 13–2). The vocal folds of prosimians show a progressive evolution from the nearly membranous structures of tarsiers to those of lemurs, which are underlined by muscle. Most monkeys have intralaryngeal outpocketings for resonance and some even have extralaryngeal air sacs. Capuchins *(Cebus)* and spider monkeys *(Ateles)* from the New World have very long, sharp-edged vocal folds, which enable them to produce loud noises. Howler monkeys *(Alouatta)* likewise have extremely long, sharp vocal folds, supported by long, deep ventricles and reinforced by a great bony dilatation of the hyoid bone.

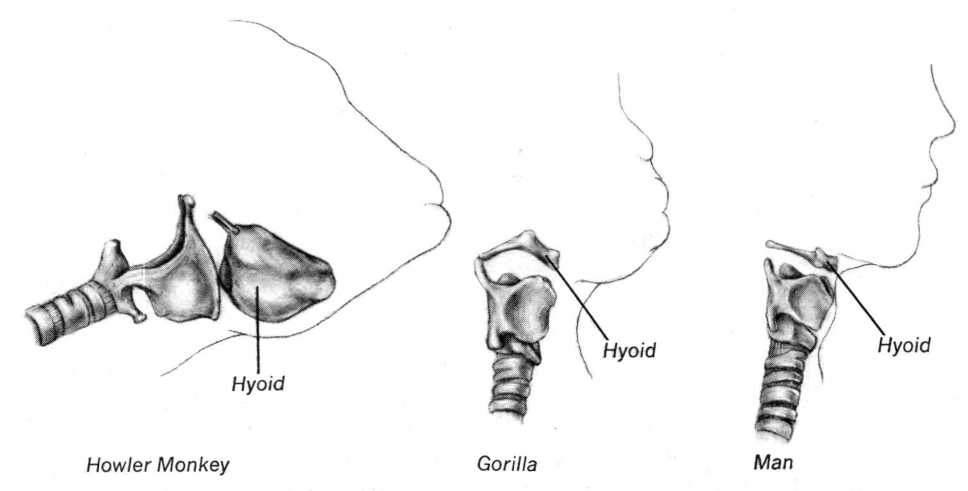

Figure 13–2. The varieties of larynges of several primates.

Equipped with such a device, these animals can emit prodigious roars. Old World monkeys have air sacs emerging from the middle of the hyoid bone. A perforation in the epiglottis serves as a passage between the larynx and membranous air sac; these recesses act as resonators. The larynx of these animals, provided with a cartilaginous skeleton, has remarkable plasticity. The pharynx is only loosely connected with the larynx; the soft palate terminates in a uvula, the first appearance of this curious structure. The sounds produced by nonhuman primates are often highly modulated and complex.

Even though superficially the anatomy of the larynx of apes appears to be similar to that of man (and some theorize that apes cannot learn to talk only because they lack a neurological base for speech), Keleman has pointed out that the larynges of the great apes differ from each other and from that of man. Among these differences are air sacs and the extremely high position of the larynx against the base of the tongue; moreover, the soft palate is almost inseparable from the epiglottis at the sides. Although the absence of a neurological basis for speech in the great apes cannot be denied, one cannot overlook the fact that their vocal apparatus also lacks the ability to produce many sounds.

Voice Production

The larynx produces sound when air, exhaled from the lungs, passes the vocal folds and causes them to separate. This at once triggers the contraction of the intrinsic laryngeal muscles, the thyro-arytenoid and vocalis, which cause the vocal folds to come together again. As this process is repeated over and over with great rapidity, the escaping air stream vibrates and produces a tone. The pitch of the tone varies according to the tension of the vocal folds: the tighter the folds the higher the pitch. When, as in singing, one attempts to produce a pure tone, the entire edge of each fold separates; in the production of speech sounds, only a part of the length of the apposed fold separates. It is impossible for the larynx to produce a pure tone; actually, a series of overtones are produced at the same time, and the final tone is the result of the effects of various resonators below and above the larynx.

The resonators—the thorax and the column of air in the bronchial tree and trachea, the pharynx, the nose, the mouth, and the paranasal sinuses—increase the magnitude of the sound and alter the mixture of tone and overtones produced at the larynx. Their importance is clearly shown when a heavy cold blocks the nasal cavity and air sinuses and causes a characteristic change in the voice. Should the infection spread to the larynx, the voice further alters and becomes hoarse, and if the vocal folds are so involved that they swell and cannot vibrate properly, the voice is reduced to a whisper or is completely "lost." A whisper can also be produced anatomically by keeping the glottis partly open all the time, partially closing the "false" vocal or vestibular folds that lie just above the true vocal folds, and by using the resonators to pick out and fuse the sounds produced.

It should now be clear how the various consonants and vowels are produced. There are two main kinds of consonants: one is "breathed" with the vocal folds abducted, for example, *F*; the other is "voiced" with the vocal folds producing the sound, as in *V*. In the glottal stop, the consonant is neither voiced nor breathed. Consonants can also be classified according to the anatomy of the mechanism that articulates them, whether the lips, the teeth, the lips and teeth, the tongue and hard or soft palate, the pharynx, or the glottis. How a consonant is produced—whether it is rolled, plosive, nasal, and so on—provides another method of classification. Vowels, on the other hand, are classified anatomically according to the position of the tongue, although the position of the lips is equally important.

Because the neural control of speech is complex, the central mechanisms taking place in the brain are not as yet clearly understood. The utterance of words is the result of the sending out of a great many impulses along numerous nerves. Before a sound can even be made, the speaker must breathe, since speech can occur only during the expiratory phase. Thus, there must be a preliminary correlation with those levels of the spinal cord from which the phrenic and intercostal nerves arise. The laryngeal muscles responsible for adjusting the vocal cords are supplied by the tenth and eleventh (vagus and accessory) cranial nerves through branches of the vagus, especially its recurrent laryngeal branch, which innervates the intrinsic muscles of the larynx. The ninth (glossopharyngeal) cranial controls the movements of the pharynx and palate, and the twelfth (hypoglossal) cranial controls the tongue musculature.

The position of the lips and the grimaces or facial expressions that often accompany speech are brought about by the facial muscles, innervated by the seventh (facial) cranial nerve. One cannot possibly speak clearly without moving the jaw, which is controlled by muscles supplied by the mandibular branch of the fifth (trigeminal) cranial nerve. Since it is more effective (and considered to be more polite!) for the speaker to look in the direction of the hearer, the eleventh (accessory) cranial nerve and nerves from the cervical spinal cord carry impulses to the trapezius, sternomastoid, and other muscles that move the head and neck.

The recognizable difference among the voices of individual human beings is brought about by many variables, such as minor variations in the way the muscles are used and their actions correlated, variations in the size of the rima glottidis, and the sometimes quite marked variations in size of the resonators. Similar differences in muscle action enable one to pronounce phones of languages other than his own. So subtle are the minor variations produced in the voice by the articulatory mechanism that the terms *phoneme* and *morpheme* have been introduced to include families of sounds. For example, in most languages the way a consonant is pronounced depends on the following vowel.

Some prosimians are usually silent, and their vocal utterances are limited to occasions of distress, mating, or the proclamation of territorial rights. Some lemurs, particularly the ring-tails, have a complicated and expressive vocabulary. Some monkeys are quite vocal and "chatter" frequently; over thirty vocal sounds have been recorded for Japanese macaques. Other monkeys bark, screech, or grunt. The gibbons emit astoundingly loud, almost metallic hooting calls that rise in pitch and rate of repetition. They also have alarm calls and calls associated with group activity. Howler monkeys roar in a deafening way. These choral utterances are emitted to assert territorial rights, to convey alarm, to indicate troop location, and sudden changes in climatic conditions. In spite of the deafening sounds they produce, males often roar when they are stretched comfortably on a limb! Orangs rarely say anything but are occasionally heard to groan or roar. Gorillas have an impressive roar but usually are content to grunt, murmur quietly, or say nothing at all. In contrast, chimpanzees are loquacious, expressing their emotions readily, often, and usually in an unrestrained manner. More than thirty sounds or "words" have been recognized in their vocabulary. In all, however, even the most articulate primates are chasms apart from man vocally.

Changes with Growth

Changes in the character of the voice take place as an individual grows older. In children, although the caliber of the larynx is relatively larger than that of the trachea, the vocal folds are shorter and more lightly built than in adults. The resonators are also smaller, and some, like the paranasal air sinuses, are hardly developed. Hence, the voices of children are characteristically high-pitched. At puberty the larynx and the resonators enlarge rapidly. Changes in the voice are particularly marked in young men as the front part of the thyroid cartilage (*pomum Adami,* Adam's apple) and the vocal folds become thicker and heavier. Boys at puberty often have an initial inability to compensate for this, and a corresponding lack of muscular control leads to "breaking" of their voice. In women, and in men castrated before puberty, the larynx does not enlarge so much in the antero-posterior direction, the folds remain light, and the voice is high-pitched. The size of the larynx has little relation to physique: the voice of a small, slight man may be deep and, conversely, that of a giant may be thin and squeaky.

Aphasia

The production of voice and speech sounds depends on the correct correlation of nearly one hundred paired or unpaired muscles, all of which must also be at the appropriate tension. Not surprisingly, both organic and functional disorders occur in speech. Enunciating properly without teeth or a tongue or with a cleft palate is difficult. Damage to the nerves that supply various muscles results in difficulty in moving the speech organs. A diseased or damaged condition of those brain centers that control the motor pathways that effect or interpret speech on the sensory side is known as aphasia. Among the more common functional disturbances of speech are stammering, or hesitant and repetitive speech; dyslalia, or defective articulation; and lalling, or the substitution of sounds.

If a child who stammers is made to wear earphones through which is transmitted a low continuous note, he speaks without stammering; as long as he cannot hear himself speaking incorrectly, he does not stammer. The stammerer gets worse the more carefully he listens to what he says. This suggests that the speech sounds created by cortical activity are heard by the speaker and correlated with what he is saying. A fault, caused perhaps by incomplete coordination, is heard and interferes with subsequent speech patterns in the cortex. A person deaf from birth has obvious speech difficulties because he cannot hear his own voice to check whether his speech sounds like the phones produced by others and because he cannot make use of other people's speech as a model on which to base his own efforts.

Other speech defects arise from different causes. Lallation, for example, is an infantile type of babbling, heard mostly among retarded persons. Dyslalia, or imperfect articulation, is chiefly caused by imperfect distribution of, or by damage to, the nerves of the speech organs.

Anatomy of Hearing

The human auditory apparatus has an external ear for collecting sound and an external auditory tube, or meatus, at the inner end of which is stretched, like the head of a drum, a delicate parchmentlike tympanic membrane (Figure 13–3). This membrane forms the outer wall of the air-containing middle ear cavity, which communicates with the pharynx.

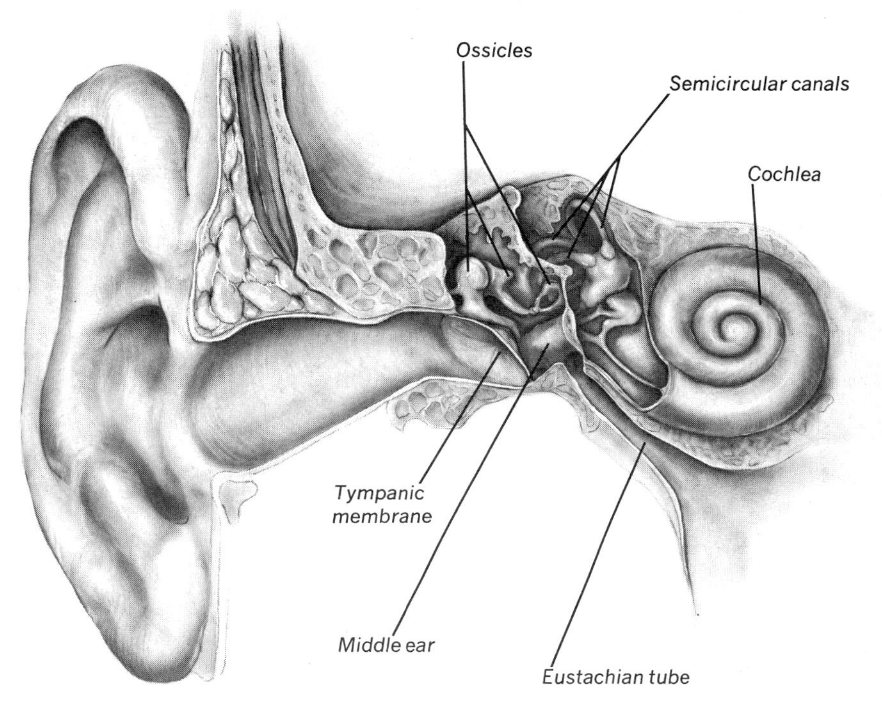

Figure 13–3. The major parts of the auditory apparatus.

Inside the middle ear are lodged the three small auditory ossicles, which convey vibrations of the tympanic membrane to the inner ear, which contains the cochlea, or organ of hearing. The vestibular apparatus also forms part of the inner ear, detecting changes in movement, such as rotation and inclination of the head, and responding to very loud noises.

The sound waves coming from the speech organs strike the tympanic membrane, causing it to vibrate. The cone-shaped membrane, with the apex, or umbo, directed inward, is set within the end of the almost circular bony tube made by the tympanic ring of the temporal bone (Figure 13–3). The long process, or handle, of the outermost of the three auditory ossicles, the malleus (hammer), is attached to the upper part of the membrane. The malleus articulates with the incus (anvil), which in turn articulates with the innermost ossicle, the stapes (stirrup). The base of the stapes, the footplate, is held by a ligament in an oval window on the inner wall of the middle ear. On the other side of the footplate and within the inner ear lies the fluid to which the ossicles convey vibration from the tympanic membrane.

Two muscles operate the ossicles of the middle ear. The tensor tympani is attached to the handle of the malleus. Its contraction preserves the cone shape of the tympanic membrane and keeps it gently taut. The tiny stapedius muscle is attached to the stapes and ensures an even articulation of the ossicles. Together with the tensor tympani the stapedius has a "damping" effect, protecting the auditory mechanisms from vibrations of too great amplitude. Damage to these muscles or to the nerves that control them results in "chattering" during movements of the ossicles, giving rise to "noises in the ears," or tinnitus. The action of the tensor tympani can cause some unevenness in the movement of the ossicles. The muscle delicately stretches the tympanic membrane and causes it to move more easily in certain directions under certain pressure conditions. Thus, when two notes of different frequency are being transmitted simultaneously by the membrane and ossicles, "difference" and "summation" tones are set up and may well be perceived by the hearer.

The three ossicles are not loose in the middle ear cavity but are held together by elastic ligaments and covered by a mucous membrane surrounded by the air that enters the middle ear cavity from the nasopharynx, through the pharyngotympanic, or Eustachian, tube (described by anatomist Bartolommeo Eustachius in 1562). For the rapidly vibrating ossicles to function properly, the middle ear must contain air with a pressure equal to that of the air outside. Should the tube or the middle ear cavity become filled with fluid or pus, the ossicles cannot vibrate. The pharyngotympanic tube is opened during swallowing to allow an equalization of pressures on each side of the tympanic membrane. The tube consists of three parts, each part varying in the composition of its confines. The inner membranous part communicates with the back of the nasopharynx through a slitlike opening above the soft palate just behind the lowest of three scroll-like conchae on the outer wall of the nasal cavity. Certain pharyngeal muscles that partly arise from the second, or cartilaginous, part of the tube normally keep the slit orifice closed, except during swallowing. The third part of the tube passes through the temporal bone to reach the middle ear cavity.

When no difficulties exist, the ossicles transmit the vibrations of the tympanic membrane to the oval window with great fidelity and amplify the movements of the membrane about twenty times. The vibrations are transmitted to the fluid filling the inner ear, in particular to that part of it called the cochlea. The inner ear, which lies within the petrous portion of the temporal bone, consists of a number of complicated fluid-filled membranous chambers and ducts called the membranous labyrinth, which in turn is enclosed within the bony labyrinth (Figure 13–4). The fluid with the membranous labyrinth is known as endolymph; that which lies outside it, but still inside the walls of the bony labyrinth, is called perilymph. Neither of these fluids, however, has any continuity or relationship with the fluid of the lymphatic system. The membranous labyrinth consists of the coiled duct of the cochlea, the saccule and the utricle, and the three semicircular canals.

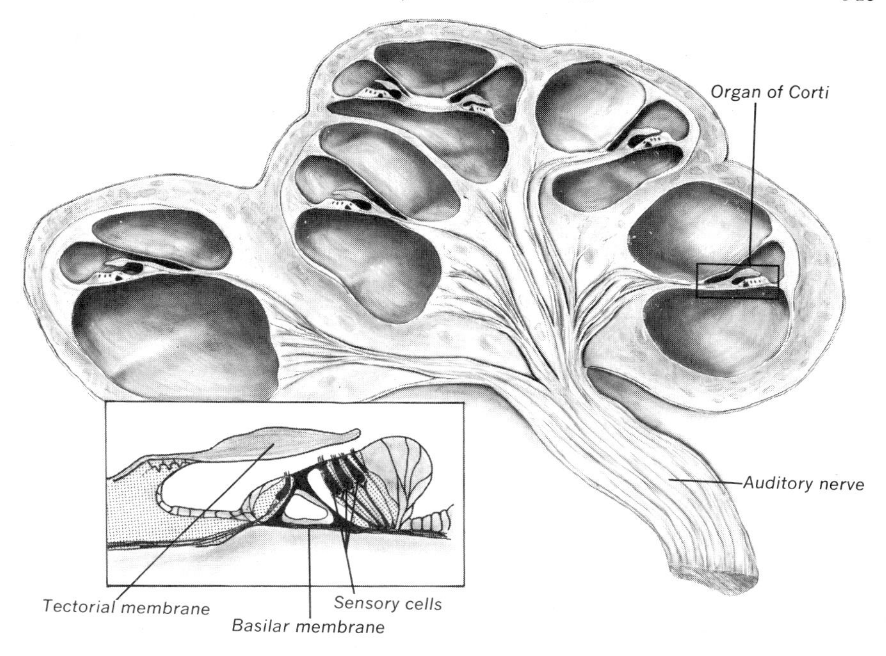

Figure 13–4. The tortuous path of the human cochlea near the middle. Insert: the organ of Corti.

The duct of the cochlea (Figure 13–4) is lodged within the two and a half coils of the bony cochlea and contains the organ of hearing (of Corti) first described in 1851 by the much-travelled "amateur" histologist, the Marquis Alfonso Corti. The tunnel within the bony cochlea is larger than the duct of the cochlea; thus there is room for one canal above and another below the duct. Two membranes divide the cochlea: the basilar membrane forming the base of the duct of the cochlea and separating it from the scala vestibuli above, and the delicate organ of Corti, which lies on the basilar membrane.

The footplate of the stapes fits into the oval window, making a watertight seal with the scala vestibuli. Vibrations of the footplate are transmitted through the perilymph to the vestibular membrane, then through the endolymph to the basilar membrane, and finally into the scala tympani to cause a corresponding vibration at a membrane-covered round window opening on to the middle ear cavity.

In prosimians and in New World monkeys, the floor of the middle ear cavity balloons out to form a structure called a tympanic bulla. In tree shrews and lemurs only, the tympanic ring is enclosed within the tympanic bulla. A bony external auditory meatus is present in tarsiers and in Old World families, but in the Lorisidae and New World families the bony meatus is absent or much shortened, and the tympanic ring is exposed on the side of the skull. These features are best verified by handling the skulls of various primates and learning how to distinguish them.

The Organ of Hearing

The organ of Corti consists of a series of pillars, or stiffened rods, arranged in a sloping fashion over a tunnel with specialized "hair cells" aligned on both sides (Figure 13–4). Supporting cells and other cells that "weight" the basilar membrane also constitute the organ of Corti. Some 3,500 hair cells are found on the inner side of the tunnel of Corti and some 12,000 on the outer side. Electron microscopy has shown that the hairs of each hair cell are arranged in rows.

These hairs project upward into a strange and almost structureless tectorial membrane. As the basilar membrane vibrates, the hairs are driven against the tectorial membrane, presumably stimulating the hair cell to initiate a nerve impulse. Early theories of acoustic physiology suggested that particular sounds were specifically transmitted in some distinctive manner by the various nerve fibers. Now we know, however, that each nerve fiber transmits discrete impulses of fairly fixed magnitude, but varying in the rate at which they follow one another. Neither the basilar membrane nor the organ of Corti has a uniform structure throughout their two and a half turns around the central cone-shaped modiolus of the cochlea. The basilar membrane is shorter, more strongly attached at its edges, and more lightly loaded with supporting and other cells at the end nearest the footplate than at the apex, or cupola, of the cochlea. Several investigators have suggested that there is a definite gradation in membrane length and tension and in the mass of cells it bears at any point along its length, so that the organ can respond to different vibrations in different regions.

The hair cells are innervated by peripheral fibers originating from bipolar neurons in a spiral canal lying within the central modiolus of the bony cochlea. The fibers pass out eventually to reach the inner and outer hair cells as naked axons. Electron microscope preparations show them ending by enclosing the base of a hair cell in an exceedingly thin, cuplike structure. Supposedly more nerve fibers pass to hair cells of the first and second coils of the cochlea duct than to the apical.

Nerve impulses from the organ of Corti travel along the peripheral fibers of the cell bodies of the auditory division of the eighth cranial nerve. They pass to the brain stem along central processes collected to form the cochlear nerve in the internal auditory meatus. The central processes relay in the medulla at the cochlear nuclei, and the majority of the ascending fibers originating from these nuclei cross over and pass up in the brain stem as a bundle (the lateral lemniscus) to relay in the inferior colliculus and the medial geniculate body. Impulses are relayed from the latter cell station to the so-called auditory-sensory center by the auditory radiation, which passes through the middle of the cerebral hemisphere in the posterior limb of the internal capsule to reach the cortical auditory reception area in the upper surface of the temporal lobe. An area of cortex immediately around the reception area helps to interpret the "meaning" of the stimuli received. Damage to one temporal lobe does not result in complete deafness on the opposite side. Another complication in analyzing the auditory pathways is the evidence that part of the vestibular apparatus is concerned with hearing, especially of loud noises.

Hearing in Man and Animals

The human auditory apparatus has considerable sensitivity and can tolerate a very wide range of intensity. It is sensitive to tones ranging from 20 to 25,000 cycles per second (now referred to as Hz, for Hertz), but young children can go as high as 30,000. The loss of high frequency sensitivity begins early; normal adult sensitivity reaches 16,000 and 18,000 Hz. The greatest sensitivity is from 2,000 to 3,000 Hz and has been measured at sound pressures as low as:

.000079 dynes/cm^2 at 2,000 Hz
.000063 dynes/cm^2 at 3,000 Hz
.000158 dynes/cm^2 at 5,000 Hz

If the intensity of a sound increases to about 1,000 dynes per sq cm, a painful sensation develops in the ear. Up to a point, the loudness of a sound is a subjective quality judged by the hearer; hence apparently normal individuals vary widely in their tolerance of loudness. A decrement in the perception of the upper limits begins at five years and continues from then on.

A number of mammals have an upward extension of the auditory range. Dogs can hear higher frequencies than man. Taking advantage of this, police and hunters employ whistles that are inaudible to man but audible to dogs. Bats can receive frequencies up to 120,000 Hz for echolocation. Some of these animals probably hear at still higher levels, but available instruments cannot calibrate accurately beyond this point. The ultrasonic echo-sounding instruments used to determine the depth of the sea are heard by cetaceans and pinnipeds. In behavioral testing, dolphins can hear up to frequencies of 150,000 Hz. Electrophysiological data indicate responses up to 250,000 Hz. However, the speed of sound in water is four times greater than in air, and we do not know how this interacts with information carried by high frequency emissions.

Despite their varieties and powers of song, the auditory sensitivity of birds is limited in frequency. However, they appear to have a remarkable ability to discriminate intensity. Whereas most of them cannot hear above 8,000 Hz, they can utilize information within that range better than we can. In place of the middle ear ossicles of mammals, birds have a columella, or bony rod, that makes contact with the drum and inner ear. The cochlea of birds is shorter and broader than that of mammals, but the nerve density is about the same. Amphibians, whose cochlea is considerably reduced in size, have a columella and many fewer hair (sensory) cells in their inner ear; hence they have an extremely limited frequency range. Fish, which have neither cochlea nor the organ of Corti, possess two auditory systems. (1) The lateral line organ, composed of thousands of individual sense organs, apparently is sensitive to frequencies below 1,500 Hz, and to near field movements. (2) The second auditory system is the Weberian ossicles attached to the swim bladder with frequencies to about 3,000 Hz. Some fish produce a number of sounds through vibrations of their swim bladder, and by tapping the fin rays against the wall of the swim bladder and grinding their teeth or jaws. The sounds produced are extremely varied, ranging between the frequencies of 20 to 4,800 cycles per second. Anecdotal reports claim that some fish can be taught to respond to specific sounds and even to discriminate pitch. Insects also have special auditory organs; in cicadas the sounds produced by the large tympanal organs in the abdomen are supposedly used to find members of the other sex.

Considerable information on the hearing of nonhuman primates has been gathered from behavioral studies. Apparently chimpanzees can hear up to 30,000 Hz, and the upper limits of several monkeys tested are known to be above those of human range. Prosimians, for example lemurs (Figure 13–5) and galagos (Figure 13–6), have superlative hearing acuity. A rich store of data is available on chimpanzees *(Pan)*, baboons *(Papio)*, rhesus monkeys *(Macaca mulatta)*, pig-tailed macaques *(M. nemestrina)*, crab-eating macaques *(M. irus,)* vervets *(Cercopithecus aethiops)*, capuchins *(Cebus)*, spider monkeys *(Ateles)*, squirrel monkeys *(Saimiri)*, marmosets *(Callithrix)*, galagos *(Galago)*, pottos *(Perodicticus)*, lorises *(Nycticebus)*, lemurs *(Lemur catta)* and tree shrews *(Tupaia glis)*. Nothing is known about gibbons *(Hylobates)*, orangs *(Pongo)*, and gorillas *(Gorilla)*.

Man is at the low end of the frequency scale; his sensitivity to high frequencies drops out first. Chimpanzees are somewhat better with an upper limit of approximately 30,000 Hz. Crab-eating macaques and pig-tailed macaques do better still, with an upper sensitivity of about 45,000 Hz. All of the other Anthropoidea tested are scattered between the last two ranges. The prosimians tested all fall within higher frequency ranges. Ring-tailed lemurs, for example, hear up to 75,000 Hz, and tree shrews 90,000 Hz. Thus, among the more primitive primates there is a gradual

Figure 13–5. The large, mobile ears of a tarsier.

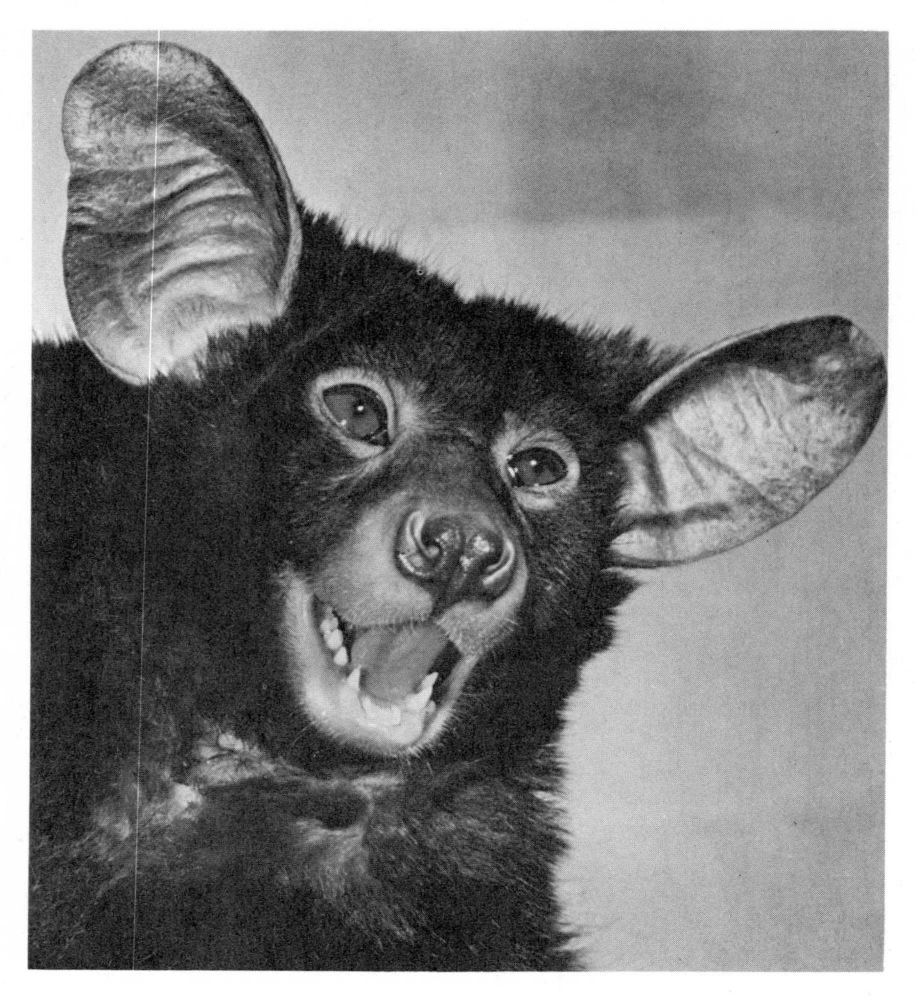

Figure 13–6. A black galago with fully extended, batlike ears.

increase in sensitivity at the upper range. Man distinguishes himself by displaying the best low frequency sensitivity and discrimination. Nocturnal prosimians, such as tarsiers (Figure 13–5) and galagos (Figure 13–6), have large ears that can be oriented toward special or startling sounds without any movement of the head in the same direction. When alert, they keep their ears erect and fully extended; moreover, the margin of tarsier ears is in constant undulating motion. One can surmise that the large external ears of these animals, like those of bats, are best adapted for the reception of high frequency sound.

Deafness

The two major anatomical causes of deafness are impairments or obstructions of sound conduction and defects in the perception apparatus (nerve deafness). Both types can occur in one individual. In nerve deafness, some damage is done to the receptors in the organ of Corti, the auditory nerve, or the hearing centers of the brain. Sometimes the bone surrounding the ear becomes involved by a little-understood affliction known as otosclerosis. This condition causes the footplate to become immovably fixed in the oval window and so effectively prevents the conduction of sound that deafness ensues.

No doubt deafness occurs in nonhuman primates as it does in man and other mammals. However, this condition can be studied only in captive animals. If they are deaf, feral animals have little chance of survival.

14

OTHER SENSORY SYSTEMS

Our senses perceive no extreme. Too much sound deafens us; too much light dazzles us; too great distance or proximity hinders our view. Too great length and too great brevity of discourse tend to obscurity; too much truth is paralysis.
(Blaise Pascal)

Like all other animals, man is aware of the world around him through his senses. Without them he could not see, hear, taste, smell, touch, feel pain or temperature, nor know whether he were standing or prone. He could experience neither happiness nor sadness, nor could he even be aware of his own living being. Such a creature would be no more sentient than a tissue or organ culture in a laboratory flask, if indeed it could be kept "alive" at all. An animal's working mechanisms, its desires, volition, and all of the other attributes of living functional beings, are dependent upon its senses. Not all senses are equally developed in all animals or even in the same animal. In fish, smell is the most highly developed sense; in birds and many mammals including man, sight; in bats, hearing. The degree of development of the senses is always reflected by those parts of the brain that perceive and interpret sensations and coordinate responses. Hence, in fish the olfactory lobes are the largest elements of the brain; in birds and most mammals, the optic centers; in bats, the acoustic centers.

Since sensory acuity is mainly subjective, most of our information about it pertains to ourselves. This is understandable because to obtain precise information from animals about their sensations is hardly possible. The paucity of laboratory data on the performance of sense organs in non-human primates is unfortunate, not only because such information would be interesting in its own right, but also because it might afford better insight into human sensory capacities. However, although exact laboratory tests are lacking, several keen observations have been made in the field. Ultimately, good field studies will guide the course of laboratory studies to be pursued. The empirical data gleaned from field observations indicate the survival value of sensory capabilities. The following extensive quotation from C. R. Carpenter on his studies of gibbons in their native habitat illustrates these points:

> The lack of sufficiently acute visual perception in judging size and distance would readily lead to falls resulting in broken bones or perhaps even death for the gibbon. The wide range of sound production must be correlated with suitable auditory reception. The arboreal gibbons like other arboreal animals depend heavily, also, upon proprioceptive reception and kinesthetic sensitivity.
>
> Visual capacities are apparently very highly developed in the gibbon. There are great demands made upon vision when a swift moving gibbon makes a 40 or 50 foot jump downward. The judgement, based on visual cues of distances and size of branches before and during jumping, must be reliable. The field worker soon learns by their alert reactions to him that distance vision in gibbons is very highly developed. In open forests or across clearings, one may disturb a group of gibbons when 400 or 500 yards away, presumably through visual processes. These primates are quick to respond to movement and the skillful observer moves very slowly when stalking groups, especially when certain animals are looking in his direction. Sometimes the reflections of light from the lenses of field glasses and camera have apparently produced flight reactions in groups being studied. Gibbons do not seem to respond as readily to form as to movement. Several times, while I lay on my back and was perfectly quiet, gibbons came near and almost directly over me in the trees at close range. At times they seemed to be otherwise occupied and did not anticipate the presence of an observer and then, with proper precautions, they could be approached quite closely or they might come within about 30 feet of the observer. Nevertheless, in several instances, newly constructed blinds caused avoidance reaction, showing that gibbons tend to avoid strange objects in their environment.
>
> The vocalizations produced by gibbons are well within the range of

the human voice; the pitch usually falls under 2,000 vibrations per second. The volume is such as to make all sounds audible to the human ear. The variety and complexity of the sounds produced may also indicate a highly developed auditory receptor capacity in gibbons. There are, however, probably fine sound quality differences such as those which lead to the recognition of familiar or strange individuals, which make considerable demands on the gibbons' auditory discriminations and these qualities are largely missed by field studies.

It has been suggested that gibbons have a poor sense of balance. There is some support for this opinion. Being brachiators, they swing beneath supports and hence do not need the same proprioceptive and kinesthetic sensory capacities as are needed by some other mammals or by birds. When they walk on two legs, the gibbons' arms are usually held up and extended to aid in balancing. Other primates, including the chimpanzee, do not show quite this form of behavior but it is also true that most other primates show a higher degree of quadro-manual rather than bi-pedal or bi-manual (brachiating) locomotion. The demands made on the balancing mechanisms of a walking gibbon are great, particularly since the visual-motor coordination habits differ for walking and brachiation and brachiation is its principal form of locomotion.

Certain kinds of contact sensations produce in gibbons, as in other types of primates, states of complete submission and relaxation. Tame gibbons may be almost put to sleep by grooming through their fur and rubbing parts of their bodies. Wild specimens can be calmed by gently brushing or currying their bodies. Through the contacts involved in making vaginal examinations or taking vaginal temperatures, adult captive specimens may readily be trained to present for these examinations without restraints of any kind. The social greeting and grooming responses show the significance of contact reception for this primate.

There is little information to be had from field studies as to the gibbon's olfaction and taste. Gibbons sometimes smell food before eating it. Nevertheless, with almost every food eaten by gibbons, I have experimented to see what it tasted like. About 80 percent of the foods are too bitter and sour for human taste. Being desperately thirsty once, I tried eating the same type of grapes on which gibbons had been eating. The fruit looked luscious and juicy, much like the southern muscadine grape. I ate several; a sharp sour flavor turned slowly to a strong bitterness and this to a burning sting which tormented my parched throat for hours. However, several fruits eaten by gibbons, such as mangoes, plums, some types of figs and a tough-skinned yellow fruit may be eaten by humans. In general the natural foods eaten by gibbons tend to be sour and bitter to human taste.

As for the gibbons' olfactory sense, too, I can report only that I have seen individuals when temporarily lost from their groups behave as if trying to find the direction which the group had gone by sniffing branches. This sensory modality may also stimulate social interactions.

From such detailed notes, we can see that gibbons possess keen visual perception, distance vision, ability to focus and to judge spatial relationships, and awareness of movement. However, they do not appreciate the significance of shape although they can determine the presence of a strange object. Hearing is probably well developed; as in other primates, the sense of balance less so. Cutaneous tactile sensibility also seems to be acute. Little is known specifically about their olfactory and taste reception, but both these sensory capabilities are present in primates.

The Senses

Environmental disturbances continually bombard all living organisms. Certain cells (sensory cells), "tuned in" to particular changes in the environment, respond by converting them to stimuli. Aggregates of similar sensory cells on the surface of the body form the receptor or sensory organs, which convert the excitations from the stimuli to nerve impulses. The nerves attached to the receptor organs convey these impulses to the central nervous system, which in turn coordinates the proper responses of the body to the stimuli and interprets the impulses as sensations. Protected deep inside the body, the central nervous system keeps in contact with the environment by these surface "feelers," the major sensory organs serving as the windows of the brain.

The constant shower of information that reaches the human brain can be classified into two main categories: one tells man about himself—the position of his limbs and the state of his organs; the other, about his surroundings. Only some of this information reaches his consciousness; much of it goes directly to the lower centers in his brain. The information from man's environment is channeled through the skin, which records the sensations of touch, pressure, temperature, and pain, and through the special senses of sight, hearing, smell, taste, and balance, perceived by the eye, ear, nose, tongue and organs of balance, respectively. The proprioceptive sense organs (L. *proprius,* self) found in tendons, muscles, and viscera send the brain information about the body's orientation in space. Nerve impulses originating in the sense organs are conveyed along pathways of nerve fibers to certain regions of the central nervous system that are capable of perceiving and interpreting the signals.

Innervation of the Skin

Most of the surface of the skin can distinguish all the modalities of pressure—touch, pain, and temperature—but these sensations are most keenly felt in such areas as the fingertips, lips and perioral areas, perigenital surfaces, axillae, and external genitalia. For many years there has been wide acceptance of the German physiologist M. von Frey's suggestion that there is a punctate distribution of sensory "spots" on the skin, each "spot" being equipped with one of four specific types of cutaneous nerve endings, each of which is capable of recording one of the four primary modalities of cutaneous sensation—touch, cold, warmth, and pain.

Tactile spots are said to be equipped with encapsulated oval corpuscles inside which a knot of nerve fibers comes together into the spiral nerve that emerges from it. One type of sensory end-organ to be described in the skin is called Meissner's corpuscle. Even though tactile spots are present nearly everwhere on the skin, these endings are found only on the friction surfaces and are particularly numerous on the tips of the fingers. Numerous studies of Meissner corpuscles with histological and histochemical methods, electron microscopy, and wax reconstruction have shown that at birth they are more or less elliptical and profuse. They become progressively fewer from birth to old age, and the surviving ones become longer. As a result of this lengthening process, the corpuscles become variously coiled and lobulated. The shape and size of these organs reflect the type of work performed by the hands and are larger and more complex in hands that do great amounts of manipulative work. Certain nerve fibers in the skin, believed to terminate as knobs and discs on epidermal cells, have been called Merkel's corpuscles after the German anatomist who first described them in 1880.

"Cold spots" on the skin were said to be subserved by groups of minute cylindrical bulbous corpuscles called Krause's end-bulbs. These are surrounded by a thick capsule, inside which a complicated plexus of thin fibrils surrounded by a semifluid substance converges into a relatively thick fiber that emerges from the bulb. Somewhat different corpuscles, described by an Italian histologist, A. Ruffini, in 1898, are thought to be associated with the sense of warmth. These oval or spindle-shaped, closely woven networks of nerve fibrils are intermingled with small bundles of connective tissue. Ruffini observed so many other encapsulated endings of intermediate form that he deemed it futile to attempt to classify them all.

The recent application of more refined techniques suggests that many of the end-organs painstakingly described in the past are artifacts. Clearly recognizable encapsulated end-organs are found in mucocutaneous junctions such as the border of the lips. In the genitalia they are known as genital corpuscles. Curiously, none are found in the buccal and oral mucosa or on the genital mucous membranes. A variety of "tactile" corpuscles, encapsulated or semiencapsulated, are found on the tongue and gums. As a rule, all encapsulated or organized end-organs are found only in genuinely glabrous skin. Hairy skin has only the end-organs around the hair follicles (described in chapter 7) and an apparently simple system of nerve networks. In the nipple of female breasts, numerous sensory end-organs are found only at the tip; there are more at the sides of the nipple or on the areola. The lactiferous ducts, however, are richly entwined by nerves, which no doubt record deep sensory stimuli. These regions, though commonly known as erogenous, are sensitive only in a limited and special way.

There is no concurrence about the specific number and types of cutaneous sensory organs and their function. Impelled by various degrees of zeal and abetted or hampered by the techniques they used, different investigators have found different "types" of organs. In the end emerges the basic architectural plan of the cutaneous sensory mechanism: in addition to the organized end-organs just described, numerous superimposed nerve nets exist throughout the dermis and become progressively finer and denser in the higher levels until they terminate just beneath the epidermis. Sensory nerves stem from a rich variety of end-organs that differ structurally and for each of which a specifically different function has been hypothesized. Today, fortunately, investigative emphasis is on the similarities, not the differences, of these organs, similarities that are evident at every stage—during development as well as in adult tissues, in functional properties, and in phylogeny.

The basic mechanism of the cutaneous sensory nerves is relatively simple. Nerve fibers from neurons in the dorsal root ganglia go to the dermis to form the complex, interconnected networks mentioned above. Thus, a stimulus applied anywhere on the skin triggers not a single response but a pattern of them. Such networks are present in all skin, glabrous or hairy, in cutaneous or mucous surfaces. Oddly enough, mucous membranes and the cornea, though highly sensitive areas, contain no recognizable organized sensory organs. Hence, the dermal nerve networks present must be the principal sensory receptor mechanisms. In hairy skin, these networks are less apparent since they are associated with hair follicles around which they are modified into end-organs of different complexities. All of these end-organs also represent modifications of the basic network. Instead of being loosely associated with epithelial structures, the free nerve filaments are rolled up into coils or balls and are surrounded by specialized capsules. Thus, notwithstanding apparent structural dissimilarities, all end-organs are seemingly interrelated and may be structural modifications of the basic network. The encapsulated end-organs are found in glabrous areas of skin, where the understructure of the epidermis is always complex. They are not usually present in hairy skin or where the epidermis has a flat undersurface.

Skin receives so many modalities of sensation that it would seem incredible for each of them to be subserved exclusively by a specific end-organ, anatomically distinct from the others. The organized end-organs found in glabrous skin are probably attuned to the reception of acute touch. If special nerve endings were fashioned to respond only to specific stimuli, they should be present on the skin wherever those specific sensations are recorded and not in the few localized areas described. Since all sensory nerve endings are related to the sensations of touch, perhaps the same nerves transmit other sensations such as temperature and pain. Hence one may, with some caution, conclude that sensory end-organs are modified structurally by the surrounding region rather than by the function they subserve.

When cutaneous sensory nerves have been damaged, the ability to transmit certain sensations, such as pain, reappears first during recovery and afterward gradually follows all other modalities of sensation. Each sensory organ in the skin and the nerve network is equipped with more than one nerve fiber. The multiple innervation of these end-organs comes from the intricate network of nerve fibers that lies beneath the epidermis. This network allows a spatial summation of stimuli that affect the skin and assist in localizing the stimulus and in grading its intensity. During the early phases of the recovery, there are periods when only one fiber has reinnervated an end-organ, which thus lacks much of its normal discriminative potential.

In nonhuman primates cutaneous sensory endings are essentially similar to man's. The distinctive end-organ characteristic of all primates is Meissner's corpuscle on the friction surfaces of the hands or feet. The homologous structures in nonprimate mammals are twisted, sausage-shaped, often elongated elements similar to those found in mucocutaneous surfaces, which R. K. Winkelmann has called mammalian end-organs. In a few primates, Meissner's corpuscles are found in certain other specialized areas: in the howler, woolly, and spider monkeys they are numerous on the ventral friction surfaces of the prehensile tail, which is glabrous and sculptured with dermatoglyphs. In gorillas and chimpanzees, they are also found on the dorsal surface of the third phalanges of the middle three fingers. These animals, being quadrupedal, walk on the backs of their fingers, the knuckle pads, which are also glabrous and have distinct dermatoglyphs. The naked patch of skin around the ischial callosities of macaques and baboons also contains organized sensory end-organs that resemble mucocutaneous bulbs. Some primates, among them baboons and macaques, have on the skin of their face and chest a curious mixture of innervation that combines the features of glabrous and hairy skin. Unfortunately, very little is known about cutaneous sensibility and discrimination in primates.

Pacinian Corpuscles and Other Proprioceptive Endings

A specialized cutaneous nerve end-organ, about whose function there has been less controversy, is a capsulelike lamellated corpuscle in the deeper parts of the skin. Observed first in the eighteenth century by

A. Vater and then by F. Pacini, the lamellated corpuscles are now known as the Vater-Pacinian corpuscles. These oval or lobulated bodies consist of a central core surrounded by a series of concentric layers arranged like coats of an onion. The central body is a dense plexus of nerve filaments converging into a nerve that emerges at the base of the corpuscle. Vater-Pacinian corpuscles are usually found in the deep parts of the dermis, particularly of the hands and feet and near joints. They are also numerous in some of the viscera and especially in the mesenteries. Good experimental evidence has revealed that stimulation of these structures gives the sensation of pressure and thus subserves the perception of movement. Vater-Pacinian corpuscles are already well developed by the eighth month of fetal life, but whether they function sufficiently for the fetus to be aware of the compression to which it is subjected at childbirth is not known.

Special sensory end-organs in voluntary muscles supply information about the tension exerted by the contraction or stretching of a muscle. These endings help the body to achieve and maintain balance and equilibration and supply information about the necessary degree of power to exert behind movements. Known as neuro-muscular and neuro-tendinous spindles, these end-organs play an important role in the maintenance of the complex neuro-muscular coordination of the body. They consist of a multilayered connective tissue capsule around a core of specialized muscular (intrafusal) and tendinous fibers. These fibers resemble embryonic muscle fibers but are smaller than the surrounding muscle fibers and have more cytoplasm and many nuclei, besides being less clearly striated. The spindles are supplied by sensory and motor nerve fibers that respectively convey impulses away from and toward the end-organs. The sensory fibers originate as fine nerves in the form of a spiral or annular plexus around the intrafusal fibers.

End-organs, then, constitute the mechanisms whereby the body receives information from the outside and from within. Though believed for a long time to be relatively stable structures, they change in response to mechanical stress, trauma, and aging. When a sensory nerve is severed, the end-organs degenerate but form again when reinnervation occurs. End-organs are the feelers or tentacles maintained on the surface and in the deeper parts of the body to keep the brain informed of the body's condition.

Visual Apparatus

Eyesight is probably man's most important sense organ. Without it, despite his large brain, he is nearly incapable of surviving unaided. Yet, the brain is the organ that distinguishes the various visual images and assigns a meaning to each. Man's two eyes are sensitive to light, discriminate intensity and color, and allow visual space perception. However, the stereoscopic superimposition of two images, which characterizes the eyes of nearly all primates, is a function not of the eyes but of the brain, which must integrate ocular functions.

In hypothesizing about the evolution of eyes, we must remember that sensitivity to light is not confined to any one major group of animals; all organisms, even unicellular ones, are thus sensitive. The development of screening pigment in invertebrates resulted in the isolation of particularly light-sensitive cells, found as raised patches, or pits, on the animal's surface. Such primitive eyespots respond only to shadows that pass across the sensitive area. Later in evolution, a transparent lens that focuses light onto the eyespot developed from ectoderm. In insects, which are unable to move their eyes, the development of compound mosaic structures allows a wide range of visual modalities. In vertebrates, the basic arrangement of the eyes evolved early and became modified with changes in habit and environment.

Man's freely movable eyeballs are set inside complete sockets (orbits) in his skull (Figure 14–1). The closure of these sockets from the temporal fossae behind by a postorbital bar in all living primates and by complete bony closure in all living anthropoids is associated with both eyes being directed forward. This arrangement serves structural and protective functions: stereoscopic vision improves, for example, as the visual axis of one eye approaches that of the other or as the angle between the visual axes is reduced. As one studies the location of the eyes in mammals, he becomes impressed with the fact that in predators the eyes are located in the front of the head, whereas in "hunted" animals they tend toward the sides. The forward position of the eyes sharpens attention, concentration, and coordination and allows better judgment during attack. The side position gives the animal a wide range of vision without having to turn the head, an adaptation that seems necessary for survival when an animal is being pursued.

Thus, the position of the eyes seemingly first subserves survival and only then, specific tasks. But if this is true, how explain the predatory characteristic of the forward position of the eyes in all primates? Among these, only man is a true hunter. Tarsiers, which are also predatory, have their large eyes fixed in their orbits like those of the owls and, like owls, must turn their heads in the direction of objects. The forward placement of the eyes must therefore serve other still unknown ends besides that of depredation; or perhaps it is a fortunate association with the reduction and downward displacement of the snout.

Six stout extrinsic voluntary muscles move each eyeball (Figures 14–1, 14–2): four recti, and two oblique. Three of the rectus muscles and one oblique are supplied by the third (oculomotor) cranial nerve. The superior oblique, supplied by the fourth (trochlear) cranial nerve, causes

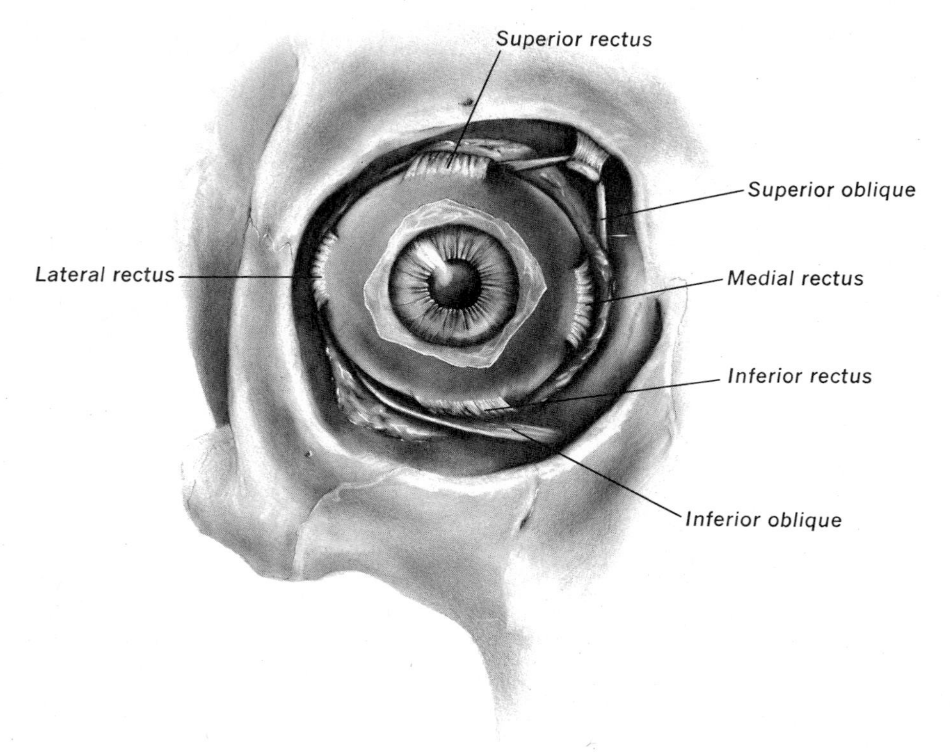

Figure 14–1. A right human eyeball in place in the bony orbit showing the six extrinsic muscles.

the eyeball to look downward and outward, giving a pathetic expression. The lateral rectus, innervated by the sixth (abducent) cranial nerve, allows the eyeball to look laterally. It makes possible the sidelong glances of lovers and was at one time given the frivolous name of *musculus amatorius.*

The human eyeball is an irregular spheroid averaging about 24 mm in diameter and composed of three coats: an outer sclera that is transparent in front at the cornea; an intermediate choroid modified in front at the iris with its central opening, the pupil; and an inner neural coat, the retina (Figure 14–3). The sclera is the tough, fibrous, protective covering of the eyeball. In lower vertebrates, particularly reptiles and birds, the sclera may be further reinforced with cartilage or bony plates. The cornea is a transparent multilayered structure equal in area to one-sixth of the total surface area of the human eyeball. Consonant with the fact that man is a purely diurnal animal, this means that only seventeen percent of the ocular globe in man is specialized as the transparent

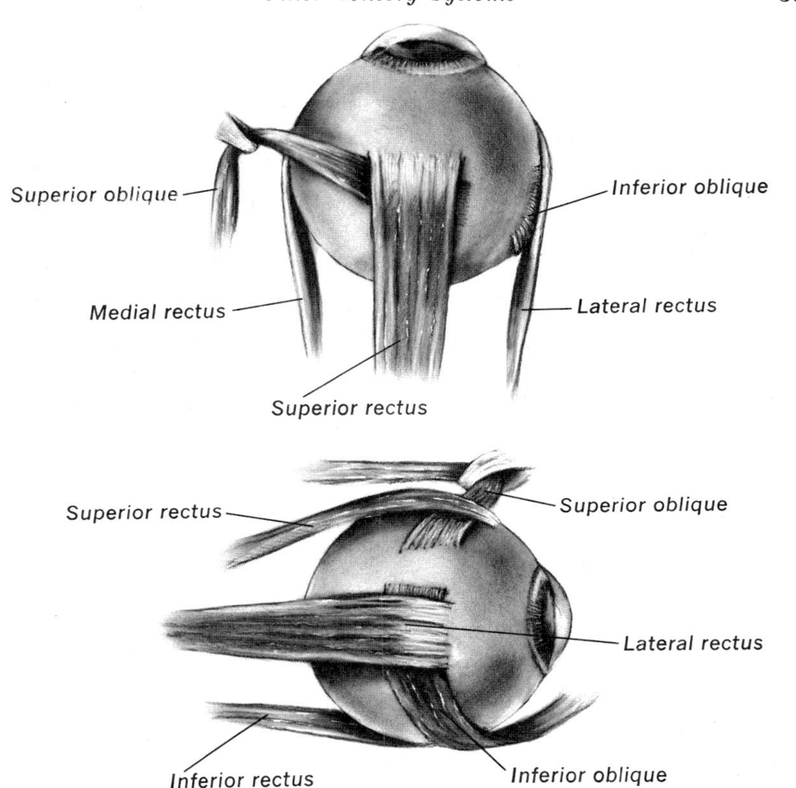

Figure 14–2. A right human eyeball drawn in two views to show the arrangement of the six extrinsic muscles.

cornea. On the other hand, in nocturnal primates, mostly the prosimians, as much as thirty-five percent of the globe is specialized into cornea. However, even the diurnal primates have a relatively larger cornea than man. Being convex, the cornea forms a powerful lens. Abnormal inequality of the curvature of the cornea is called astigmatism. A marked natural astigmatism is found in whales and seals. Birds, sheep, chimpanzees, and a few other animals have a visibly pigmented cornea. No blood vessels are found in the cornea.

The middle coat, the choroid, is deeply pigmented and well supplied with blood vessels. Fish have a "silvery membrane" in the choroid that gives their eyes a metallic luster. Many mammals, but not man, have a brilliantly colored layer on all or part of the choroid, the tapetum, which causes the green reflections from the eyes of nocturnal animals. In carnivores, cattle, horses, and whales, the tapetum produces different-colored reflections of strong intensity. The function of the tapetum is to reflect back onto the receptor organs in the retina the light that passed through them without being absorbed. Without a tapetum, this light would be absorbed by the choroid pigment. Among the primates only the nocturnal douroucouli, or night monkey (*Aotus trivirgatus*), from South America and the nocturnal prosimians have a well-developed tapetum.

The front part of the choroid forms the iris, a contractile diaphragm around a central pupil that may be round as in man, oval as in seals, or slit as in cats. The different colors of the iris are the result of the arrangements of pigment within it: blue eyes have pigment only on the back surface, brown and black eyes have pigment all through it. Smooth muscle in the iris can constrict or dilate the pupil. Most primates have a circular pupil, except the lorises, lemurs, and tarsiers. In some lemurs and galagos the pupil is vertically oval; in tarsiers it is horizontally oriented. Lorises have vertically oriented slit pupils.

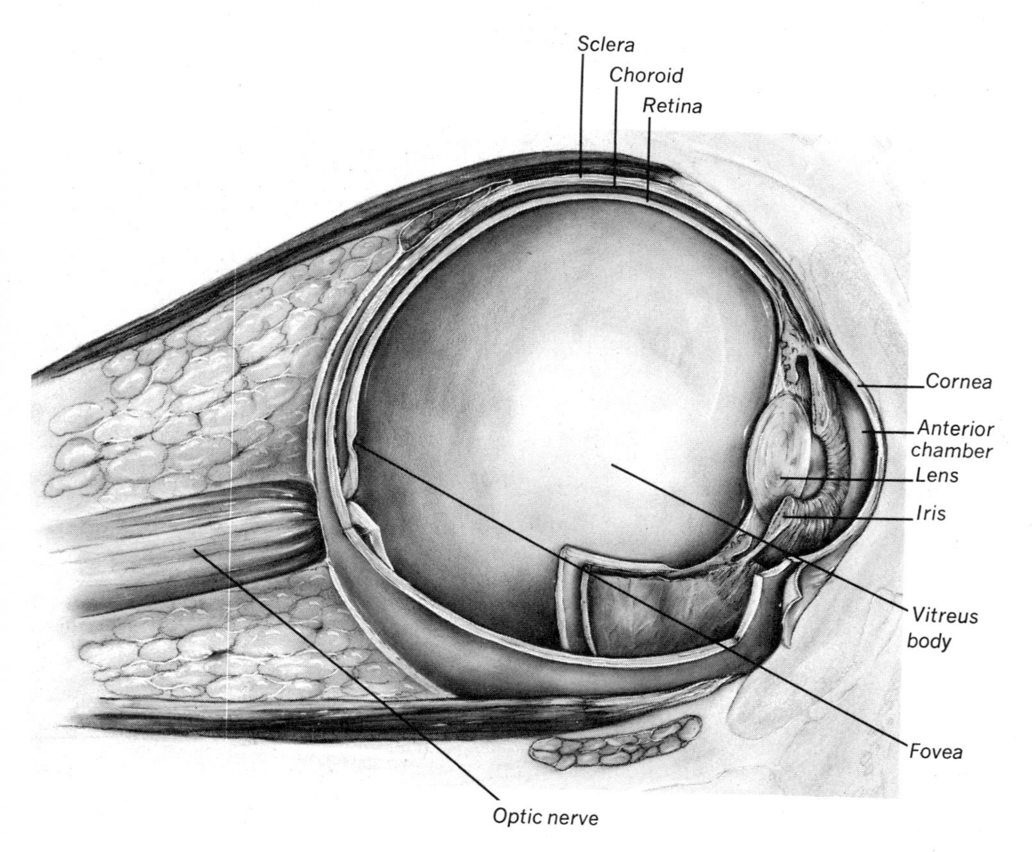

Figure 14–3. A human eye dissected to show its three coats and some of the other structures.

The iris forms a diaphragm immediately in front of the vitreous lens, which is kept in place by fine suspensory ligaments attached to the ciliary body (Figure 14–3). Ciliary muscles hold both the iris and the lens in place. The contraction and relaxation of these muscles control the tension exerted by the suspensory ligaments. The lenses of fishes, like those of rats and mice, are almost spherical. In primates the posterior surface of the lens is convex but the anterior is flattened. The human lens, about 10 mm in diameter, develops from a "placode" of ectoderm, becomes enclosed in a capsule, and thickens as successive generations of lens fibers are added to its surface. The lens divides the inside of the eyeball into an anterior and a posterior chamber in front, the two separated by the iris, and a vitreous body behind. In cephalopods (squid and octopus) the cornea is perforated and the anterior chamber filled with sea water. The two front chambers of mammalian eyes contain a clear aqueous humor, which in man consists of ninety-eight percent water and has a refractive index of 1.34. The aqueous humor, coming from ciliary vessels behind the iris, circulates through the pupil and is drained at the edge of the anterior chamber by a canal in the corneoscleral junction named after Friedrich Schlemm, a Berlin anatomist who described the canal in 1830. Birds have a very large anterior chamber and fishes a very small one.

In the space between the lens and the retina is the vitreous body, a translucent, jellylike substance. Its composition is similar to that of the aqueous humor, but it contains a mucoprotein that gives it its jellylike consistency. During fetal life the vitreous body is traversed by the hyaloid vessels that supply the developing lens. These vessels later degenerate and the lens is nourished by the aqueous humor.

The Retina and Color Vision

The retina is the neural, light-sensitive layer of the eye, which, early in embryonic development, grows from a cup-shaped outgrowth of the forebrain vesicle. This means that the optic "nerve," which conveys impulses from the retina to the brain, is actually a part of the brain and not a nerve in the usual sense. The retina lines about four-fifths of the inside of the human eyeball and consists of an outer pigmented layer against the choroid and an inner light-sensitive layer composed of neurons modified into rods and cones. The rods contain a special pigment called visual purple (rhodopsin). Processes from the rods and cones relay impulses stimulated by light to processes of bipolar neurons that compose a bipolar retinal layer. The distal neurons from this layer reach neurons in a surface ganglionic layer. Nerve fibers from this layer collect impulses from all over the retina at the optic disc where they enter the optic nerve. The optic disc is the point of exit of neurons on their way to brain centers and, having no receptors over it, is called the "blind spot."

Strangely, the light-sensitive elements of the retina are directed away from the light, which must traverse all of the retinal layers before reaching the receptors. This has been referred to as an inverted retina. Cones are more numerous in diurnal animals, rods in nocturnal ones. Retinae possessing rods only are present in all Lorisidae, in some Lemuridae, in *Tarsius* and in the night monkey. Estimates indicate that some 7 million cones and 125 million rods are found in a human retina. Yet only about one million nerve fibers are in the optic nerve. This means that many receptors, particularly those at the periphery of the retina, share a common pathway to the brain.

A yellowish area near the posterior pole of the retina, the macula lutea, has a depressed central part, the fovea centralis, which contains closely packed cones and no rods. The macula and fovea are the areas of keenest vision and subserve functions in binocular and stereoscopic vision. They are well developed only in man, apes, and some monkeys; prosimians, except *Tarsius,* have no true fovea.

The visual sense is relatively acute in most primates. G. B. Schaller believes that in the gorilla, keenness of sight, as well as of hearing and smell, is roughly similar to that of man. The great American primate behaviorist Yerkes believes that the orang has rather poor eyesight, but he did not have experimental data. Observations on the visual apparatus have been made mostly in the macaques, which are communicative and make good experimental animals. Macaques enjoy colored objects and have a superior quality of vision and visual attention. Their discrimination of brightness and color and their visual acuity and form vision are perhaps as good as those of man. The receptive elements of the retina display various degrees of size and shape in different animals. As a rule the larger they are, the coarser is the image they see. Primates and birds, which have very fine receptors, have a higher quality of vision than those animals that have coarse ones.

A human eye supposedly can distinguish 160 different colors. Color vision is possessed by some bony fish, turtles, lizards, birds, and by all primates except the nocturnal prosimians. It may be present in some snakes and marsupials, but is slight and poorly developed in cats and dogs. Color vision is usually attributed to the cones of the retina. Thomas Young propounded the theory, subsequently developed by H. L. F. von Helmholtz in the nineteenth century, that three primary receptors are found in the retina for red, green, and violet. A second theory postulates six types of receptors. Experiments have confirmed that several different types of receptors exist in a mammalian retina, each with certain spectral responses. No cytological evidence has been revealed, however, of different types of cones, although the presence of three different pigments has been demonstrated in single monkey cones.

Primates depend so much upon their sight that they can be designated "visual animals." Only man, apes, and monkeys have a well-developed true central fovea in their retina, specialized for refined vision. Apparently, this makes possible the careful manipulation of objects with hands and feet under visual guidance. The optic centers of the brain develop precociously in human and other primate embryos and are considerably larger than those for hearing and speech. This accords with the fact that refined vision in man is superior to his auditory sense. Perhaps hearing is the most important auxiliary sense for human survival.

Blood supply to the retina The retina of lower vertebrates has no blood vessels; true retinal vessels are found only in mammals. Blood supply may be derived from the choroid (certain rodents, ungulates, and bats), or from a pleated vascular membrane (pecten) in the vitreous (birds), or it may ramify close to the surface (snakes). In primates the vessels leave the optic disc from a central artery that traverses the substance of the optic nerve. Numerous branches spread over the surface of the retina and penetrate it. In man and the other primates, the entire retina (holangiotic) is supplied with vessels. When the inside of the eyeball is examined with an ophthalmoscope, the back looks red because of the blood in the choroid. The tapetum in nocturnal primates alters the color.

Visual pathways The nerve impulses that leave the retina follow somewhat complex pathways in the brain. The two optic nerves pass back to the base of the brain where they come together at the optic chiasma (the Greek letter chi is cross-shaped) and leave by two optic tracts. All vertebrates have a chiasma. In submammalian vertebrates and marsupials, each optic nerve crosses over completely to the opposite side. In lower mammals only a small proportion of the fibers remains on the same side, and in man and other primates only about one-third or less cross over; those from the lateral part of the retina do not cross. About one-half of the fibers from the macula cross. Most of the fibers in the optic tract pass to the lateral geniculate body of the brain, which, together with the medial geniculate body, has developed from the thalamus and lies close to it. These are the relay stations for sight and hearing.

In primates the lateral geniculate body is large and layered. Six layers receive fibers from sharply defined retinal areas. Fibers from the fovea have wide representation in that they pass to three laminae. Some fibers (fewer in man than in most animals) pass from the optic tract to the nucleus of the oculomotor nerve. These connections allow reflex eye movements—movements of accommodation—and alterations in the degree of constriction of the pupil.

Fibers pass from the lateral geniculate body to the visual cortex through a conspicuous tract, the optic radiation, located mainly on the inner side of each occipital lobe of the brain. The visual cortex, or the striate area of the brain, is largely and finely convoluted in man, with a surface area of up to 4,000 sq. mm. This is the true visuo-sensory area around which lie the association areas called visuo-psychic and visual memory centers. As images from each eye are superimposed in the striate area, they give rise to binocular vision. Association and other types of fibers pass from the striate area forward to all parts of the cortex in a most complex and so far undetermined pattern.

Taste and Smell

In mammals the receptors for taste are primarily localized in the tongue, where aggregates of sensory cells form taste buds. These are congregated in the so-called vallate or circumvallate papillae of the tongue, but many are also found singly or in groups in the foliate and fungiform papillae and on the soft palate, the epiglottis, and the pharyngeal and laryngeal walls.

The distribution of taste buds in the various classes of vertebrates shows some interesting adaptations. In larval lampreys they are found on the pharyngeal lining, but in adults they are outside, on the surface of their heads. In most fish taste buds are located on the outer surface of the body, but in sharks and rays they are found in the lining of the mouth and pharynx. Amphibians have them confined to the lining of the mouth, tongue, and pharynx. Among the reptiles, snakes and crocodiles have taste buds far back in the mouth cavity, the only part of the mouth that is not highly cornified. Birds usually have them on the floor of the mouth along the lower jaw.

On the mammalian tongue, the vallate papillae are found at the back of the tongue usually separating the posterior third from the anterior two-thirds. The number of these papillae varies in different species; in man about nine of them are arranged in a V-shaped row, the apex of the V pointing posteriorly. In other primates each species has a characteristic number arranged in a unique grouping. Each vallate papilla is surrounded by a moat, at the bottom of which open the ducts of certain glands whose watery secretion keeps the moat flushed clear of food particles and prevents the taste buds from becoming encrusted with debris.

Taste buds are barrel-shaped organs composed of spindlelike

gustatory cells surrounded by palisades of supporting cells. The free surface of each taste cell ends in a fine hair; these hairs are all pulled together in a small brushlike fashion inside the pore of the taste bud. Gustatory cells are in contact with delicate nerve fibers that converge at the base of each taste bud, where they leave in a bundle to join branches of the seventh, ninth, and tenth cranial nerves. The ability to discriminate between sweet, bitter, salt, and acid has generally been attributed to specific taste buds, whereas the corresponding sensations are said to be mediated by certain enzymes in the gustatory cells. No evidence supports the existence of four kinds of taste buds, but different patterns of enzyme systems in the taste buds may discriminate the different modalities of taste. The color of foods, the consistency, temperature, moisture, texture, and other factors are undoubtedly important in taste appreciation.

Man's preferences for certain foods and total rejection of others is to a large extent acquired. Even in our occidental civilization such foods as raw oysters, fish, or eggs, relished by some gourmets, are emetic to others. In older people the taste buds at the top of the papillae perish. The few that remain tend to collect at the sides of the papillae. Perhaps this explains the preference or tolerance of older people for more highly seasoned foods than is shown by children.

Gustatory impulses are conveyed up the brain stem in the tractus solitarius, which relays them to some still undetermined part of the cortex. The full appreciation of gustatory perception is intimately related to olfactory sensation and the two often cannot be separated. Southern Mediterraneans, Orientals, and Latin Americans use great quantities of garlic, hot peppers, and outspoken spices, whereas more conservative northerners consider such delicacies as "poisonous." Similarly, one cannot equate human taste preferences with those of other animals or even of higher primates, as was shown by the observations on gibbons quoted early in this chapter. Whereas some foods may be acceptable to both, many are acceptable to one but not to the other. The sweetness of sugar is a favorite of most animals and man, who also responds favorably to most nonnutritive compounds that are sweet. Other animals, however, have definite preferences for specific compounds and not for others. Primates are, as a group, omnivorous. Most will eat fruit, roots and tubers, nuts, leaves, flowers and parts of tender shoots, even bark and grass, birds and their eggs, lizards and frogs, crustaceans, and various types of insect. Baboons are known to eat meat. A few forms exist in the wild on a restricted diet. Asian langurs, African colobus monkeys, howler monkeys, and indris are leaf-eaters by preference. Some prosimians are largely insectivorous. Just how much taste is involved in preferences and selection is difficult to assess.

An interesting hereditary characteristic of man is his ability to taste phenyl-thio-carbamide (PTC), an organic substance that some persons (tasters) find insufferably bitter, and others cannot detect at all (nontasters). A single gene probably produces this characteristic, and the taster gene appears to be dominant to the nontaster one. Although the proportion varies in different populations and races, about seventy percent of human beings are tasters. Monkeys and apes can also be divided into tasters and nontasters, as B. Chiarelli has shown. When all primate species have been tested in significant numbers, the results should be revealing.

Smell is probably one of the most ancient of the primary senses. Fish, with gigantic olfactory organs, have olfactory lobes that make up the largest part of their brain, whereas birds, with practically no sense of smell, have very small olfactory lobes. The olfactory sense is highly developed in mammals, among which carnivores and ungulates seem to have the largest and best developed olfactory organs. These sensory receptors (olfactory epithelium) are located in the back of the nasal passageway. Man, with a relatively small expanse of olfactory epithelium, is called microsmatic. Dogs and many other mammals that live more by their keen, discriminatory sense of smell than by visual perception, are called macrosmatic. Some cetaceans, commonly said to be devoid of a sense of smell, are anosmatic. Olfactory sensitivity in macrosmatic mammals, and even in microsmatic man, is remarkably high. Quantities of certain substances so minute that they cannot be detected by physical instruments can still cause a strong sensation of smell. The ordinary human nose can readily detect one part of vanillin by weight in 10 million parts of air, or one twenty-three millionth of a milligram (23 nanograms) of mercaptan in a liter of air.

Since man is a microsmatic animal, the many subtle odors in his environment are seldom explicit to him, or drift in and out of his consciousness, and are mostly out of reach. Perhaps we have been indifferent to the science of chemical systems of communication, highly developed among some animals, because we are so insensitive to odors. Such macrosmatic animals as deer, dogs, rodents, and others have systems of glands that release into the environment chemical substances that convey general or very specific information to other members of their species, and sometimes to those outside their species. These substances in complex are known as pheromones (Gk. *pherein,* to carry, *hormánein,* to excite). The most exquisitely developed systems of the pheromones are found among the insects in which they serve as attractants, phagostimulants (Gk. *phagein,* to eat), alarm and trail markers, and primarily sexual attractants. Male moths can track down females miles away by following their scent. Most insect pheromones are so species-specific that they have been identified by the name of the species.

In many animals the olfactory discriminative accuracy reaches the level of molecular differentiation. The specificity and accuracy of response are based not only on the animals' selective reactions to a particular signal, discriminated against a background of other active signals, but also on the all-or-none efficiencies of these chemicals on highly selective receptors.

The existence of individual and species-specific odors has been shown by the ability of some animals to discriminate between members of the same species and between members of other species, including predators. It seems clear that the individual and species-specificities of these odorous signals are related to a biochemical individuality that can be demonstrated by immunologic comparability tests. For example, although dogs can discriminate between the odors of individual human beings, they cannot tell identical twins apart. In most mammals pheromones are secreted by sebaceous glands, to some extent by apocrine glands, and by specially modified glands, which are usually under the control of sex hormones.

Substances secreted by different glands in the same animal are said to give rise to an olfactory language. In mice, for example, several odors have been identified from different sources with different physiological significance, and with different effects upon receptors. Mice have an individual odor, an alarm odor, a male urine odor, an alien odor, an individual and group odor from the sweat glands in the foot pads, and one from the coagulating glands of the male reproductive glands. Such a chemical patterning of information has unlimited possibilities for coding both qualitative and quantitative physiological events.

There is good correlation between the development of chemosensory messages and the life style of animals. Most nocturnal animals have a highly developed olfactory sense, and odors are their principal guides; this is a good adaptation because, regardless of environmental circumstances, chemical stimuli are always present and available. The same is true for animals that live in burrows.

In spite of his microsmatic nature, man must emanate many odorous substances, well equipped as he is with sebaceous glands, the major source of his characteristic species and individual odor. Man also has the best-developed axillary organ, an area of skin abounding in apocrine and sebaceous glands and eccrine sweat glands. All of the secretions emanating therefrom are poured onto hairs encrusted with colonies of microorganisms that break down the secretions into odorous substances. If these odors are strong and easily detected by our microsmatic nose, what volleys of messages they must convey to highly macrosmatic animals!

Sex-linked, all-or-none, or differential sensitivity to certain types of odors has not yet been extensively studied in mammals. However, there is no doubt that the modulation of sensitivity levels is under the influence of neuroendocrine states.

Many of the prosimians, as part of their social and reproductive behavioral repertory, mark objects with the secretion of special scent glands, which may be sebaceous or apocrine glands, and sometimes both. In these animals the sense of smell has a very important social and survival function. The possession by certain Lorisidae and Lemuridae of a naked, moist rhinarium (like a dog's muzzle), together with curved nostrils and a hairless central groove or philtrum (p. 13) to the tethered upper lip, has led to the expression "olfactory" muzzle for such features. Monkeys and apes, when given unfamiliar food, survey it first visually, sniff it with great deliberation, and then either reject or savor it with caution, only to sniff again. Some time ago in a colony of ring-tailed lemurs at the Oregon Regional Primate Research Center, observations were made that point out the importance of odors in these macrosmatic animals. Two gravid animals gave birth to young at the same time, one naturally, the other, because of previous pelvic injury, by Caesarean section. When the operated mother rejected her young, it was offered to the normal mother who had her own baby (twinning occurs now and then in these animals). When she also rejected the foster child, the infant was rubbed against her vulva and perineal area, whereupon she sniffed it, licked it clean, and accepted it. Both young were reared.

In the back of the nasal passageways, the sensory olfactory epithelium lines delicate scrolls of bone. The epithelium, yellowish in color, contains specialized olfactory sensory cells and supporting cells. The surface of the bipolar sensory cells terminates in a process that runs between supporting cells to end in a fine olfactory hair on the surface epithelium. Olfactory cells are really nerve cells. At their base they continue as nonmedullated nerve fibers that converge into many bundles. These go through holes in the cribriform plate of the ethmoid bone and end in the two olfactory bulbs of the brain. Relatively small in man, olfactory bulbs are large in some primitive mammals. Neurones in the olfactory bulbs send fibers along the two olfactory tracts to the brain. Through very complicated pathways, the majority of the olfactory fibers finally reach the cortex at the uncus on the underpart of the temporal lobe. In macrosmatic mammals the uncus forms part of a larger piriform lobe. In a human brain this is very much reduced. Macrosmatic mammals have a relatively large part of the olfactory tracts passing to an olfactory tubercle on the base of the brain. This tubercle, generally absent in man, may be seen in other microsmatic animals as a small eminence in a region called the anterior perforated substance.

In many mammals an accessory olfactory area is present in the lower, front part of the nasal cavities, called the vomeronasal organ or organ of Jacobson, after Ludwig Jacobson, a nineteenth-century physician to the French army, who described it. In man this appears only during embryonic development and disappears almost completely by birth.

Auditory and Vestibular Perception

Although one naturally thinks of hearing as the proper function of the ear, equilibration was the basic ancestral function of what is now the auditory organ. The paired internal ears of most lower vertebrates consist of a series of sacs and canals (membranous labyrinth) filled with fluid endolymph, and surrounded by an optic capsule of cartilage or bone. Certain areas in the labyrinth (maculae) contain specialized receptors with sensory hairs (neuromasts). Calcium carbonate deposited in gelatinous material forms bodies (the otoliths) that cover the tips of the hairs. The shape of otoliths varies from species to species. Such a static balancing organ provides information about the degree of the tilt of head or body in a gravitational field.

Information about movements of the head is also provided by another set of membranous (semicircular) canals. Three of these exist on each side of the head, two in each set vertically placed at right angles to a third horizontal one. Thus, one lies in each of the three spatial planes. Each canal has a dilated enlargement (ampulla) that contains specialized sensory hair cells on a raised crista. Any movement of the head stimulates the receptors by setting up vibrations in the endolymph within the canals. Evidence suggests that the vestibular apparatus evolved partly from the primitive "lateral" line system of fish, which is sensitive to vibrations in water. Marine mammals do not have any such lateral line systems, but dolphins at least have numerous encapsulated endings under the snout skin. Whether in water or air, mammals require an effective balancing mechanism. Nerve fibers from the receptor organs travel along the vestibular division of the eighth cranial nerve to nuclei in the midbrain. Extensive connections occur with postural mechanisms and the flocculonodular lobes of the cerebellum.

Other Receptors

Various types of sensory endings within viscera and blood vessels provide information about the affairs of these organs. Whether the structural or functional characteristics of these endings in man differ from those of other mammals is not known for certain; hardly anything is known about them in nonhuman primates. They include generalized endings in the gut and other viscera, and certain somewhat specialized ones in relation to the heart, large arteries, and veins. Baroreceptors, present in heart chambers and large veins near the heart and in the carotid sinus, detect changes in blood pressure. The carotid sinus is a dilatation at the junction of the common and internal carotid arteries. The vessel wall is more elastic at this point and is innervated by the glossopharyngeal nerve. Structures such as the carotid body and aortic arch body may function as chemoreceptors that detect changes in the composition of blood. These structures might be expected to show special modifications in deep-diving aquatic mammals and in those mammals that run particularly fast, but little is known about such matters. For example, it is not fully understood what "tells" a cetacean that it has dived for long enough and must surface to breathe. Recent research has shown that certain vessels are more extensively innervated with receptors than was hitherto believed; but with the slowed circulation and the redistribution of blood that occur on diving, it is difficult to see how such receptors work.

15

SOME DISADVANTAGES OF BEING MAN

. . . genetic evolution, if we choose to look at it liverishly instead of with fatuous good humor, is a story of waste, makeshift, compromise, and blunder.
(Sir Peter Medawar)

Man has reached a point where he can begin to look back to his origins and, with some hesitation, even forward to his future. We can now attempt to evaluate *Homo sapiens* as well as to label and classify him. At the outset of this assessment, however, we must contemplate how much man had to give up to become man. With the attainment of his unique characteristics—an upright posture, a big head, extraordinary mobility and longevity, consciousness, intelligence—the pinnacle of the order Primates, man also acquired defects in construction and in function. Most of the foregoing chapters have already alluded to his weaknesses, and by now it should be obvious that despite the general belief in his perfection, *all* is not well with *all* of his biological particularities. As John Napier has written, "The 'scars of evolution' as they have been called can be seen daily in their thousands in outpatient departments all over the world."

However, part of man's genius is that the very distresses and troubles caused by defects in his structure and function have spurred him on with endless zeal not only to correct them but even to postpone, perhaps indefinitely, his inevitable demise. Nearly all the recorded history of man witnesses to his interest in medicine and to his commitment to search out the causes of his maladies and to devise reparative measures. Yet, as man learns more about himself, the number of unsolved problems, unanswered questions, and unrepaired defects seems to increase rather than decrease.

Like all other animals, man suffers the almost constant trauma inflicted by external agencies and by the wear and tear of his internal organs. He carries within himself numerous microorganisms, some of them pathogenic, and is often infested with parasites. He is especially susceptible to certain diseases, many of which, as far as we know, are unique to him. His cells, particularly late in life, tend to undergo neoplastic and malignant changes. His longevity makes him prone to degenerative diseases, especially of his nervous and cardiovascular systems. Some result from dietary or other excesses and deficiencies, but the causes of other diseases still elude us. Ironically, some of the afflictions that beset modern man exist precisely because as *Homo sapiens*, the "wise man," he has managed in some degree to insure the survival and reproduction of individuals whose genetic constitution is so abnormal that they would have perished without his intervention. Some abnormalities alter the development of parts or of whole organs and affect their function. Basic (congenital) disturbances in vital biochemical processes are caused by what are known as "inborn errors of metabolism." The failure of man's anatomical construction to meet the demands made on it by his physical pursuits often ends in disaster to the very physical endowments that equipped him to pursue such activities! No doubt, these pursuits too have a genetic background because man's genes make him the particular creature he is. Nonetheless, we consider them to be particular attributes that spring from man's particularities. Other mammals too, including nonhuman primates, occasionally exhibit certain "abnormal" conditions similar to ours, but about which we have as yet too little information. Since malformed or incompetent feral animals, whether primates or other mammals, have practically no chance to survive, the only way we can learn about their congenital defects is to keep them captive and thus protect them from their own kind as well as from their natural enemies.

Man's Faulty Skin

Although man has learned to live in external environments with which his nearly naked skin is ill-equipped to cope, he is almost constantly harassed by numerous, sometimes nameless, skin disorders. These are so common that an entire medical specialty, dermatology, is devoted to them. Although one of the oldest of the medical arts, it has until recently made little progress in establishing a solid scientific basis for therapy, largely because of the difficulty in understanding its functional complexities. Even so, by carefully examining the skin, a trained observer can obtain much information, not only about local afflictions of the skin itself, but also about conditions such as anemia, jaundice, shock, heart failure, certain endocrine and metabolic disorders, and contagious diseases.

In Chapter 7 we discussed the abundance of large sebaceous glands, which, influenced mostly by androgenic hormones, secrete large quantities of sebum on the head and around the anogenital areas. Yet, very little is known about the function of sebum other than that it is a pheromone. Man alone is afflicted with acne, a distressing condition that centers around these glands. Largely a disease of adolescence, in some unfortunates it continues into adult life. Generally only the glands of the face, neck, shoulders, and chest are involved; these glands elsewhere, particularly on the scalp and anogenital areas, though as large as those in the affected areas, are rarely involved. This singular disturbance of man's skin, which occupies a prominent place among his biological particularities, though not crippling is disfiguring and causes untold anguish.

As a result of poor hair cover, skin can undergo serious damage. For example, in cold damp climates, such as England, Ireland, and northern Europe, it is often afflicted with a blistering condition known as chilblain, which usually affects the hands, feet and legs, and sometimes the face and disappears during the summer. In some individuals direct contact with certain substances, whether natural—plant, animal, or mineral—or synthetic, such as drugs, detergents, and soaps, can cause hypersensitivity or even severe immunological responses.

Even the blessings of essential sunlight can be damaging and sometimes fatal to man. Lacking a protective hairy coat, man's skin burns if exposed too suddenly to intense ultraviolet irradiation. Gradual exposure allows the epidermis to form a thicker protective horny layer. The complement of melanocytes that reside at its base also forms appropriate quantities of melanin pigment, which, when distributed to epidermal cells, absorbs ultraviolet light to protect the living cells from the damaging effects of sunlight. Some individuals are so "photosensitive" that sunlight not only is pernicious but can be fatal. Fair-skinned people who have been exposed to the sun for many years often develop virulent carcinomas. Ulcerations of several sorts can also develop in response to exposure to sun. The absence of pigment results in albinism, which is due to a recessive gene. It occurs sporadically in all races of man, and has been recorded in many mammalian species. The discovery of a true baby albino gorilla, "Snowflake," in the Congo, now in the Barcelona zoo, has stirred considerable worldwide emotional interest. In true albinism, pigment is also lacking in the iris; hence albinos have to take special precautions against exposure to bright light.

Minor incidents, such as ingrown hairs caused by shaving, ingrown toenails, fungal infections (athlete's foot), and corns caused by the wearing of shoes, can be disabling. The causes of exfoliative conditions of the epidermis—ichthyosis and psoriasis, blistering diseases of many sorts, neurodermatitis, the precipitous loss of hair in the many forms of alopecia areata, neuralgias, itching, and other such torments—are either only partially understood or not understood at all.

The sweat glands, which pour water on the surface of the skin for thermo-regulation, when overstimulated can become blocked at the surface and cause water retention and small blisters (prickly heat). Under extremely warm conditions, man can lose so much sweat, together with salt, that the resulting muscular weaknesses and cramps can cause serious difficulties. Conversely, very cold conditions can lead to frostbite and even a loss of fingers and toes because of a failure of adequate peripheral circulation. In a lesser degree, relatively short immersions in cold water, which cause the skin to be pale and drawn, lead to failures in the peripheral circulation bringing on the muscle cramps so dangerous to swimmers. Evidently, the characteristics of man's skin limit the range of conditions that he can tolerate in his environment.

These are only a few examples of skin disorders in man. Infinite varieties of stressful situations that daily assault the skin are known. No doubt many of these woes are caused by the contacts of man's relatively naked skin with deleterious materials in his environment, by an increasingly complex diet (including drugs), and by defects in cutaneous construction.

Disadvantages of an Erect Posture

Man's skin is not his only liability. In addition to his constant proclivity to fall on his face, his upright posture entails many disabilities. Balancing on two feet has always proved a hazardous feat, from the time a man learns to walk and sustains countless cuts and bruises from tumbles, to the time he shuffles into an unsteady and tottering old age. In becoming upright, man underwent a series of associated and interrelated changes in his framework, perhaps none of which has been perfectly completed.

Man's characteristically large head derives its egglike appearance from his big brain. Man's head is not so huge as that of the gorilla, and it lacks the bulky masses and bony crests found on the skull of this great ape; the size of man's jaws is also much reduced. The bony braincase has expanded so much that most of it lies immediately below the scalp. Thus, a blow on the head that could easily cause damage to a human brain would leave a male gorilla unscathed. All scalp wounds are potentially dangerous: blood vessels in this tissue are firmly held and do not contract when cut; thus considerable hemorrhage can seep unnoticed into a pillow or mattress. Bleeding can be easily stopped, however, by pressing the scalp over the vessel against the skull. Notwithstanding the obvious advantages of cerebral development, man's brain is damaged relatively easily and is his greatest anatomical liability. Blows to the head can cause many different lesions, from mere headaches and contusions to depressed fractures of the skull, in which fragments of bone enter the brain. A severe blow can jerk the brain against the opposite side of the skull and cause cerebral damage a good distance away from where the blow fell *(contre coup)*. All fractures of the skull are dangerous because they can damage the bones of its base and cause a loss of cerebrospinal fluid and possible brain damage.

Moreover, head injuries can affect the blood supply to the skull and brain. Should the nutritive vessels to the skull itself, the meningeal vessels, be damaged, hemorrhage will occur on the inside of the bone but outside the dural covering, and the clots formed may press on the cerebrum. In various degrees, the compression will paralyze the motor region of the cortex supplying the opposite side of the body. Sometimes after a blow the meningeal vessels remain unharmed, but those on the surface of the brain hemorrhage within the meninges (subarachnoid) and lead to severe vascular damage of the brain. Man's large, rounded head, supported by a slender, usually not too muscular neck, is easily damaged in sudden jerky movements, as in car and airplane accidents.

As man's head became balanced on top of his vertebral column, he lost a snout and acquired a flat face. His most important sensory receptors, the eyes, became more exposed to danger than they are in other creatures, especially since he has neither heavy overhanging bony eyebrows nor long vibrissae on the forehead. Hence, abrasions of the eye surface and inflammation of the conjunctiva frequently occur. Fortunately, tears, with their lubricating and mild bacteriostatic action, reduce the potential damage and keep the eye surface translucent. From their origin in the lacrimal gland, tears traverse the eye and drain from the nasal side of the eye through the nasolacrimal duct into the nasal cavity. This mechanism usually suffices to rid the eye of foreign substances, but when overworked, the ducts become inflamed or blocked, and the eyelids, bathed in a sea of tears, become swollen, puffy, and liable to infection (blepharitis). Under normal circumstances the tarsal glands of the eyelids secrete a substance that prevents tears from overflowing the lid margins and also coats the thin film of tears on the front of the eyeball with an oily layer that slows evaporation.

Man's upright posture has affected the paranasal air sinuses too. The relative positions of the openings have become completely altered from those of quadrupedal animals. The exit from the maxillary sinus at the top of this large space causes fluids to accumulate in it when it becomes inflamed. Nasal mucous membranes are so close to bone that when inflamed and engorged they reduce or obstruct the air flow and drainage from the air sinuses. Even though man responds automatically to these discomforts by sneezing and snorting violently through his air passages, he is seldom able to improve the congested situation. Because the passage (infundibulum) draining the frontal sinus is narrow and curved, it is easily blocked; the sinus fills with fluid or pus, which causes the so-called sinus headache. Pus running down from the frontal sinus can also be directed, by the arrangement of a slit known as the hiatus semilunaris, into the opening of the maxillary sinus, thus involving it as well.

With the recession of the snout of ancestral forms, the jaws became shorter and the teeth smaller. Man was thus forced to choose food that is easily portioned and chewed. Man's earliest artifacts were designed to compensate for the deficiencies of his jaws and teeth. An overcrowding of teeth in too small jaws leads to faulty eruption, irregular arrangement, and poor occlusion and bite patterns. These irregularities, aggravated by the incidence of caries and conditions affecting the gums, make man's dentition one of his chief weaknesses. Defects in dentition also affect the movement of the mandible and lead to discomfort, pain on biting, and such conditions as "clicking jaws."

These are not just local conditions, of interest only to dentists. Diseased teeth and gums have general effects on the health of the body. Poor chewing can result in gastric disorders, and, with the teeth "set in a sea of pus," toxic conditions and infections can spread to other organs. Another common dental defect in man is associated with the development of the last set of molars (wisdom teeth) late in life, when his jaw growth is usually complete. Forced to grow in too small a space, wisdom teeth are often tipped, cannot erupt, and become impacted.

The peculiar construction imposed by bipedalism makes it easier for man to eat sitting down, with head and trunk held vertically. Thus, ingested food and fluid are aided on their way by gravity; provided that one stops breathing as one swallows, all goes well. However, there is very little room for maneuver, and mistakes often occur. One of man's worst disadvantages is to have his airway and food pathway cross over in the laryngeal region. It is imperative that one pathway be closed for the other to function properly. During breathing, the soft palate hangs down, partially shutting off the back of the oral cavity and leaving the nasopharynx open. The esophagus is collapsed, and the opening of the larynx moves forward beneath the covering flap of the epiglottis and is occluded while the esophagus is opened. The soft palate is raised upward and backward against the back wall of the pharynx and prevents food or fluid from passing into the back of the nose. Should man cough or snort while swallowing, to his social dismay, food can be forced up through the nose. When small particles of food are aspirated into the larynx and trachea, their sensitive mucous membranes immediately stimulate forceful expulsion by coughing.

Aside from convenience, physical and psychological advantages, and social grace, sitting to eat has much to be said in its favor. This position concentrates attention on the orderly alternation of swallowing and breathing, reduces the risk of inhalation at the wrong time, slows the respiratory rate and excursion at each breath, and allows a redistribution of blood from the limbs to the viscera. It also increases the abdominal capacity by relaxing the abdominal muscles. Sitting provides a more stable platform from which fixative and synergistic muscles can hold and move the head, aiding the mechanism of swallowing. We hesitate to emphasize what every child knows, that one swallows more easily with his arms on the table. This fixes his shoulder girdles and thorax and gives advantage to the muscles of the neck and throat. Man walks upright, but it is much to his advantage to sit to eat.

It might seem that bipedalism, which frees the upper limbs from involvement in locomotion, would be associated with a shoulder girdle and arm well adapted for prehension. The human upper limbs and pectoral girdle are indeed highly maneuverable, but their function depends on the integrity of a slender strut of bone, the clavicle, all too easily broken. This bone is the only anterior bony protection to the structures at the side of the root of the neck. Here pressure affecting the brachial plexus of nerves passing out to the arm can cause paralysis of certain arm muscles. A sudden wrench or unnatural movement can cause dislocation of the shoulder joint, in which the articular surfaces are modified for mobility rather than for bearing weight.

Man's arm and hand have other disadvantages. Granted, he has freedom of movement, superlative ability to hold, grasp, point, and manipulate; indeed, this last word indicates manual versatility. Furthermore, with instruments of his own making, man can extend this precision to microscopic levels and can use the fingers to adjust things so small that he cannot see them with unaided eyes. Yet, with this obvious usefulness the upper limbs can cause trouble, even beyond that of being "all thumbs." The development of manual dexterity came about with the adoption of an upright posture, but when man falls he subconsciously reverts to a tetrapodal form: he puts out his hands to break the fall and save himself from damage. His upper limbs, however, are not too well adapted for this function. Many pages in orthopedic textbooks describe the fractures and dislocations suffered by hands and arms in such circumstances, e.g., fractures of one or more of a series of bones from the first metacarpal of the thumb, the scaphoid of the wrist, the head of the radius at the elbow, as well as the shafts of the long bones and those of the strutlike clavicle. The constant use of the hands exposes them to continual risk of infection and minor damage that can have far more disabling consequences for a working man than similar afflictions elsewhere.

Unlike quadrupedal animals, man carries his abdominal viscera in front of him, rather than below. Because of his spinal curvatures, the viscera are not all supported within and above the pelvis but are retained in the body cavity by the corsetlike action of the abdominal musculature. Having no bone in its front wall, this musculature also plays an important part in respiration, trunk movements, and in any function such as parturition or defecation that is aided by increasing intraabdominal pressure.

The construction of the abdominal wall is not perfect either. Man is a placental mammal and during fetal life has an umbilical cord, which is essential only during intrauterine life. However, the site of attachment of the cord to the body wall marks a potential weak spot in adults. As in scrotal mammals, man's testes descend before birth through an inguinal canal in the lower part of each side of the abdominal musculature. After birth testicular vessels and the vas deferens pass through this canal within the spermatic cord, establishing still another potential weakness in the abdominal wall. Pressure by viscera against these weak places may cause them to herniate (protrude) through these regions. Umbilical and inguinal hernias of abdominal contents tend to occur more often as man ages. The floor of the pelvic cavity is formed for the most part by the levator ani muscle (elements of which are derived from muscles that moved the tail in lower forms), which also helps to support the viscera. Damage to this muscle can occur at parturition, and, if not repaired, the resulting weakness leads to prolapse of the uterus, perhaps of part of the bladder, and even of the rectum.

Except when completely prone or supine, man's vertebrae exert various degrees of pressure upon the cartilaginous discs (intervertebral) that separate them, particularly upon the nucleus pulposus in the middle of the discs. Pressure upon the nucleus is obviously greatest when one is in an erect position. When under more pressure than it can normally support, the nucleus, being composed of semifluid substances, may bulge or burst through its fibrous surrounding (anulus fibrosus) and cause what is erroneously called a slipped disc. When such herniations occur posterolaterally, they exert pressure upon spinal nerves, which results in pain and discomfort; sciatica is one such disorder. These conditions seldom occur in young people, whose discs are strong, but the incidence increases after age twenty-five. Although derangements can happen at any level, they are commonest in the lower lumbar and lower cervical regions.

The downward pressure of the upper part of the body on the sacrum at the base of the vertebral column overtaxes the ligaments of the sacroiliac joint. This and unknown other conditions cause lower back pains, one of the common afflictions of middle-aged people. In all individuals the vertebral column shows some slight side-to-side curvatures, called scoliosis. This normal deviation is accentuated to the left in right-handed people, and vice versa. Mild scoliosis is not normally serious but can become troublesome when exaggerated as a result of shortness of one leg, injury, or disease. The absence or reduction in the size of the bodies of thoracic vertebrae aggravates the dorsoventral curvatures and results in hunchback (kyphosis). The compensatory lumbar curvature is known as lordosis, which can be abnormally increased by defects in lumbar vertebrae, such as the absence of their arches.

The disadvantages of bipedalism do not end with the upper portions of the body. The narrow and elongated neck of the femur, for example, is set at an angle of 120° to the shaft. When, through accident or any other cause, the neck is absent or its angulation with the shaft is not exactly right (coxa vara), walking becomes an awkward waddle. The construction of the knee joint is a constant potential hazard to the semilunar cartilages between the tibia and the femur. These cartilages move with the tibia and, being resilient, change shape during the rotation of the joint, acting as shock absorbers. If subjected to sudden violent and abnormal movements, the edges may be caught between the femur and tibia and a cartilage ripped from its attachments. Only surgery can restore the effectiveness of the knee.

The patella, the large sesamoid bone inside the tendon of the quadriceps muscle where it crosses in front of the knee, gives man a knobby-knee appearance. With the leg extended straight the patella rides above the condyles of the femur. When one kneels, the patella is in front of the femoral condyles but does not come in contact with the ground unless one leans well forward. The patella, in fact, protects the cartilage of femoral condyles from the rubbing of the quadriceps tendon. This is an adaptation for an upright posture and does not do well when one is working on the ground. After prolonged laborious genuflexions the bursa between the patella and the skin can become irritated and give rise to "housemaid's knee." Moreover, after injury acute irritation can occur at the joint membrane of the knee.

The knee joint is one of the strongest in the body and dislocation is rare; it is, however, one of the least secure. It is not a simple hinge; rotation of the femur on the tibia or vice versa occurs to some extent throughout the hingelike action. It is formed between the two longest bones of the human skeleton and considerable leverage can be brought to bear on its structure. The articular surfaces are poorly adapted and were it not for the strong ligaments binding the bones and the powerful musculature, the joint could not withstand the weight-bearing stresses and the range of movement to which it is subjected. Either by accident or because of excessive exertion, twisting strains can cause tearing or detachment of the intraarticular (semilunar) cartilages and can also injure the external (collateral) and the internal (cruciate) ligaments causing sprains or even rupturing them. In general, ligaments limit the range of movement at a joint and when torn result in loss of stability; this is particularly true of the knee joint when the strong anterior and posterior cruciate ligaments are torn. These ligaments restrain the femur upon the tibia, and when they are severed an excessive gliding movement develops between the two bones, which is incapacitating indeed. Sudden movements, or abrupt checking and changes in direction, can demand more than man's locomotor apparatus can tolerate. Torn muscles and tendons can severely incapacitate the activity of the knee. The locomotor anatomy is reasonably efficient for ordinary life, but being what he is, man strives for perfect performance, for sudden action, for athletic records, and then finds out all too soon the disadvantages of his construction.

At the ankle joint, the processes of the tibia and fibula (malleoli) prevent the ankle from slipping sideways. This adaptation allows easy to-and-fro rocking movements, but sudden and forcible sideways twisting of the foot can result in fracture of one or both malleoli (Pott's fracture) or in dislocation of this joint. The arches in man's feet, designed to give spring to his gait, have inherent weaknesses in construction and often collapse. Weak arches may be inherited; but regardless of their strength, they can "fall" under the burden of excess weight that stretches beyond their capacity the tendons and ligaments that hold the bones together. Fallen arches result in a distortion of the tarsal bones that reduces the efficiency of the feet in walking or running. Flat feet tend to splay outward and may lead to knock-knees and outward deviation of the big toe (hallux valgus), a crowding of the second toe, and hammer toe and bunions.

Disadvantages Related to Man's Long Life

Ironically, longevity is one of man's chief disadvantages. A long life in a society where individual members do not have to struggle constantly for survival could have its rewards. Since, as a rule, aging people no longer need to provide for themselves, they finally acquire time to enjoy the "good things" in life—travel, companionship, contemplation. However, this blessing is more easily relished by the affluent and by those whose intellectual attainments have better prepared them for the contemplative life. "Retired" people of means can travel, live in luxury, and purchase those pleasures they could not afford when young. Scholars can bury themselves in their books, sensitive persons can observe, integrate, and reflect; and all can enjoy companionship and philanthropy. This stage of man's life could be nearly utopian; free from care and conflict, he could attain serenity. And, indeed, some fortunate ones do.

But, how sad! Lacking financial, intellectual, or physical means, men often find the last decades of a long life endless and agonizing. Physical comeliness wanes inevitably and inexorably, as do social and personal graces. Old people become increasingly self-centered and solitary. With the deterioration of their brains, they progressively lose their memory, perception, sensitivity, and the ability to integrate and associate ideas and experiences. Together with all this fades what we have come to call man's dignity, the very essence of what makes him man. The twilight of a long life is, too often, a perpetual bout with boredom, accompanied by mental and physical torments.

The frame steadily and insidiously weakens as man ages. The primary reproductive organs of women cease to function altogether and thereby cause disturbing changes in hormonal balance, and atrophy and fibrosis occur in some of the secondary sex organs. Disturbances in the progressively useless vagina are many, and both benign and malignant tumors of the uterus are common. Breast cancer, not necessarily limited to older women, is more frequent in them. Age changes are less precipitous in men, who experience a gradual waning of libido and physical sexual function and suffer a number of unique discomforts. Most usual of these are disorders of the prostate gland, an organ essential only for reproductive purposes, which becomes enlarged in nearly all men past their physical prime. Even if the prostate undergoes only a simple enlargement, its cufflike location around the urethra tends to strangulate the latter, inhibiting the free flow of urine from the bladder and making urination increasingly difficult and painful. Even more serious, however, is the tendency of the prostate to become cancerous in some old men. More amenable to surgery and chemotherapy than other cancers, prostatic carcinoma is a rampant tumor that metastasizes exuberantly.

The cardiovascular system begins to undergo alterations as soon as a man is full grown, or sooner. Being progressive and cumulative, these disorders worsen steadily. Arteriosclerosis, more common in men than in women, is almost universal in old age. With the impairment of their vessels, the brain and the heart gradually lose their efficiency and deteriorate. Heart muscle fibers accumulate waste substances like fat and fibrous tissue. The complicated heart valvular system becomes worn and damaged, rendering the closures imperfect and allowing a backflow of blood. Some details in the architecture of the vascular system are not too well tailored for an erect posture; think, for example, of the long path that blood must traverse to return from the legs to the heart. Even in a short-statured individual, blood from the legs must be returned through thin-walled veins against gravity. The delicate valves of the superficial veins may be inefficient or become damaged, and the vessels dilated and engorged with blood (varicosed). It is likely that every person over twenty-five has at least some varicose veins. In addition, blocking of the myriads of tiny lymph vessels in the legs causes a poor return of lymph, which accumulates in the tissues and produces swelling.

Faults of Communal Living

Man pays a high membership fee to live in a social community. Studies of population dynamics in other animals show that the success of any given community flowers during the early stages of its establishment, when intragroup aggression is at a minimum. Females breed at an early age, producing large and healthy litters, and the population increases rapidly. When the population reaches a critical number, an inexplicable rise in competitive fighting occurs and the females becomes less fertile and less fecund. They have smaller litters and the young are of inferior quality. At this stage, the population stabilizes. Although what factors cause these negative conditions are not known for certain, crowding is known to cause stress, which could, in turn, trigger chains of deleterious phenomena.

Under chronic stress, the adrenal glands become enlarged and secrete great quantities of hormones to prepare the animal to cope with the situation. If, however, the stress should continue beyond tolerable limits, the adrenal glands atrophy and the animal succumbs. Man as man is singularly subject to stress. His crowded living conditions inevitably breed tensions, but oddly enough he seems to prefer them even when he no longer needs to fear unduly the menace of human and nonhuman predators. But more basically, man by nature would seem almost predestined to be the prey of continual anxiety, fear, longing, anger, jealousy—the very passions that generate stress. The consequences of stress in an animal, human or subhuman, are many and puzzling. Under stress man is much more susceptible to both internal and external parasitic infestations. He is an easy victim of disease; even those organisms that his body normally hosts without discomfort may become pathogenic. Finally, he is subject to many cardiovascular disorders, stroke, and the common but dramatic hemorrhagic stomach ulcers; several authorities consider that the suicide rate also increases significantly. Although population crowding no doubt greatly increases man's stress, a gross oversimplification would be to explain the complex concatenation of human ailments by one single factor. We do not know what this kind of stress has done to man's mental health, but the number of his psychic disturbances is increasing alarmingly. No doubt, social and environmental pressures stamp their imprint on man's constantly taxed consciousness.

A large number of diseases, especially venereal, thrive only in communal environments. Leprosy and poliomyelitis, both peculiar to man (no other animal is known to be afflicted by them or has been successfully infected with the organisms) will be mentioned here. Both diseases are ancient and both have caused untold misery. Curiously, leprosy thrives in populations where hygienic standards are very low or nearly nonexistent, whereas poliomyelitis does best in a society where the standards are high. Leprosy poses practically no problems under clean conditions; in fact, in aseptic conditions it is difficult to infect an individual with leprosy organisms. Why poliomyelitis does best in hygienic environments is still not known. Perhaps young children reared in such an environment are protected from subclinical contacts with the disease at an age when they might have developed an immunity to it and so fall victims later. Fortunately, both leprosy and poliomyelitis are controllable today—the former by cleanliness, the latter by immunization with vaccines.

Comments

Considered here are only a few of those disadvantages that plague man because he is man. Some of these reside within his framework, causing internal disruption of his bodily functions; others are external agents, preventing him from adapting to his environment. As man learns to recognize these disadvantages, he also learns to compensate and at times indulge himself in the fantasy of "creating" the perfectly equipped man. Indeed, writers of science fiction, films, and television have been surpassing themselves in devising "new" creatures with superhuman attributes. This is an amusing game, but we think that these "creators" ought to be made to take examinations on the evolution, embryology, anatomy, and nervous systems of their creations.

Pollution, which is the result of man's population explosion, his technology, affluence, avarice, and carelessness, is now a ubiquitous threat. It is possible that it will never be cleaned up. Could prolonged pollution exert an effect on the anatomy of man? Arthur Kretchmer (1971) thinks it could, perhaps encouraged by modern geneticists in the process, and visualizes the emergence of a new species, *Homo effluviens.* Polluted Man would have larger eyes to see through smog; protective transparent inner eyelids; elongated, spatulate index fingers to work machinery; protective flaps to his small ears; hypertrophied nasal filtering structures and detoxicating pouches in his cheeks; much enlarged and more efficient lungs; spindly limbs with wasted muscles and flabby cushionlike buttocks; testes much enlarged in a vain effort to maintain fertility. A creature of fantasy perhaps, but pollution is still here for man to cope with.

To understand man as he is or as he will be we must establish his past, which is immutable, and realize that he is the result of countless genetic mutations. Throughout millions of years these mutations have for the most part been self-eliminating; thus, the chances of new mutations that would bring forth a *Homo sapientior* (a "wiser man") are infinitesimally small. Any mutation that is likely to occur will doubtless only repeat the "failures" of the past. More probably behavioral changes will distinguish our successors from us, changes that are long overdue. We conclude by restating that the disadvantages of being man arise as a direct result of his past, of the material of which he is made, and of the pattern of his form ordained by his genes. Yet, man still has one ultimate advantage over all other animals; he is the only vertebrate capable of altering the conditions of his environment to his advantage by his own efforts. Furthermore, he knows that he can.

16

THE AGES OF MAN

Yes, it is useful to prolong human life.
(I. Metchinikoff)

To speak of the "human family" may seem to imply that all men belong to one great, friendly, interrelated group. Obviously, however, this is not the case. One of the behavioral characteristics of *Homo sapiens* from time immemorial has been a proclivity for gathering in large social groups. As a highly social animal, each individual depends on others for his needs, protection, and emotional gratification. The evolution of group society, tribalism, and "togetherness" lies in man's long dependence on others, beginning even before birth. As an infant, he requires the attention, positive help, training, education, and experience generally given first by his parents, then in childhood and adolescence by other elders. Throughout life, despite his sometimes noisy assertions to the contrary, man never becomes wholly independent.

It is almost inevitable then that man should be characterized by those very traits that derive from his group existence in society. Although this strong social orientation does not immediately influence his anatomy, except when certain fashions demand that he deform some parts of his body, it does affect his biology. Social influences leave their imprint on his behavior, physical and mental abilities, nutrition, growth, reproduction and longevity, the diseases to which he succumbs, and on his resistance to them.

Large social groups probably broke down into smaller isolated ones that eventually formed separate populations and resulted in the appearance of new forms that influenced genetic drift. Any social group, either of man or of other species, is composed of individuals with a basically similar genetic makeup that still admits of differences of sex, physique, temperament, leadership, and age. The age range in a human population is more extensive than in any other social group of mammals.

Although man lives longer than any other mammal, his body cannot resist indefinitely the onslaught of senescence; when he can no longer withstand the pressure of his environment, he succumbs, As Alex Comfort sees it, senescence represents an increasing likelihood of death; it is not an "inherent" property of multicellular animals, "but one which they have on several occasions acquired as a potentiality, probably through the operation of evolutionary forces directed to other biological ends."

The human child is frightened by the knowledge that his parents will grow old and die. Yet, acceptance of this fact marks an important advance in the child's own maturation. Aging involves deterioration in various, sometimes detectable, often subtle, ways. The steady progression of this decay eventually causes death with or without the intervention of a genuinely pathological cause. Not surprisingly, medicine, having made outstanding advances in the cure and prevention of disease, is now examining with increasing vigor the phenomenon of senescence and the feasibility and advisability of lengthening man's life span. Advances in transplantation techniques have enabled surgeons to equip an otherwise doomed patient with a "new" heart taken from a donor immediately after death. Surgically the procedure is not especially difficult, but grave dangers exist that the transplanted heart will be rejected by tissue, cellular, and chemical responses of the host to foreign material. Although these responses can be somewhat reduced or even eliminated by radiological, chemical, or immunological means, the host patient is left largely devoid of normal resistance and at the mercy of pathogenic organisms. Some enthusiasts have gone so far as to suggest that healthy human hearts be "stored" in large primates for later transplantation, when needed, into suitable human recipients. Other organs, such as liver and kidneys, gonads and endocrine glands, have been transplanted in animals and man with varied degrees of modest success. No two organs appear to be accepted, or rejected, in precisely the same manner. The more alike the genetic makeup of the cells of donor and host, the more likely the acceptance of the graft under the conditions devised.

Although designed solely to prolong life, transplantation experiments are performed in a society that is still vexed with moral considerations. Has the donor really "died"? Whose life should be prolonged and who shall decide? Does it matter whether white men will have "black" hearts? Shall we accept the use of nonhuman primates, and probably other mammals, as "stores" for human organs? If, as seems possible, surgeons learn to transfer organs from animals to men, shall we find that literally bearing "the mark of the brute" alters the host in any way? When the foreseeable possibility of using transplants to provide deformed or abnormal children with a better life or to increase their expected life span becomes a reality, how shall we decide which children are to be the fortunate hosts?

Attempts have been made to divide life into a succession of discrete growth phases and to analyze them in terms of alteration, not only in the tissues of the body but especially in the skeleton. Unfortunately, not all changes proceed simultaneously or at the same rate, and many are not easily correlated because in human populations each is an expression of genetic, hormonal, nutritional, and environmental variables.

Age is calculated from the date of birth, although the Chinese and the Japanese apparently add another year to allow for prenatal life. In the first twelve to fourteen years, man goes through childhood into puberty, when the onset of maturity becomes obvious. The word "puberty" refers to the appearance of hair on the pubic region just above the genital organs where at puberty a certain amount of fat accumulates. Critics have pointed out that the word is not particularly apt because hair sometimes appears normally in the pubic regions of children; it is, however, too much a part of the language to be altered now. Puberty is marked by numerous anatomical, physiological, and behavioral changes, mostly related to the increased activity of the gonads.

The reproductive phenomena associated with puberty have been described earlier. These are accompanied by an enlargement of the ovaries, uterus, and mammary glands in girls (Figure 16–1) and of the testes, seminal vesicles, and prostate in boys. The size of the gonads increases slowly for some years before puberty, but the active growth of mammary glands in girls and of the prostate in boys is particularly dramatic at this period. While the endocrine organs enlarge, their functions accelerate.

Until the age of 9 or 10, boys and girls grow at an almost equal rate. But around 12 most boys gain an ascendancy, which they maintain except for a brief period around 14 when girls overtake boys. After 14 boys again take the lead. By this age, girls frequently show certain adult feminine features: broadening of the hips, enlargement of the breasts, and a relative shortness of the legs (Figure 16–2). In boys the most striking pubertal change occurs in the larynx. Under the influence of endocrine activity, this organ enlarges in both sexes but more so in boys,

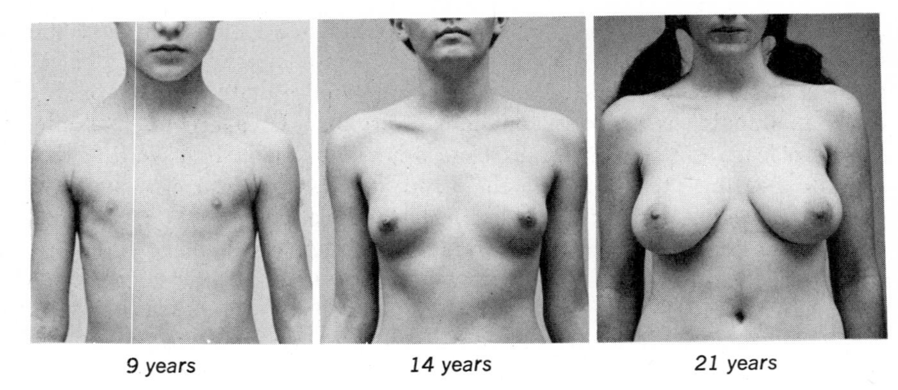

Figure 16–1. The development of breasts. Left: 9-year-old girl; middle: 14-year-old girl; right: 20-year-old woman.

in whom it expands markedly in one year and develops the angular form known as Adam's apple. The lengthening of the vocal cords, together with other changes in size and shape, brings about the "breaking" of a boy's voice. The larynx of girls grows one-third longer at puberty but retains its smooth form.

Apart from the fusion of three elements in the hip-bone (ischium, ilium, and pubis), few ossificatory changes occur during the early part of puberty. Changes in the bone structure occur rapidly, however, during adolescence, a period that lasts until about age 18 to 20 in girls and 20 to 23, or even later, in boys. At the end of this period of consolidation, the skeleton in both sexes reaches full length. Girls usually add about two inches to their stature during adolescence, boys about four to five inches. Changes likewise occur in hair growth: the beard and moustache appear in young men, but full efflorescence of the beard and body hair does not take place until adult life. Bone growth then ceases at the end of adolescence, when the skull, vertebral column, hip-bones, and clavicle reach their final form. Growth of the clavicle gives the young man of 25 to 27 years noticeably broader shoulders than those of the "stripling" of 18 to 20.

After a man becomes an adult, his biological nature attempts in various ways to maintain him in the fullest effective anatomical and functional condition. But, at the same time, the first signs of the aging process become apparent in nearly all the systems of the body, although not at an equivalent rate or to a similar degree. Certain systems are more specifically and more dangerously assaulted than others, and an individual's life expectancy is in direct ratio to his ability to withstand the ravages of the various aging processes. Some people appear to maintain an amazing equilibrium for many years.

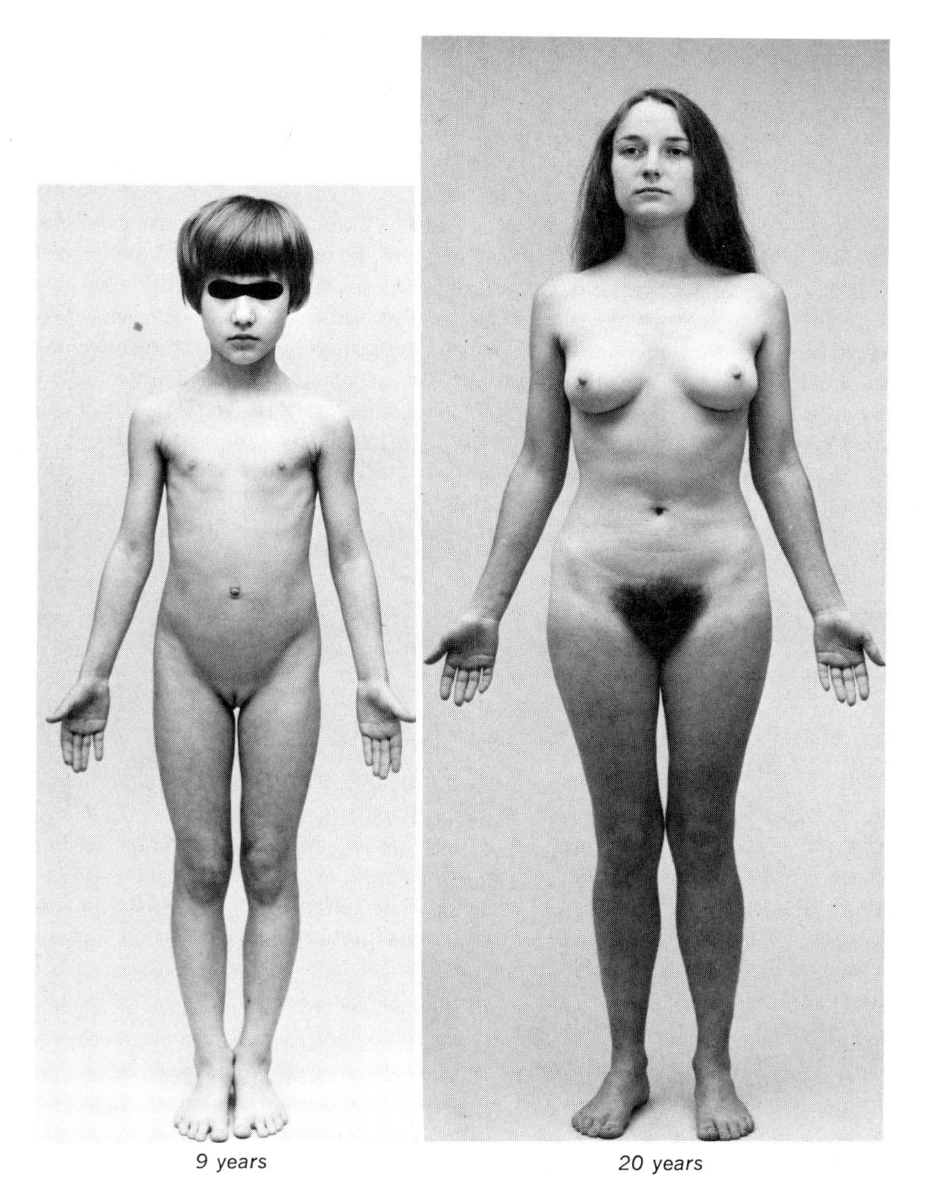

Figure 16–2. The attainment of an adult female body.

Longevity in Man and Animals

Man lives on an average longer than the elephant and probably longer than the allegedly long-lived perch and carp. It is foolhardly to quote even authenticated maximum longevity records as evidence of longevity in animals. Little can be added to Francis Bacon's trenchant observation written over three hundred years ago: "Concerning the length and brevity of life in beasts, the knowledge which may be had is slender, the observation negligent, and tradition fabulous; in household beasts the idle life corrupts, in wild, the violence of the climate cuts them off." Despite the inherent pitfalls in interpreting longevity records, they provide, at least in man, some indication of how long the hardier ones, given freedom from accident or attack by disease, can survive. The brief list in Table 16–1 gives maximum longevity records for some animals.

The greatest reasonably authenticated age reached by man is 120 years. But Christen Jacobsen Drakenburg (1626–1772) is said to have lived under seven Danish kings during the 146 years of his life. After

Table 16–1. Maximum Longevities in Some Animals

	Years		*Years*
Snail (*Helix*)	6–7	Cat (*Felis*)	20–30
Ant (*Formica*)	7–15	Seal (*Phoca*)	15–20
Queen bee (*Apis*)	5	Finback whale	
Lobster (*Homarus*)	33	(*Balaenoptera*)	25–50
Carp (*Cyprinus*)	40+	Pig (*Sus*)	12–20
Frog (*Rana*)	12+	Elephant (*Elephas*)	60–70
Tortoise (*Testudo*)	152+	Horse (*Equus*)	40+
Eagle owl (*Bubo*)	68	Tarsier (*Tarsius*)	12
Goose (*Anser*)	35+	Bush baby (*Galago*)	14
Pigeon (*Columba*)	30+	Lemur (*Lemur*)	27
Bare-eye cockatoo		Rhesus monkey	
(*Cacatua*)	75	(*Macaca*)	29
Shrew (*Blarina*)	1½	Baboon (*Papio*)	29
Bat (*Rhinolophus*)	7+	Gibbon (*Hylobates*)	31
Rat (*Rattus*)	4–5	Chimpanzee (*Pan*)	41
Rabbit (*Oryctolagus*)	10–15	Gorilla (*Gorilla*)	33
Dog (*Canis*)	15–20	Man (*Homo*)	109 (England)

many adventures at sea and fifteen years in slavery at Tripoli, he was reported to have lived "quite a respectable life after the age of 141." In England a man had a birth certificate that showed him to be 109 years, and another, without a birth certificate, was shown to be over 111 years.

Certain geographical areas, among them Bulgaria, the Caucasus, and Russia (Abkhasia and Daghestan), claim to have large numbers of old men. However, these claims have not withstood critical examination. The list in Table 16–2 gives the average age at death of some categories of eminent men. For what it is worth, it provides a few facts for reflection.

Table 16–2. Average Longevity of the Eminent in Various Fields of Activity

Field of activity	*Average age at death in years*
Members of the president's cabinet (United States)	71.39
Entomologists	70.99
Inventors	70.96
Historians	70.60
American college and university presidents	70.11
Geologists	69.79
Chemists	69.24
Educational theorists	69.06
Educators, all kinds	68.98
Economists and political scientists	68.68
Contributors to medicine and public hygiene	68.57
Botanists	68.36
Philosophers	68.22
Historical novelists	67.89
State governors (United States)	67.02
Authors of words to church hymn tunes	66.94
Mathematicians	66.62
Composers of grand opera	66.59
Composers of choral music	66.51
Composers of chamber music	66.26
Naval and military commanders (born from 1666 to 1839)	66.14
Authors of political poetry	64.47
Painters in oil	64.22
British authors and poets	63.91
Hereditary European sovereigns	49.14

Expectation of Life

The value of estimating a man's life expectancy from tables with a fair degree of accuracy is especially relevant to those who "insure" against dying too soon (life insurance) and to those who "insure" against living too long (endowment insurance). Senescence has already been defined as the increasing liability to succumb to the external environment. John Graunt and later Sir William Petty in England first demonstrated during the seventeenth century that the risks and chances of perishing from disease can be quantitatively expressed. The latter, realizing the importance of population studies, incidence of diseases, fertility, and mortality rates, urged the creation of a government institution for collecting such data. Demography, or the statistical study of populations, received its start when the eighteenth-century Prussian clergyman, J. P. Süssmilch, sought to demonstrate a divine design in the numerical relationships of vital statistics. Whatever his intent, his work did reveal the need for such investigations and emphasized that large numbers must be examined if the results are to be meaningful. The value of such calculations has increased steadily since the census system was introduced in England in 1801. Several studies have shown how the physical and intellectual characteristics of man can be numerically expressed and compared with those of the "average" man by the application of probability theory.

One may wonder whether the anatomical characteristics of man are in any way related to demographic statistics. In one way or another the peculiarities or generalities of his structure have some bearing upon vital statistics. Biological factors that influence mortality and longevity range from risks to the embryo, fetus, and child before and at birth to changes that inevitably appear in his structure as he matures and grows old. Among these factors are the blood groups of his parents, the harmony of his development, and the ability of his endocrine system to regulate his internal environment. His large brain enables him to attain an intellectual level that in turn depends on the intrinsic quality of the brain. To be a part of society he must act on information received from his senses and must communicate. His anatomy is also peculiarly sensitive and susceptible to certain diseases, and his long life makes him a prey to the chronic exposure of various irritants and poisons.

Nevertheless, human longevity has increased steadily (Figure 16–3). Estimates indicate that prehistoric man had only a slim chance of surviving until age 40. In 1850 one-sixteenth of all babies born in England died during their first year; nearly one hundred years later (1945), the infant mortality in England and Wales was still 46 per

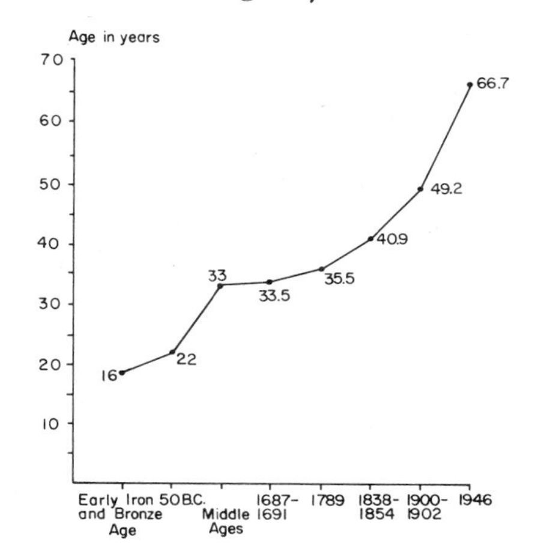

Figure 16–3. Average length of life from ancient to modern times. *(From Dublin, Lotka, and Spiegelman, 1949. Ronald Press Co., New York.)*

1,000 live births. The life expectancy in 1871 for an English boy was 41.4 years; in 1971 it was 68.6 years. Having reached the age of 10, an American girl in 1911 could expect to live another 50.7 years. If she had reached 10 in 1947, she could then look forward to another 62.3 years. A girl born in 1960 has a life expectancy of 74.1 years; but if she survives 60, her life expectancy is 79 years. The numerous statistics available, most of them somewhat confusing, all point to the fact that the longer one survives, the longer he is expected to live beyond that point. The life expectancy of men has consistently been 5 years below that of women. These figures account only for the average; what happens to the individual depends on a bookful of variable factors: his constitution, his reaction to his environment, and the nature and nurture of his peculiarities.

Age Changes in the Skeleton

The most reliable and practical way of determining the age of an individual is to examine his skeleton and teeth. No other part of the body, however carefully examined, can divulge quite as much information. The younger the person, the easier it is to estimate his age accurately from the bones, especially from the incremental lines in the teeth. To make these estimations, it is not always necessary to have the dried bones; much useful information can be obtained from X-ray films. When problems arise about the alleged ages of living refugees or of children thought to be changelings, X-ray films of the skeleton have been the only available means of identification.

The first important thing to ascertain in the development and aging of the skeleton is whether well-defined stages can be identified into which we can with some assurance categorize unknown material. The American anthropologist W. M. Krogman has divided the major skeletal changes during a man's life into seven stages, which, however, do not exactly correspond to the seven ages of man in Shakespeare's *As You Like It.*

Krogman's first period extends from birth to the age of five. During this time all but one of the primary and most of the secondary centers of ossification appear, earlier in girls and in better-nourished individuals. The actual time of appearance varies according to sex, nutrition, and genetic factors. These centers can be discerned with X-rays, as Figure 16–4 clearly shows.

About eight weeks after fertilization some ossification begins at specific centers in each bone. Bones are mainly formed by intramembranous or by intracartilaginous ossification. The first occurs chiefly in the bones of the vault of the skull, in the facial skeleton, and the clavicles. Intracartilaginous ossification is characteristic of the bones of the base of the skull (basi-cranial axis), of the vertebrae (axial skeleton), and of the limbs (appendicular skeleton). The essential difference is that intramembranous ossification takes place in a soft, mesenchymal tissue, whereas intracartilaginous occurs inside a cartilaginous model of the future bone (Figures 16–5, 16–6).

All bones start to ossify at primary centers in the middle of the membranous or cartilaginous tissues. In long bones the primary centers are in the shafts; later secondary centers appear at the ends (epiphyses) of each bone. Much later still the epiphyses fuse with the shaft to form the adult bone.

At birth most bones have small primary centers of ossification. In long bones the primary and secondary centers are separated from each other by cartilage plates. In flat bones several primary centers are separated by membranous tissue. Long bones are completely formed only when bone replaces all the intervening cartilage (Figure 16–5). Since cartilage itself continues to grow, some years are required before it is completely assimilated. Each center of ossification grows in two ways: by an increase in thickness, or girth, caused by a deposition of bone on the periphery of the growing center of ossification, and by an increase in length caused by an extension of the growth center along the axis of the bone, according to its shape. Internal modeling occurs during the entire growing period; as a result, though the bone increases in thickness, it does not become a nearly solid rod with only a small central cavity.

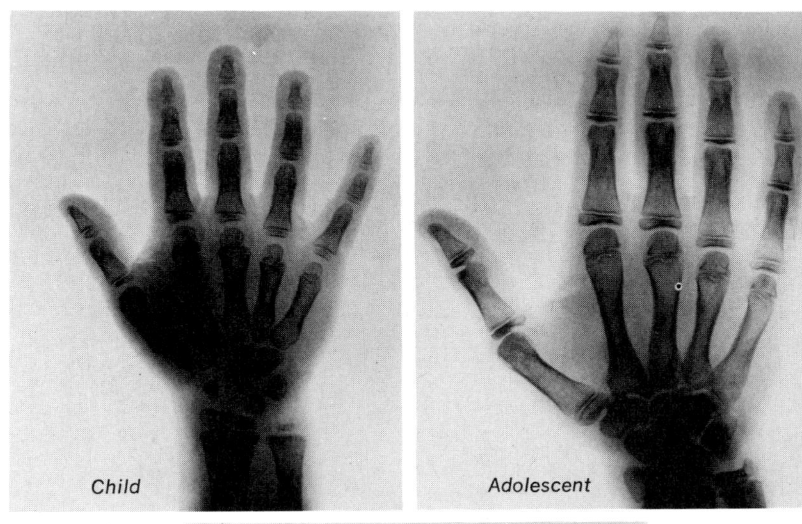

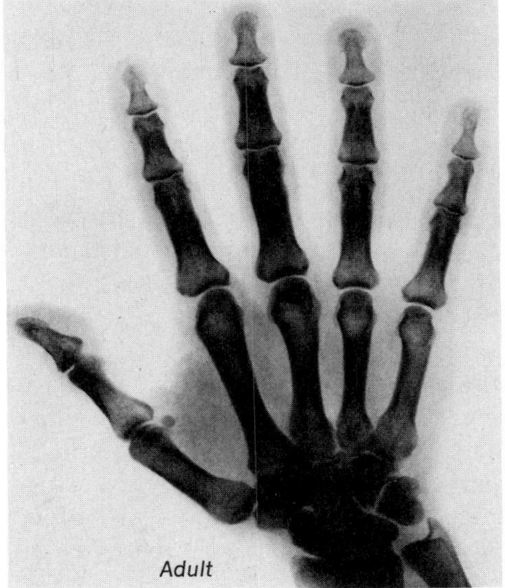

Figure 16–4. X-ray film of the hand of a child, an adolescent, and an adult.

Concomitant with the removal of bone from the marrow cavity is the laying on of bone on the outer surface so that the cavity increases in size as the bone increases in diameter. The cartilaginous plates of the epiphyses remain for some time between each shaft and the ends of a bone and can be seen on an X-ray film as a narrow, dark line. As long as this plate is present, the bone can continue to grow in length.

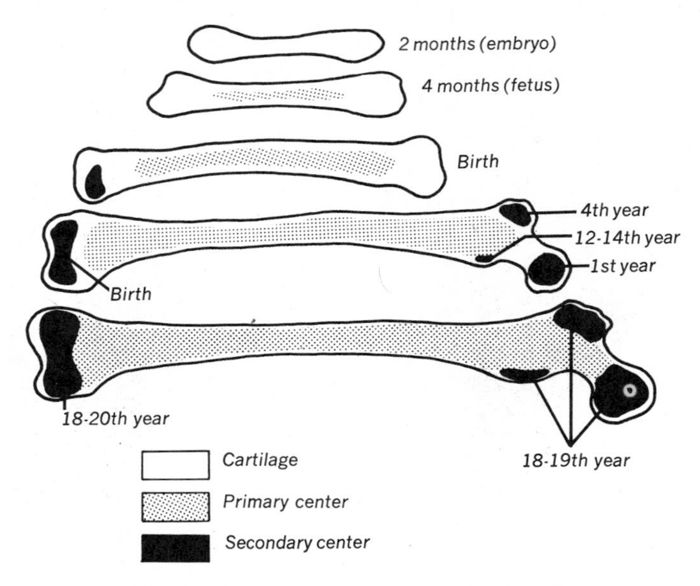

Figure 16–5. The development and growth of the long bones in man.

In the rare condition known as achondroplasia, the growth of epiphysial cartilage stops prematurely and the limbs and some skull bones are not as long or as large as they ought to be. Various factors affect bone growth, e.g., how much calcium, phosphorus, vitamin D, and hormones are included in the diet.

Krogman's second period extends from 5 to 12, during which appear the last primary centers of ossification, that of a little bone in the wrist called the pisiform, and those of a few secondary ones. The main changes now are in alterations and growth in the ossificatory centers that appeared before birth or during the first period.

The third period lasts from 12 to 20 years, extending from puberty to late adolescence. During the first few years, growth and change in the size of the ossification centers continue. Toward the end of the period, new and important series of events take place. The epiphysial plates gradually disappear, the primary and secondary ossification centers fuse, and the length of the bone is completed. The fibrous tissue that covers the bone, the periosteum, retains considerable osteogenic potentiality, seen when a bone is fractured any time before, during, and after the end of the third period. Complete union between the shaft and the ends of bones occurs earlier in girls, who reach their adult height before boys. Those epiphyses that were the first to appear are the last to fuse with the shaft. This means that in some bones one end goes on growing for a variable period of time after the other end has stopped. Just as certain factors affect the time of appearance of ossification centers,

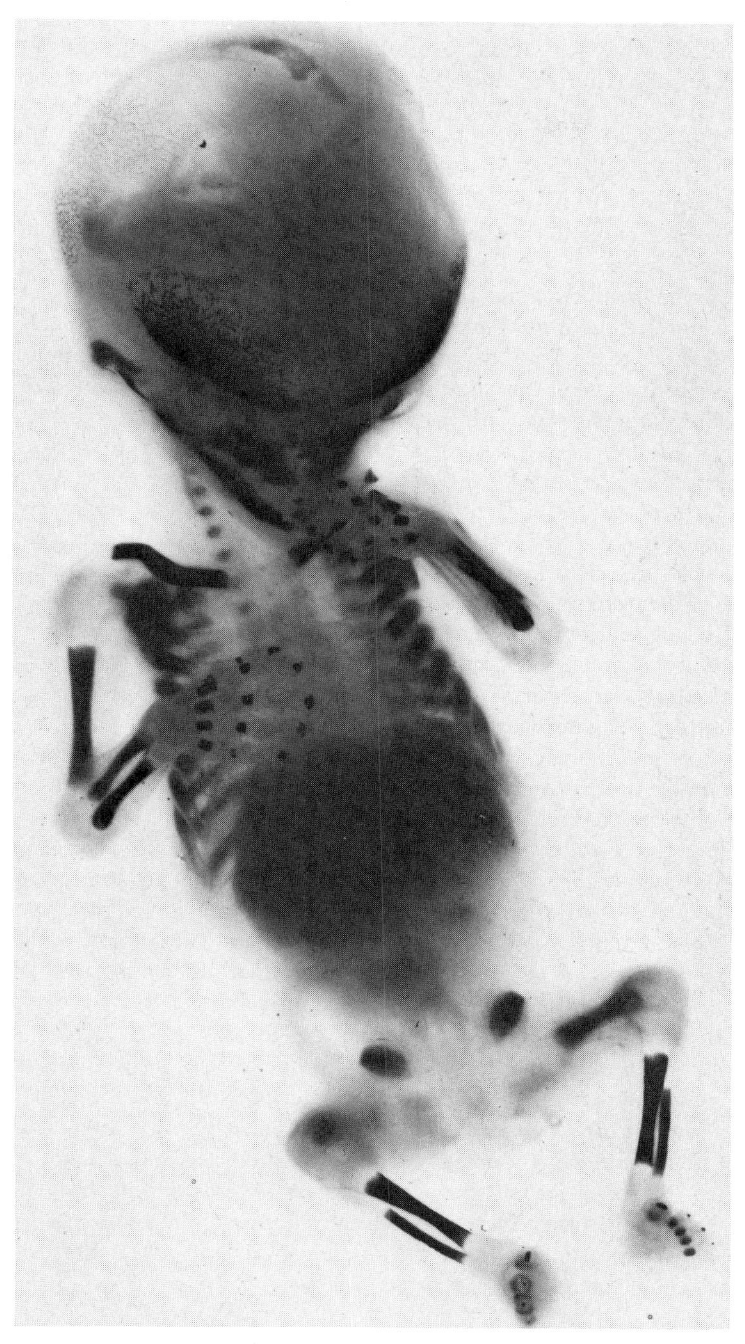

Figure 16–6. Four-month-old human fetus stained with alizarin and cleared to show the centers of ossification.

so also similar and other factors affect the time of fusion. Sex and nutrition, deficiency of certain vitamins (D), and lack of certain hormones from pituitary, thyroid, and parathyroid glands all affect the time of union.

The fourth period of bone growth, a short one lasting only from 20 to 24, is really a continuation of the third period. By age 20, the union of most of the epiphyses of the body has occurred, except in the clavicles, the base of the skull, the pelvis, and the vertebral column. A secondary center of ossification, variable in its appearance, is present on the inner or medial end of the clavicle. This usually appears at about 21 and closes a year or two later, though it can appear as early as 16 and fuse as late as 28. The activity of this center on the inner end of the clavicle broadens the shoulders of a young man after the age of 18.

Bony replacement of cartilage in the base of the skull starts at about 17 in both sexes and is completed by about 22 in girls and 24 or later in boys. Although it is not quite correct to say that ossification of the base of the skull is not completed until the third molar (wisdom) tooth has erupted, the statement does at least indicate that the base of the skull is one of the last parts of the skeleton to be completely ossified. Bony changes in the skull continue until quite late in life. In the pelvis a marked adolescent burst of growth in the region of the pubic symphysis is responsible for the enlargement of the pelvis that is particularly noteworthy in girls. The ends of the vertebral bodies take the form of a series of flattened rings covering the upper and lower surface of each body. These ringlike epiphyses become linked with the body of the vertebra at age 25. The vertebral column, then, goes on growing much longer than the limbs.

By the end of Krogman's fourth period all the bones of the skeleton have reached their adult size and shape. In long bones the cartilage has been replaced by bone, except at the articulating surfaces, and in membranous bones the only fibrous tissue not replaced by bones is that lying in the space between the individual bones, the sutures.

At the sutures of the skull, the main changes take place during the fifth period, which extends from 24 to about 36. These sutures gradually close as bone invades the fibrous tissue between them; the process commences on the inner aspect of the skull. The time of closure of different sutures varies considerably: scarcely any two skulls are similar in the timing or order of closure, and occasionally one side of the skull shows a more advanced degree of sutural closure than the other. These and other factors make it difficult to determine the precise age of a particular skull bone; the teeth usually provide a better indication of the age of a skull.

The sixth period, extending from about 36 to 50, the early part of middle age, shows further progress in the fusion of cranial sutures.

Toward the end of this period, early degenerative changes start in the encapsulated joints between two cartilage bones (synovial joints). Lipping, or extension of excess bone over the edge of the articular surface, results in the limitation of movement at the joint. Moreover, a loss of cartilage on the articular surface narrows the distance separating the bony elements. Areas of rarefication of bone, resulting from localized withdrawal of calcium salts, can be seen by X-ray examination, particularly near the articular surfaces.

Determining the age of a particular bone during the final or seventh period, from 50 to the time of death, is even more difficult. Some skeletons just look older than others. Sutural closure continues at different rates, marked joint changes are more noticeable, and the rarefication of bone becomes more pronounced. The deposition of calcium salts appears in some places where it is not normally found, e.g., in the cartilages of the larynx and in those at the anterior extremity of the ribs. All the bones become more brittle, and the breast-bone (sternum) is less pliable.

W. M. Cobb writes that whatever the influences that determine "the formal pattern of bone architecture, the pattern itself is an age character developing in childhood and youth, maintained during mature life and lost as age advances or infirmity intervenes."

Age Changes in Soft Tissues

All tissues and organs undergo various deteriorative changes with age, many of which are well known and have become part of our colloquial language but only a few of which are really understood. Sooner or later gray hairs appear on the head, usually first on the temple, so named to indicate that graying of the hair there reflects the ravage of time (L. *tempus*). Gray hair results from an imperfection that develops in the enzyme system (tyrosine-tyrosinase) in melanin-producing cells, the melanocytes. Although some melanocytes are still present in the hair follicles that produce gray hairs, they fail to synthesize melanin. The onset of grayness on the scalp is uneven, varying with the age of occurrence and with certain genetic factors. The hair on the axillae and pubes often remains pigmented even when that on the rest of the body is all white.

The belief that scalp hair becomes thicker with advancing age is false. The diameter of the hair shafts there does not increase after the age of 15. A steady decrease in the population of long hairs on the scalp occurs as one becomes older. The thicker hairs are more resistant to these changes and in the decimated hair population of a venerable scalp most of the extant hairs are the coarser ones. Moreover, baldness (see chapters 1 and 7) is superimposed on other aging changes.

The main change in the skin is a loss of elasticity, which causes it to become loose and wrinkled. Attempts to estimate aging by ascertaining the degree of loss of elasticity are mostly futile since such loss is too variable to be correlated with any general mechanism. Existing skin creases become more marked and new ones appear partly because of the loss of elasticity and partly because of continued creasing in a particular region. Furrows develop on the forehead, crow's feet appear at the outer angle of the eye and mouth, and creases are particularly noticeable on the face. The skin gradually becomes drier, more parchmentlike, and wizened in old age (Figure 16–7). The mechanisms responsible for these devastations are unknown, none of the theories having withstood scientific scrutiny.

The decline in reproductive capacity with the aging of the gonads has considerable biological importance. In mammals with large litters (polytocous), this decline is first manifested in a decreased litter size. In women the onset of menopause is relatively sudden and is followed by a long postreproductive period. Theories have been advanced, although without evidence, that menopause precedes final follicular exhaustion and is brought about by the failure of ovarian response to pituitary hormones. On the contrary, evidence indicates that the output of pituitary gonadotrophin does not decrease at the time of or even after menopause. Perhaps the ovary is responsible for the aging of women, since in elderly women the ovaries are often totally deficient in glandular elements. Neither the removal of gonads nor the onset of menopause has any definite relation to longevity. The cessation of gonadal activity is not, therefore, a cause of senescence. Ovarian hormones are often used for substitution therapy

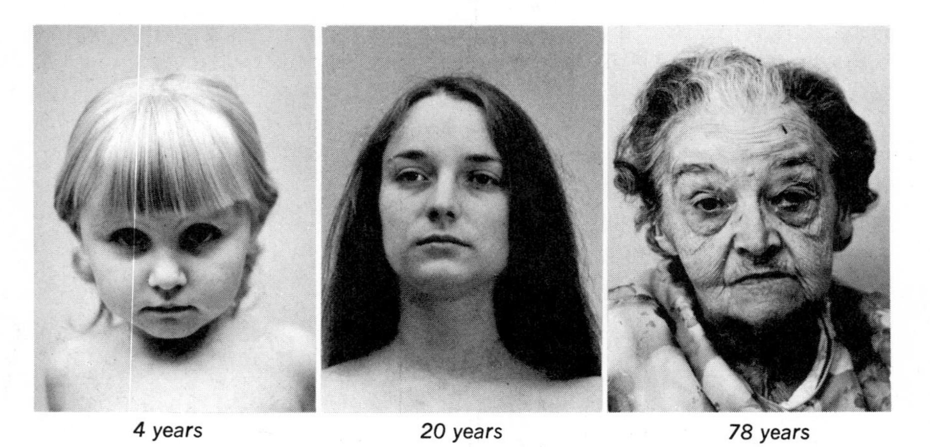

Figure 16–7. Aging.

in postmenopausal women. These hormones bring about changes in the skin and vaginal epithelium that are as near to a "rejuvenation" as any other known substance can produce, but they do not increase longevity.

Age Changes in Cells

The aging of individual cells has recently been the object of increased attention. Certain types of cells undergo active division throughout a person's lifetime. These are found in the basal layer of the epidermis, in bone marrow and lymph nodes as precursors of blood cells, in the lining of the entire gastro-intestinal tract, and as germ cells in the testis. Such cells exist only between one cell division (mitosis) and the next, and can be called "intermitotics." They exist in any one place for such a short period that hardly any evidence of aging has been demonstrated in them. Some intermitotic cells remain behind as seeds for more cell division, whereas others steadily differentiate into more specialized cells and lose the ability to divide. The basal cells of the skin, for example, differentiate into keratin-containing cells of the stratum corneum. Megaloblasts of red cells gradually acquire more of the respiratory pigment, hemoglobin, as they mature into adult red cells (erythrocytes), and spermatogonia give rise to a number of cell lines that eventually transform into spermatozoa. The intermitotics that no longer divide and that can be called differentiating mitotics have a relatively limited life span.

A third type of cell, known as reverting postmitotics, is found in such organs as the liver; once the organ has attained adult size, its cells scarcely ever divide. The usual fate of these cells is to accumulate pigment and other substances and eventually die. In the event of injury, however, even in relatively aged animals, these cells "revert" to type and undergo a series of division that give rise to cells that differentiate into functional cells until the approximate number of cells lost by injury has been restored.

The fourth group of cells, fixed postmitotics, cannot revert and rejuvenate. These highly differentiated cells, once completely defined, are incapable of further division. In this group belong red blood cells, voluntary and heart muscle cells, and neurons.

Such a classification suggests the hypothesis that a cell that undergoes or has the potential to undergo division is considerably less prone to the damaging effects of aging than one that can no longer divide. In other words, cell division maintains cell youthfulness. The ability of certain cells to synthesize such substances as deoxyribonucleic acid in more than adequate amounts apparently endows them with the ability to divide.

Some specific alterations in cells appear to indicate aging. Fragmentation, vesiculation, or reduction in the number of mitochondria, the accumulation of phospholipids within the cell, particularly in neurones, and fragmentation of the Golgi apparatus are indicators of aged cells. In nuclei, vesiculations of nuclear contents into the cytoplasm and alterations in their staining reactions can be found. Many tissues increase their accumulation of calcium, iron, and certain pigments, known loosely as wear-and-tear or "senility pigments."

Attempts to relate enzyme alterations in cells with aging have not yielded much information. All cells contain small bodies in the cytoplasm, known as lysosomes, to which are attached a number of hydrolytic enzymes. If the activity of these enzymes were not restricted or inhibited by other agents within the cell, they would destroy it. This is what happens when a cell dies and undergoes autolysis. Thus, each cell holds its death at bay by its ability to restrict or inhibit the destructive tendencies of lysosomes. Perhaps the activity of lysosomes is responsible for the steady decrease in the number of neurons in the aging brain.

Aging processes could be initiated or speeded up or both by an insufficient supply of vitamins and other vital substances or by inefficient body utilization of them. Vitamin C deficiency, in particular, has been shown to be somehow involved in cellular aging. Excessive intake of certain simple food substances, such as amino acids, fats, and carbohydrates, reduces the longevity of some animals, but little information is available concerning their effect on man.

Theories of Aging

One can conceive that under ideal circumstances tissues could remain unchanged and animals live forever. This "foreverness" seems to be man's goal in studying the aging processes. Had this goal been achieved in the past the numbers of each species would eons ago have exceeded the limits of their natural ecological niches. The total inhabitable surface of the earth and the oceans, lakes, and streams would have long ago been overpopulated, and the competition for survival would have been magnified to such an extent as to destroy life. But, neither tissues nor organisms are indestructible. In mammals, at least, the reproductive organs have a limited functional span, and with the waning of gonadal activity the other organs undergo a steady decay. Continuous use and attrition leave their stamp upon all tissues, and the agents that control the life processes gradually deteriorate until the body machinery becomes progressively less able to cope with its own functional requirements and with the hazards to which it is constantly exposed. When they can no longer withstand the onslaught of natural competitive forces, organisms succumb. It is singularly true of animals with a circumscribed reproductive function that when this function ceases, the individual dies, as if nature had ordained that organisms that are no longer useful in genetic succession are ipso facto useless and must be eliminated. Some justification can be found for this belief when one analyzes the situation in all vertebrates except man. The perpetuity of each species, after all, can be assured only by reproductively vigorous individuals. Hence, the elimination of those no longer able to reproduce establishes a natural order of things.

The theory that the deterioration characteristic of the aging man begins at a specific time and can be retarded or circumvented is specious. Aging processes in man are insidious, inexorable, and cumulative, many beginning even before maturity is reached. Man's hopes that these phenomena can be reversed or avoided are understandable since he does not want to die, but his chances of warding off old age and death are meager.

Many theories of aging have been proposed, critically discussed, and discarded. Senescence is accepted as an inherent property of multicellular animals, a product of natural selection, arising by chance and perpetuated as a beneficial attribute. Senescence can be the deteriorative result of the pace of living or of toxic or pathological processes; perhaps it is the consequence of changes in tissue fluids or the end of a declining rate of cell growth and replacement. It has been postulated that each cell must eventually die when certain vital substances, like the enzyme and replication systems, are exhausted and can only be replaced by a cell division that does not occur. Cells progressively lose the power to replace themselves. This steady dissipation of "growth energy," known as Minot's Law, has been expressed by Sir Peter Medawar as "development, looked at from the other end of life." Again, senescence may be a measure of declining selection-pressure, the living system having persisted after its "biological program has been exhausted."

Some evidence shows longevity in man to be inherited in the same way as liability to certain diseases. For obvious reasons more is known at the moment about the heredity factors involved in a relatively short life span. It has been said that if a man wants a better chance of longevity he should choose long-lived ancestors. But how much selection has had to do with the increased expectation of life during the last fifty years is not clear. There is little warrant for arguing about the inheritance of general health or bodily vigor when so little is known about the nature of senescence or the way it is being modified by man's increasing success in adapting to his environment.

Male animals are, as a rule, shorter lived than females. The human male is more vulnerable than the female beginning with the fertilization of the ovum, after implantation, and in childhood—times when claims of greater occupational hazards are hardly possible. In some animals the age of the mother at the time of conception, but not so much that of the father, influences adversely the longevity of the offspring, but little information is available for man. Children of women in their late thirties and early forties have slightly greater chances of being in some way abnormal, physically or mentally. Some people also advocate sexual continence or abstention to increase longevity, but no scientific data support the claim that ascetics live longer than sexual athletes. Regular sexual indulgence and a normal number of pregnancies in women apparently do not affect their expected life span.

The interesting, fortunately rare, phenomenon of progeria should be mentioned here. This is the premature aging of children or adults, which has generally been attributed to endocrine disturbances, particularly of the pituitary. The condition, however, is too generalized to be explained simply in terms of hormonal activities. Progeric children cease to grow, their skin becomes wrinkled, making them look like little old men or women, and they seldom survive the third decade. In adults the hair suddenly grays and falls off, and the skin and other organs undergo abrupt deterioration.

Man's attempts to increase his life span have been many, sometimes bizarre. Restricting the caloric intake in the diets of young animals can cause a prolonged suspension of growth and a delaying of maturity without any apparent aging changes. However, the value of the experiment is limited by the need to allow weight increase to avoid death. There is little evidence that starving a man during his childhood or adolescence will affect longevity. These are interesting attempts, but the methods seem too heroic and uncertain to be useful in increasing human longevity. Malnutrition, certain diseases, and other factors can postpone the onset of puberty in man, but such deficiencies usually shorten, not prolong, life. Conversely, some believe that too rapid an increase in body growth during childhood shortens the life span. Such shortening could be mediated by some pituitary exhaustion or even overfeeding during childhood, but too little is known about this too.

The death of an organism seems to be traceable to the individual cells that make up each of its organs. With the passing of time, cells gradually accumulate obstructive substances or become irreparably damaged by any of the innumerable hazards to which they are constantly exposed. Such a process continues insidiously until the essential organ to which they belong is impaired in its role of maintaining the organism, and death ensues.

The inevitable conclusion of this chapter should make one reflect briefly on the wider implications of increased life expectancy in man. Medicine has not yet increased the specific longevity of man although it has reduced the number of people who die before reaching the expected age. The efforts of preventive and curative medical treatment are directed largely toward aiding everyone to reach his "three score years and ten." Some people, but not necessarily all, will live longer.

By his own efforts man is creating a situation in which large numbers of people will remain alive in each of the first seven decades, with a sudden drop in the eighth. The number of old people in a static population will increase. Man has to face up to the fact of a preponderance of old people in the world of tomorrow. Despite savage remarks to the contrary, no valid biological or philosophical reason can be adduced why this should be detrimental to the species. Admittedly, however, this ratio will necessitate considerable social readjustment.

17

Reflections on his nature

If then in outward manner, form, and
mind,
You find us a degraded, motley kind,
Wonder no more, my friend! The cause
is plain,
And to lament the consequence is vain.
(J. Hookham Frere)

In the foregoing chapters we have discussed most of man's unique biological attributes. *Homo sapiens* has emerged within and from the order Primates endowed with such a complex nature that even those properties that he shares with other animals are masked in him by numerous, often infinitely subtle, modulations. His anatomical differences should, and occasionally do, provide a clue to his special nature. As he attained a disproportionately large brain, an upright posture, remarkably skilled control of his hands, and increasing longevity, he also developed a society structured more upon group effort than upon rugged individualism. Perhaps the complexity of his society developed because alone man might not have survived the constantly menacing environments from which he had to wrest his food and shelter. The vagaries and inclemencies of his surroundings forced him, a nearly naked creature, to construct shelter and don protective garments. Lacking such natural weapons as horns, fangs, claws, and hooves for obtaining food and fending off enemies, he had to outwit his enemies by banding with others and constructing weapons for defense, aggression, and most important, for obtaining more than his share of food by baiting, snaring, and killing his prey.

By nature omnivorous, man learned to control edible vegetation, particularly in those parts of the world with clearly defined seasons. As he gained an understanding of growing plants, he also learned to gather and store seeds, roots, and tubers, to plant and cultivate them, to protect the seedlings from marauding or parasitic creatures, and most important, to harvest and store the produce during times of plenty in order to survive during the lean seasons. Hunting, fishing, and farming, using the tools he had made, and domesticating animals for food, transportation, and work—all this demanded that he, defenseless and alone, should form social groups. Orderly planning, social cooperation, and the accumulation and sharing of skills became not only the necessary means of fashioning his society, but also the ultimate goal of its development. This remarkably complex structuring of an initially simple life-style was due to his anatomical advantages over all other creatures.

The integrity of human society is largely maintained by the repetition of symbolic procedures. Scholars have pointed out that through the use of symbols order and discipline are maintained in the home, armed forces, affairs of state, industry, and religious institutions. Since, despite his naturally aggressive proclivities, man has had to live in social groups to survive, he has likewise had to incorporate into his society certain regulations that restricted his individual freedom of action while insuring the integrity of the group. Laws, rules, regulations, mores, rituals, and taboos no doubt arose from the necessity of having to live together. Taken in their strictest utilitarian sense, these restrictions protect the group from disintegration and individuals from extinction. However, social restrictions promote challenge and man has rarely resisted challenge. Furthermore, social restrictions insist that the individual subserve society, and thus mask or obliterate his ego. This at once stirs up competition and intragroup aggression, since very few men can passively observe the spectacle of their ego, their most important possession, being suppressed.

In a relatively static society concerned with agriculture, animal husbandry, and the simple arts, the family is the mechanism for passing on its basic culture. Individuals of different generations within the family can pool their knowledge and experiences on matters as widely separated as childbirth, house decorating, and investment. Essentially stable and conservative, the family is a source of strength through closely knit ties and links, a refuge from the rapidly moving world outside.

The family also evolves rules of behavior, creates standards to guarantee respect, and establishes its own moral values within a framework set by society. The role of the family is repeated at higher levels in schools, institutions, colleges, committees, and governments. Everywhere it helps to foster a stable emotional background, but at the same time it generates man's greatest problem: there are too many men in the great family of Man.

Today several factors operate against the cohesion of the family unit. The increasing ease with which individuals can travel away from the family leads to new experiences, new behavior patterns, and increased realization that things are being done better elsewhere. The family is slow to accept and assimilate new knowledge and is even less adept at transmitting new ideas. Educated parents are no longer experts about all knowledge. Too often, children trying to get a clear idea of, for example, thermodynamics and electronic equipment find parents out of touch, out of date, or hopelessly uninformed.

The raison d'être of the family is reproduction. But, even the importance of this function has diminished since the extraordinary increase in population has created the attitude that, beyond a certain replacement point, reproduction is unwise. No longer can a father boast about a family of a dozen; in fact, such men may represent only a lack of social regard, carelessness, or stupidity. The importance of the family as a fit place in which to raise children is questioned daily, especially since other agencies have effectively assumed many of its functions. The earlier a child's exposure to education, the earlier all young and flexible nervous systems are given equal opportunities to develop, the sooner family influences are weakened. Whereas only some people today seriously or overtly propose the abolition of the family, the responsibility for raising children is gradually but noticeably passing from the family to society itself. This shift alone has a profound effect on the way man will behave in society.

Recently, one of the least understandable manifestations of this alienation from home and family influence has been the insouciant attitude and behavior of many young people toward the "father figures" of previous generations. Obviously, there has always been a generation gap both in human and in nonhuman families, and there has always been a difference of attitude between the men and the boys, between teacher and pupil, between authority and people, between them and us. But in recent years this gap has widened, and many of the under-thirties regard their elders as lacking in imagination and understanding, oppressive, boring, unresponsive, and outright failures. By their appearance, behavior, and attitude, these rebels make it very clear that they want nothing much to do with authority of any kind, that they favor revolution and the establishment of a new society. Adult reaction to this rebellion has been variable: considered anarchic by some, the young are lauded by others for the new vitality and outlook they have brought to the social scene. No one can deny that, largely through the explosion of our technology, family strength and unity have been seriously undermined by the ready availability of information and speedy dissemination of news, the increase in affluence and leisure time, and perhaps the introduction of radio, telephone, and television. Furthermore, the family itself is partly responsible for its own disintegration. When a man no longer needs to work all of the daylight hours to earn enough to provide food and shelter for his family, being bored, out of touch, and lazy, he indulges himself. Women with automatic washers and dryers, dishwashers, garbage disposals, vacuum cleaners, and other labor-saving appliances in their homes and enough money to pay for babysitters are no longer imprisoned at home with their children. Thus emancipated, they are free to work, to play tennis or bridge during the day or at night, and to leave the children with that universal foster parent, the television set.

Comparative studies of social behavior in nonhuman primates are sometimes useful in understanding our own society, but the number of living species is so great and their form, size, habitat, habits, and biological attributes so varied that social interactions among members of the various groups differ enormously. Many of the lowly prosimians—small, nocturnal, mostly forest-dwelling and carnivorous—are solitary. An unmistakable pattern can be traced from these small, arboreal primates to the larger, open-country, predominantly vegetarian ones. Even among the forest-dwellers, there is an obvious progression from a solitary existence, with a dispersed population, to compact family groups. For the most part, the latter exhibit defensive behavior and little aggression toward other members of the group. The largely vegetarian dwellers of the grasslands or savannas, which range over a less well-defined home territory, belong to variably large groups with a better structured social order. They display both intra- and intergroup aggression, defensive behavior, and greater sexual activity and competition than the members of the smaller groups.

Chapter 12 discussed man's almost incessant preoccupation with sex, whether inside or outside the family. This may be one of the most important factors in keeping individual members of society interested in and tolerant of other members. Sexual behavior, in fact, plays a pivotal role in human society: it is part of its self-preservation and the basis of monogamous families. Man at different times has experimented with every conceivable type of family organization. Among monogamous societies, the family is largely patriarchal with strong matriarchal undertones. Polygamous societies have not flourished, and only a few isolated groups have practiced or still practice any of the many types of polyandry. Thus, monogamy would seem to have been the basis for social success in the past. But today that institution is threatened on all sides, and if it is destroyed, as there are indications that it will be, the social order will have to undergo many drastic changes. Among other social animals, including primates, males, whether dominant or not, prefer certain females, but do not exclude the others. This, however, does not have much to do with the family unit. In a primate natural troop, the father's identity is seldom known, and it is the mother around whom the family revolves. She unifies the group, protects it, and probably passes on to it the essential instruction for survival in the society.

In addition to their generally monogamous proclivity, human primates are exogamous, rarely mating with close relatives. Consciously or not, even the more primitive extant tribes often avoid the disastrous results of intrafamily mating. Yet, among other primates interbreeding prevails within some groups. Free-ranging baboons remain near one another, and their cohesion is maintained by inbreeding. Different troops can be distinguished by the physical or behavioral similarities of their members, each troop often showing a predominance of certain color and length of hair, and length and form of tail or muzzle. However, in a semi-free-ranging troop of Japanese macaques, we have never observed brother-sister or son-mother mating, although symbolic mounting often takes place. It is difficult to know whether father-daughter mating occurs since the father is often unknown and has no role in family life. However, at the Oregon Regional Primate Research Center, in a colony of Celebes black apes, which is maintained as a harem, father-daughter mating occurs often since the male does not tolerate competitors.

Since women are receptive to the advances of men during almost the entire ovarian cycle, copulation is a social function that transcends reproductive purposes. Except for lionesses and some nonhuman primates, the females of nearly all other mammals rebuff the male, sometimes viciously, except during estrus, a brief interval in the cycle when ovulation is most likely to occur. Indeed, most other primate males show little interest in females during anestrous periods. S. L. Washburn and I. De Vore observed that female baboons become sexually aggressive at estrus and "actively solicit the attention of the males and may mate with several in close succession. In the later stages of estrus, the female forms temporary consort relationships with the dominant adult males. These consort pairings may last a matter of hours or days and usually dissolve peacefully." Most of the solitary primates or those living in small social groups have fairly well-defined breeding seasons, a fact that greatly influences their social structure.

Although cooperative living restricts individual freedom of action to some degree, it makes individual efforts more effective and gives the individual time to relax, rest, and think. Thus, the modern monogamous man, free from the endless responsibilities of providing shelter and food and of looking for a new mate, can expend his energies on activities that give him pleasure and intellectual stimulation. Man, then, has emerged with a nature divided between devotion to the needs of the social group and orientation toward solitude where he can idle, amuse himself, wonder about his nature, and dream about conquering the world around him. Articulate symbolic speech has perpetuated his accumulated learned skills and experiences and has itself become increasingly complex as the facts that need to be recorded become more complicated. Such may have been the evolution of human culture; and this very culture, far more than man's anatomy and physiology, separates him unmistakably from all other animals.

Charles Darwin *(The Descent of Man)* proposed that man alone is capable of progressive development; he alone uses fire, makes tools, domesticates animals. Darwin overlooked the fact that some insects, like the ants, have also more or less "domesticated" other insects such as the aphids. Only man has the power of abstraction, can form general concepts, is self-conscious, and can comprehend himself. Man alone employs language, has a sense of beauty, is liable to caprice, feels gratitude, and is disturbed by the unknown. Man alone believes in God and is endowed with a conscience. Darwin maintained that man and other primates have some "instincts" in common, exhibiting the same intuitions, sensations, passions, affections, and emotions. Man is by nature jealous, suspicious, and emulative. He alone can display gratitude and magnanimity. He shares with some other primates the proclivity to cheat and deceive, to be vindictive and revengeful; but we do not know whether, like man, the other creatures are sensitive to ridicule, have a sense of humor, and feel wonder. Certainly many nonhuman primates have the faculties of mimicry, attention, deliberation, selection, memory, imagination, association, and reason of some sort; and, though less often than man, some of them may suffer from mental derangement. The mental endowment of other animals is reflected in remarkable individual variations. Everyone who has had a pet dog, cat, or monkey knows that some are dull witted and occasional ones very bright.

The immense differences between the mental capacity of the highest nonhuman animals and that of the lowest "normal" man has only rarely been called into question. In an amusing passage, Darwin imagines an anthropomorphous ape taking a dispassionate view of his own case. The ape would have to admit that though he could artfully plunder a man's garden and use stones and sticks for fighting or cracking nuts, the idea of chipping stones into tools was beyond him. Still less could the ape pretend to pursue a train of metaphysical thought, solve an equation, or reflect on matters divine. He could concede that some of his kind could and did admire the sexual skin and fur of their lady friends. By cries and grunts, postural attitudes and gestures, some of the ape's perceptions and simpler desires could be communicated to others. But either it had not occurred to him or he did not have the natural attributes to express definite ideas by specific symbolic sounds or the recording of his speech with graphic symbols. The ape might be willing to go to the aid of his fellows, adopt orphans, and risk injury and death for them, but disinterested love for other living creatures, that most noble attribute of man, was quite beyond his comprehension. Darwin's conclusion of these fantasies is that the best and highest distinction between man and other animals is his moral sense and agreed with Marcus Aurelius that man's social instincts are the prime principle of his moral constitution.

Faced with so many unknowns, all but the Skinnerian psychologists must admit that the whole story of the biological nature of any animal, and of man in particular, is not told even when his genetic, anatomical, physiological, and biochemical characteristics have been listed. In addition, man shares with other animals imprinted behavior patterns, generally called instincts, and acquired habits of behavior and thought. What he learns, remembers, and integrates constitutes the raw materials for logical reasoning. Man has emotions, a temperament, and self-consciousness, i.e., he is aware of his own personality and character. He dreams when he sleeps and continues to dream in his wakefulness. He searches endlessly for his free will and meanwhile is held captive by his fear of the unknown. He uniquely needs to communicate to others his inner experiences, and in his inability to do so satisfactorily he becomes lonely and despairs. Watching other men's achievements, man sees himself as in a mirror; he sees in actuality the very experiments he wanted to try if only he had dared, if only he had not been so afraid of failure. Man seems always to labor under stress: he is plagued by feelings of aggression and competitiveness with his fellow men without whom he could scarcely survive.

Deeply concerned with his own consciousness and with those elements not immediately related to the external world, man creates a philosophy (denied by B. F. Skinner), the search for underlying causes and principles. Henri Bergson, one of many who have struggled to define consciousness, thought of it as complex and indivisible, with all of its phases, whole or partial, merging with every succession. From birth to death, consciousness streams forward uninterruptedly, broadening its banks, expanding and deepening with maturity, and then dwindling into old age. It is probably qualitative, and its intensities are immeasurable. The apparent increase in the "quantity" of consciousness whereby one seems more "intensely conscious" is only chimerical, for it is the stimulus upon the mind, not the mind itself, that is really intensified. At the core of such a philosophy of consciousness is its continuity and duration, characterized by alterations, increasing storage of ever-lengthening past experiences, and an inexorable forward progression. Such unpredictable features as sentient experience, habit, and memory lend to consciousness the appearance of newness; truly conscious activities, linked with the passage of time, are never exactly the same from one instant to the next.

Consciousness is the ability to orient oneself in space and time with memory and habit, with judgments and decisions, and with some degree of insight into the nature of being. Wakefulness and vigilance are integral parts of consciousness. The seat of consciousness may be in the subcortical levels and that of awareness in or near the thalamus, but the cerebral cortex gives subtle meaning and even purpose to them. Clinical observations of patients recovering from confused, concussed, or comatose states, deep sleep, or anesthesia suggest that levels or layers of consciousness exist.

Man's awareness of his environment, his ability to adjust to it, and, when this is not possible, his success in adapting that environment to his own needs have contributed much to his survival. Thus, his first requisite has been a finely developed power of observation and investigation. The continued visual, auditory, gustatory, and manipulative examination of the things about him endows the human child with ever-increasing information that in time gives him a degree of mastery over his world. In different degrees, other animals also learn to adjust and even to modify their environment by familiarizing themselves with it, profiting from its abundance, and protecting themselves from its dangers. This seems to be instinctive since all animals, from insects to man, have an inherent curiosity about their surroundings, which they explore constantly to determine the least change. The difference between primates and other mammals appears to lie in the degree to which they respond to their environment and in their capacity to perceive and react to it.

Like human beings, all animals and particularly monkeys and apes learn many things from exploration and experimentation, some of which do not seem to be pertinent to their immediate survival. For example, some of the higher primates play with sticks and manipulate them, but only a few, especially chimpanzees, put these skills to useful but very limited purposes. Tactile and visual explorations are correlated by the manipulation and inspection of objects. Man's hands may have acquired increased skills through visual guidance, which in time led to progressively greater tactile sensibility and discrimination. The development of motor mechanisms responsible for the attainment of high degrees of muscular coordination of the hands and fingers in man parallels his highly developed sense of touch, and this is one of the major assets of his awareness.

The ability to store, coordinate, and retrieve information gathered through experience is to a large extent the very substance of man's intelligence. With the attainment of greater degrees of intelligence, adaptability, and fantasy, his nature becomes progressively more complex.

Sentient man needs relaxation, freedom from boredom, and physical and mental pleasures. An essential part of these requirements is his need to play. All recorded history indicates that people have played games. Not all sociologists agree that athletic games are a substitute for innate aggression, but games in general provide ritualistic and ordered patterns of interaction. The writings of Homer and Virgil witness to their importance in Greek and Roman society, to participants and spectators alike. Some have even suggested that the Olympic games were partly responsible for the relatively peaceful coexistence of independent city-states in ancient Greece. In modern societies, where man is no longer obliged to devote his full time to providing for his basic needs, increasing numbers of professional organizations—athletic, social, and intellectual—have emerged. As in the past, a kind of adoration is paid to athletic heroes, accompanied by near hysteria when international, national, or even local sporting events take place. Perhaps these are not necessarily attempts to combat boredom or substitutes for aggression. One of man's idiosyncrasies is his need for challenge as well as for pleasurable diversions. Chess and card playing, quiz shows, hunting, fishing, mountain climbing, skiing, and swimming may not be merely the simple indulgence of man's need to overcome ennui. Perhaps even the comedy of the space race among nations and the spectacular achievement of landing on the moon or orbiting unmanned spaceships around Mars expresses more than anything else man's need to challenge himself to his deepest core.

Nearly all animals play. The aerial circling of flocks of birds, the endless chasing, romping, and play-fighting of dogs and cats are common examples of animal activities that serve no apparently "useful" purpose. Free-ranging troops of baboons, macaques, and other social primates indulge in intricate patterns of mock fighting, chasing, hair pulling, splashing in water, tossing and retrieving sticks and stones, and many other obviously amusing activities. Our Japanese macaques at the Oregon Center roll snowballs each time enough snow falls to warrant the effort. Such activities, however, usually cease with adulthood; only man continues to play most of his life.

One area of modern "sport"—hunting and fishing—poses a conundrum. Whereas early man preyed on other animals for survival, modern man obviously does so for other reasons since he can more easily purchase his prey than hunt for them. Either he naturally enjoys killing, or he learns to enjoy it; both are disturbing thoughts. However, such presumably atavistic expressions of man's patrimony as hunting and fishing are so acceptable that fish and game commissions in the United States and other countries spend fortunes stocking streams, fields, and forests with fish, fowl, and game to be killed by man. Although pleasure may be the primary goal of the hunter and fisherman, perhaps his ceaseless quest for challenge is an equally strong motive. On the other hand, we are faced with the frightening spectacle of men fighting not only sharks, crocodiles, and bulls but each other as well. And we are reminded of the now outmoded public tortures and executions that in times past drew huge crowds of curious spectators who could have stayed away. The incredible atrocities committed by all people at war attest to an ugly animality in nearly all men. The puzzle remains: whether play of all sorts is an outlet for or a symbol of innate hostilities, an attempt to overcome boredom, or a physical and mental challenge to man's abilities.

Man is at the top of the order Primates as its latest model, with millions of years of tailoring behind him. This realization should make him pause to reflect on the permanency of his hold on this exalted position. Dinosaurs, moas, and Irish elk all showed a disturbing tendency to vanish almost as mysteriously as they had appeared. Other species have been replaced by forms better adapted to their environment or, like the living fossils, have remained in small, isolated populations, no longer ruling and dominant.

Man constantly tries to reassure himself about his biological presentiments. Theologians have emphasized his divine nature, his soul, and the near-perfection of his creation by divine intervention. Scientists have tried to provide an anatomical and physiological basis for these beliefs by emphasizing his conspicuous cephalization, which has endowed him with a larger and better brain than all other creatures, and by pointing out that such an "improved model" has "changed gear" and operates at a supra-animal level. Even T. H. Huxley believed that, although man is "one with the brutes," something more hopeful than brute intelligence is to be found in the nature of his brain. Whereas we concede Huxley's point, we are forced to admit that in the whole range of human brains, some are so poorly endowed as to make their affinity with brute animals more obvious. Furthermore, the American iconoclastic psychologist B. F. Skinner denies that man has attained either freedom of thought and action, a philosophy, or personal dignity. We think Skinner goes too far.

We have seen in chapter 5 that man has searched urgently in his "top model" brain for the existence of special regions that are responsible for his human attributes. But he has been better at investigating the smaller components and their functions than the whole. It is still beyond his ability to "explain" in all but the crudest neurological terms the fundamental activity of a human brain—thinking. Despite the disappointing results of his search, even the very act of searching attests to the unique nature of his intelligence.

Guided by a superior brain, man made things with his hands. He discovered fire and the use of stones and metals. He invented the wheel and later, engines, electronic devices, and computers. He discovered and harnessed nuclear power. This inventiveness suggests that man is mastering his environment and in a temporal sense is making enormous strides to delay the inevitable decay of his species. Such an optimistic view is further buttressed by his advances in space explorations. He is on the verge of freeing himself from the limitations of earth, which some people pessimistically believe is being exhausted and polluted by overpopulation and overexploitation. Ironically, during his frenzied attempts to master all manner of technical feats, man has treated his environment with contempt and has so changed and polluted it that in many places it may be beyond redemption. Smoke, insecticides, chemical pollutants of land and water, billions of unwanted objects, and streams and lakes unfit for flora and fauna have rendered much of what was habitable earth a disastrous accumulation of discarded commodities.

This is admittedly a sorry form of escapism. The human race has effectively jeopardized its existence by its increasing dependence on motorized and electrical power. Failure of electricity for more than a few hours can create a shattering dislocation of near air-raid proportions in a large conurbation such as New York City. Conceivably, a computer-managed society can come to be plagued not only by unemployment problems undreamed of by leaders of managerial revolutions but particularly by the abysmal boredom that comes from unending leisure time.

We cannot escape the fact that in addition to his unique anatomy, physiology, and biochemistry, man is distinguished from all other animals by genius, which the French naturalist Buffon is said to have defined as "an ability to be patient." Equipped with this kind of patience, an intellect, and a relentless drive to excel, literary scholars, philosophers, historians, and scientists have done filigree work in erudition and have pursued goals, which, when achieved, offer promises of recognition and aggrandizement. But can it be that there is much more to man's genius than the search for recognition? In his eternal search for self-expression, man has created the arts, those products of his culture that derive from and appeal to his sensory perception. But of all the arts, none is more singularly human than, and enjoys such wide acceptance as, music.

Furthermore, regardless of personal preferences, nothing can compare with the overwhelming achievement of a full orchestra. Only consider the many factors that must be synchronized to produce a performance. The composer, like the playwright, tells both the conductor and the players what to say and how to say it. Then, equipped with specific directions, each musician must be able not only to follow the instructions but also to translate them into sounds that, despite being played according to the composer's specification, must somehow reflect the individuality of the player. The true musician does more than follow tempi and dynamic marks; if he did not, all players would sound alike. Furthermore, no two players have identical techniques, embouchure, or, for that matter, emotional response to the music. Still another factor individualizes a musician's performance: the quality of his instrument. Since most instruments are handcrafted, no two of them sound exactly alike, just as no two voices are identical. Finally, the conductor, always aware of the composer's wishes, must respect the performers' individuality and unify all these elements through his own wizardry, his own interpretation. This is a remarkable human achievement in terms of individual and communal effort. An orchestra is, in fact, the epitome of a micro-human society.

Music must surely be an extension of man's speech, his chief means of communication. Most animals make noises of some sort. We have seen that among vertebrates sounds are generally produced by the larynx, but when that organ is incompetent or missing, animals have other means of communication. Insects rub their legs or wings against vibratile membranes to produce sound patterns distinctive of each species. Fish produce sonic vibrations in the water by gritting their teeth, rubbing their fins together, or expelling air from their swim bladder. Amphibians croak, crake, or drill by inflating and deflating vibratile pouches on the sides of the head or below the lower jaw. Most reptiles are mute; but some hiss and shake rattles at the end of their tails, and crocodiles roar. Birds produce numerous and occasionally complex sounds, and the males of some species articulate stereotyped patterns of sounds into what we call "song." In spite of their elaborate songs, most birds have an extremely limited vocabulary; the quality of their song results mostly from an astounding amount of overtones in the background. Birds that have little to say make sounds by hissing; inflating and quickly deflating their air sacs; fluttering, flapping, or slapping their wings; and drumming on hollow logs with their bills. Mammals produce infinite varieties of vocal expressions and sounds, mostly with the larynx, which is often equipped with amplifiers, dampers, and modifying devices. Actually, however, the vocabulary of any given animal is restricted to a few basic sounds. Birds and mammals have such distinctive dialects that experts can trace their origin by differences in their accents, inflections, and phraseology.

All animal noises are messages. One would think that the more elaborate and distinct the sound, the clearer the message; however, we know too little about this subject, and only the speech of some animals has been decoded with modest success. Sounds are designed primarily for intraspecies communication: to identify oneself, to attract a mate, and to warn or threaten real or potential enemies.

Man, then, is surrounded by a noisy world, but the noise conveys only a few basic messages. We have stated earlier that among our biologically unique credentials, none is more remarkable than speech. Human speech is not only a departure from all the assorted sounds produced by other animals but also a product in some way controlled by nearly every part of the brain. It is no wonder that speech is the major means whereby we communicate basic information and express subtle shades of emotion.

Perhaps all of the basic words in any language are or were originally onomatopoeic. With the passing of time, words were distorted or even replaced by other symbols and sounds. No doubt, music was born not separately from but together with speech, and their evolution has progressed concomitantly.

To understand the evolution of music, one must understand man, the adventurer and experimenter. Innately aggressive and hostile, he also experiences moments of exhilaration, joy, serenity, and love; and because he is a social animal, he is compelled to communicate his feelings. To illustrate what adventure has to do with music, let us cite an analogy. We walk, run, and jump on two legs; this is our locomotor behavior. However, why do we try to run faster than we need or should? What compels us to scale mountains, abusing legs and trunks that were not constructed for such efforts? What makes us pour out cascades of words when we feel joy, sorrow, or a sense of wonderment?

There is not a single controllable biological attribute that we have not altered or pushed beyond its normal limit of function. Man does with his voice and speech exactly what he does with legs, arms, and trunk. He experiments with every conceivable quality of his vocal apparatus, and the song he produces is probably the result of three factors: an inclination to experiment, a need to communicate emotions more meaningfully than through the spoken word, and an attempt to mimic natural sounds in nature. Centuries of natural selection, experimentation, refinement, and innovation have given rise to song, the basis for all music.

No one knows much about the genesis of musical instruments, probably because they are not basic to man's survival. However, they are in some ways man's most intriguing artifacts. The construction and playing of musical instruments was made possible only by our bipedal gait, which emancipated our hands. Musical instruments of sorts have been used by man perhaps from the time he learned to fashion tools. And, in a sense, musical instruments can be traced to objects that produce natural sounds. A dry branch firmly attached to a trunk produces audible vibrations when snapped, and branches of different size produce vibrations of different pitch and quality. Striking a hollow log produces a booming sound, and blowing through partially chewed straws sets up reedy vibrations. Records are scarce but it would appear that even back in the dawn of history, music was an integral part of man's culture. It could not and cannot be separated from man, whether the indomitable, destructive, cruel human animal, or the idealistic, gentle, godlike being.

Music is more than culture and art; it is a way of life. It is a means of communicating thoughts, experiences, and emotions that cannot be articulated in other ways. This is its basic function. It conveys man's sense of order and symbolic logic; it expresses his sense of beauty, the ultimate in human emotions. More than other forms of art, music can convey the ugly and tragic with the same medium as the beautiful and joyous and can enable us to experience vicariously one or the other.

Everyone is, to some degree, a musician just as everyone is an athlete. We all run and toss balls, but the gifted do so with such adroitness that we pay them for doing it, and as we watch them perform we identify ourselves with them and are entertained by the spectacle. Similarly, we all sing or whistle, and make what can, charitably at least, be interpreted as musical noises. When we hear a virtuoso, we become that virtuoso.

Except for those who have a "tin" ear, most sentient persons enjoy music. In our age, when what were once luxuries have become essentials, men in all strata of society no longer live by bread alone. Furthermore, ours is a violent age that needs the arts, music in particular, to mellow the acidity of our spirits and to blunt the edge of our aggressions. In this age, when our very future is uncertain, perhaps music, our common culture, will help us to live through the anxieties and stresses of overcrowding, overcompetitiveness, ennui, and lack of purpose.

Sweeping generalizations about the arts and the sciences are common, as if each occupied a separate niche. Thus, imaginary barriers have been erected between the two and have obscured the reality—that they really are the two facets of man's genius. Music belongs to everybody—rich and poor, enlightened and bigoted, brilliant and stupid.

Perfection has been the goal of man's quest. His journey through history is a record of painful ascent toward some pinnacle of greatness. Reaching for eminence, along the way man transmutes his ideas and ideals into the arts. Plagued by his real or imagined shortcomings and a desire to escape the vexing realities of the present, he creates utopias, and in his attempts to reach them often sacrifices the past and the present to a mythically better future.

The products of man's creative urge are like mirrors, reflecting his need to refer everything to himself, in whom and through whom all the elements of his life are integrated and justified. Thus, the content, meaning, and value of his artifacts are rooted in the sentient being who assesses them. Gradually man has fashioned an idealized image of himself, better than he really is, and woven it into his culture. And through this culture he has made his real impact on history. Without it, he is merely the human animal, one among many; with it, he is a human being—the idealized image of himself. Humanness, then, is achieved by man's conscious attempt to escape the tyranny of his real being. Paradoxically, he is compelled to seek the truth about himself, only to find the human animal ever ready to spring upon and rend asunder the culture he has struggled to fashion in such a way as to mask and suppress his basic animality. Herein lies at once the most mysterious and the most hopeful aspect of man's nature: with pathetic insistence, he continues to search for the ideal man, the humane being.

REFERENCES

Numbers in parentheses indicate page numbers of quoted material in text.

(22) Aristotle. (1943) *Generation of Animals.* (tr. A. L. Peck). Book V. Ch. 3. Cambridge, Massachusetts, Harvard University Press; London, Heineman.

(358–359) Carpenter, C. R. (1964) *Naturalistic Behavior of Nonhuman Primates.* University Park, Pennsylvania, Pennsylvania State University Press.

(9) Jones, F. Wood. (1929) *Man's Place Among the Mammals.* London, Edward Arnold.

(141) Napier, J. R. (1967) The antiquity of human walking. *Sci. Amer., 216,* No. 4:56.

(337–338) Schultz, A. H. (1961) Some factors influencing the social life of primates in general and of early man in particular. In: *Social Life of Early Man.* (ed. S. L. Washburn). Chicago, Aldine.

(28–29) Ward, F. O. (1858) *Outlines of Human Osteology,* 2nd ed. London, H. Renshaw.

FOR FURTHER READING

Chapter 1

Campbell, B. G. (1966) *Human Evolution: An Introduction to Man's Adaptations.* Chicago, Aldine.
Huxley, T. H. (1863) *Evidence as to Man's Place in Nature.* London, William E. Norgate.
Napier, J. R. and Napier, P. H. (1967) *A Handbook of Living Primates.* London/New York, Academic Press.
Romer, A. S. (1941) *Man and the Vertebrates,* 3rd ed. Chicago, University of Chicago Press.
———. (1962) *The Vertebrate Body.* Philadelphia/London, W. B. Saunders.
Tanner, J. M. (1964) *The Physique of the Olympic Athlete.* London, Allen & Unwin, 1964.
Ward, F. O. (1858) *Outlines of Human Osteology,* 2nd ed. London, H. Renshaw.
Young, J. Z. (1957) *The Life of Mammals.* Oxford, Clarendon Press.

Chapter 2

Buettner-Janusch, J., ed. (1963–1964) *Evolutionary and Genetic Biology of Primates.* New York/London, Academic Press.
———. (1966) *Origins of Man.* New York/London/Sydney, John Wiley.
Clark, W. E. Le Gros. (1955) *The Fossil Evidence for Human Evolution.* Chicago, University of Chicago Press.
———. (1960) *The Antecedents of Man.* Chicago, Quadrangle Books.
Dolhinow, P. and Sarich, V. M., comps. (1971) *Background for Man: Readings in Physical Anthropology.* Boston, Little, Brown.

Gregory, W. K. (1951) *Evolution Emerging.* New York, Macmillan.
Hill, W. C. Osman. (1953–1966) *Primates: Comparative Anatomy and Taxonomy.* Edinburgh, Edinburgh University Press.
Hooton, E. A. (1946) *Up from the Ape.* New York, Macmillan.
Morris, D. (1962) *Biology of Art.* New York, Alfred A. Knopf.
Napier, J. R. (1970) *The Roots of Mankind.* Washington, D.C., Smithsonian Institution Press.
——— and Napier, P. H. (1967) *A Handbook of Living Primates.* London/New York, Academic Press.
Simpson, G. G. (1945) The principles of classification and a classification of mammals. *Bull. Amer. Mus. Nat. Hist.,* 85:1–350.
———. (1962) Primate taxonomy and recent studies of nonhuman primates. *Ann. N.Y. Acad. Sci.,* 102:497–514.

Chapter 3

Abbie, A. A. (1968) The homogeneity of Australian aborigines. *Archaeology and Physical Anthropology in Oceania,* 3:223–231.
Comer, J. P. (1967) The social power of the Negro. *Sci. Amer.,* 216:21–27.
Coon, C. S. (1963) *The Origin of Races.* New York, Alfred A. Knopf.
Dobzhansky, T. (1962) *Mankind Evolving: The Evolution of the Human Species.* New Haven/London, Yale University Press.
Dolhinow, P., and Sarich, V. M. (1971) *Background for Man: Readings in Physical Anthropology.* Boston, Little, Brown.
Edmonson, M. S. (1965) A measurement of relative racial differences. *Curr. Anthrop.,* 6:167–198.
Hooton, E. A. (1931) *Up from the Ape.* New York, Macmillan.
Napier, J. (1970) *The Roots of Mankind.* Washington, D.C., Smithsonian Institution Press.
Washburn, S. L. (1963) The study of race. *Amer. Anthrop.,* 65:521–531.

Chapter 4

Adelmann, H. B. (1966) *Marcello Malpighi and the Evolution of Embryology.* Ithaca, New York, Cornell University Press.
Chauvois, L. (1957) *William Harvey: His Life and Times: His Discoveries: His Methods.* London, Hutchinson Medical Publications.
Cole, F. J. (1949) *A History of Comparative Anatomy.* London, Macmillan.
May, M. T., ed. (1968) *Galen: On the Usefulness of the Parts of the Body.* Vols. 1 and 2. Ithaca, New York, Cornell University Press.
O'Malley, C. D. and Saunders, J. B. de C. M. (1952) *Leonardo da Vinci on the Human Body: The Anatomical, Physiological and Embryological Drawings of Leonardo da Vinci.* New York, Henry Schuman.

Oppenheimer, J. M. (1967) *Essays in the History of Embryology and Biology.* Cambridge, Massachusetts, M.I.T. Press.
Sherrington, C. S. (1940) *Man on His Nature.* London, Cambridge University Press.
Singer, C. (1957) *A Short History of Anatomy and Physiology from the Greeks to Harvey.* New York, Dover Publications.
——— and Underwood, E. A. (1962) *A Short History of Medicine.* Oxford, Clarendon Press.
Whitteridge, G. (1964) *The Anatomical Lectures of William Harvey.* Edinburgh, Livingstone.

Chapter 5

Campbell, H. J. (1965) *Correlative Physiology of the Nervous System.* New York, Academic Press.
Connolly, C. J. (1950) *External Morphology of the Primate Brain.* Springfield, Illinois, Charles C Thomas.
Dodgson, M. C. H. (1962) *The Growing Brain.* Bristol, John Wright.
Harrison, G. R. (1956) How the brain works. *Atlantic, 198,* No. 3:58–63.
LaBarre, W. (1954) *The Human Animal.* Chicago/London, University of Chicago Press.
Noback, C. R. and Montagna, W., eds. (1970) *Advances in Primatology.* Vol. 1, The Primate Brain. New York, Appleton-Century-Crofts.
——— and Moskowitz, N. (1962) Structural and functional correlates of "encephalization" in the primate brain. *Ann. N.Y. Acad. Sci., 102:*210–218.
Papez, J. W. (1929) *Comparative Neurology.* New York, Thomas Y. Crowell.
Sholl, D. A. (1956) *The Organization of the Cerebral Cortex.* London, Methuen.
Sperry, R. W. (1958) Physiological plasticity and brain circuit theory. In: *Biological and Biochemical Bases of Behavior* (eds. H. F. Harlow and C. N. Woolsey). Madison, University of Wisconsin Press, pp. 201–224.

Chapter 6

Ducroquet, R., Ducroquet, J., and Ducroquet, P. (1968) *Walking and Limping: A Study of Normal and Pathological Walking.* Philadelphia, J. B. Lippincott.
Elftman, H. (1951) The basic pattern of human locomotion. *Ann. N.Y. Acad. Sci., 51*:1207–1212.
Hooton, E. A. (1931) *Up from the Ape.* New York, Macmillan.
Morton, D. J. (1922) Evolution of the human foot: I. *Amer. J. Phys. Anthrop.,* 5:305–336.

———. (1924) Evolution of the human foot: II. *Amer. J. Phys. Anthrop.*, 7:1–52.

Napier, J. R. (1956) The prehensile movements of the human hand. *Journal of Bone and Joint Surgery, 38B*:902–913.

———. (1967) The antiquity of human walking. *Sci. Amer., 216*, No. 4:56–66.

Schultz, A. H. (1961) Some factors influencing the social life of primates in general and of early man in particular. In: *Social Life of Early Man* (ed. S. L. Washburn). Chicago, Aldine, pp. 58–90.

Washburn, S. L. (1968) *The Study of Human Evolution.* Condon Lectures, Eugene, Oregon, Oregon State System of Higher Education.

Chapter 7

Fitzpatrick, T. B., Arndt, K. A., Clark, W. H., Jr., Eisen, A. Z., Van Scott, E. J., and Vaughan, J. H., eds. (1971) *Dermatology in General Medicine.* New York, McGraw-Hill.

Montagna, W. (1962) *The Structure and Function of Skin,* 2nd ed. New York/London, Academic Press.

———. (1963) Phylogenetic significance of the skin of man. *Arch. Derm.*, 88:1–19.

———. (1965) The skin. *Sci. Amer., 212*, No. 2:56–66.

Needham, A. E. (1964) Biological considerations of wound healing. In: *Advances in Biology of Skin.* Vol. 5. Wound Healing (eds. W. Montagna and R. E. Billingham). Oxford, Pergamon Press, pp. 1–26.

Pillsbury, D. M., Shelley, W. B., and Kligman, A. M. (1956) *Dermatology.* Philadelphia, W. B. Saunders.

Rothman, S. (1954) *Physiology and Biochemistry of the Skin.* Chicago, University of Chicago Press.

Chapter 8

Day, M. H. (1967) *Guide to Fossil Man.* London, Cassell.

Greene, D. L. (1967) *Genetics, Dentition and Taxonomy.* University of Wyoming Publications, 33 No. 2:93–168.

Stack, M. V. and Fearnhead, R. W. (1965) *Tooth Enamel.* Bristol, John Wright.

Widdowson, T. W. (1946) *Special or Dental Anatomy and Physiology, Dental Histology, Human and Comparative.* 7th ed. Vols. 1 and 2. London, Stapler Press.

Chapter 9

Brown, J. H. U. and Barker, S. B. (1962) *Basic Endocrinology for Students of Biology and Medicine*. Oxford, Blackwell.

Gorbman, A. (1959) *Comparative Endocrinology*. New York, John Wiley.

Harris, G. W. and Donovan, B. T. (1966) *The Pituitary Gland*. London, Butterworths.

Hughes, G. M. (1963) *Comparative Physiology of Vertebrate Respiration*. London, Heinemann.

Lisser, H. and Escamilla, R. F. (1962) *Atlas of Clinical Endocrinology*. St. Louis, C. V. Mosby.

Pincus, G., ed. (1947–1967) *Recent Progress in Hormone Research*. Vols. 1–23. New York, Academic Press.

Pitt-Rivers, R. and Trotter, W. R. (1964) *The Thyroid Gland*. London/Washington, D.C., Butterworths.

Whalen, R. E. (1967) *Hormones and Behavior*. Princeton, D. Van Nostrand.

Willoughby, D. P. (1970) *The Super Athletes*. South Brunswick/New York, A. S. Barnes; London, Thomas Yoseloff.

Chapter 10

Asdell, S. A. (1964) *Patterns of Mammalian Reproduction*, 2nd ed. Ithaca, New York, Cornell University Press.

Cole, H. H. and Cupps, P. T. (1959) *Reproduction in Domestic Animals*. New York, Academic Press.

Corner, G. W. (1963) *The Hormones in Human Reproduction*. New York, Atheneum.

Harrison, R. J. (1971) *Reproduction and Man*. New York, Norton.

Parkes, A. S., ed. (1960–1966) *Marshall's Physiology of Reproduction*, 3rd ed. Vol. 1, Pt. 2; Vol. 3. London, Longmans.

Young, W. C., ed. (1961) *Sex and Internal Secretions*, 3rd ed. Baltimore, Williams & Wilkins.

Zuckerman, S., ed. (1962) *The Ovary*. Vols. 1 and 2. New York, Academic Press.

Chapter 11

Corner, G. W. (1944) *Ourselves Unborn*. New Haven, Yale University Press.

Crew, F. A. E. (1965) *Sex-determination*. London, Methuen.

Hill, J. P. (1932) The developmental history of the primates. *Phil. Trans. Roy. Soc. London, B221*:45–178.

Mossman, H. W. (1937) Comparative morphogenesis of the fetal membranes and accessory uterine structures. *Contrib. Embryol., 26*:129–246.

Parkes, A. S., ed. (1952–1966) *Marshall's Physiology of Reproduction,* 3rd ed. Vols. 1–3. London, Longmans.

Streeter, G. L. (1942) Developmental horizons in human embryos. (Description of age group XI, 13 to 20 somites, and age group XII, 21 to 29 somites). *Contrib. Embryol., 30*:210–245.

———. (1945) Developmental horizons in human embryos. (Description of age group XIII, embryos about 4 or 5 mm long, and age group XIV, period of indentation of the lens vesicle). *Contrib. Embryol., 31*:27–63.

———. (1945) Developmental horizons in human embryos. (Description of age groups XV, XVI, XVII, and XVIII). *Contrib. Embryol., 32*:133–203.

———. (1949) Developmental horizons in human embryos. (Review of the histogenesis of cartilage and bone). *Contrib. Embryol., 33*:149–167.

———. 1951 Developmental horizons in human embryos. (Description of age groups XIX, XX, XXI, XXII, and XXIII). Prepared for publication by Chester H. Heuser and George W. Corner. *Contrib. Embryol., 34*:165–196.

Wynn, R. M. (1967) *Cellular Biology of the Uterus.* New York, Appleton-Century-Crofts.

Chapter 12

Altmann, S. A., ed. (1967) *Social Communication Among Primates.* Chicago, University of Chicago Press.

Benjamin, H. (1961) Prostitution. In: *The Encyclopedia of Sexual Behavior* (eds. A. Ellis and A. Abarbanel). New York, Hawthorn Books, Vol. 2, pp. 869–882.

Brecher, E. M. (1971) *The Sex Researchers.* New York, New American Library.

Ford, C. S. and Beach, F. A. (1951) *Patterns of Sexual Behavior.* New York, Ace Books.

Goodhart, C. B. (1964) A biological view of toplessness. *New Scientist, 23*:558–560.

Johnston, J. W., Jr., Moulton, D. G. and Turk, A., eds. (1970) *Advances in Chemoreception.* Vol. 1. Communication by Chemical Signals. New York, Appleton-Century-Crofts.

Kinsey, A. C., Pomeroy, W. B., and Martin, C. E. (1948) *Sexual Behavior in the Human Male.* Philadelphia/London, W. B. Saunders.

———, Pomeroy, W. B., Martin, C. E., and Gebhard, P. H. (1953) *Sexual Behavior in the Human Female.* Philadelphia/London, W. B. Saunders.

Masters, W. H. and Johnson, V. E. (1966) *Human Sexual Response*. Boston, Little, Brown.

Michael, R. P. and Keverne, E. B. (1970) A male sex-attractant pheromone in rhesus monkey vaginal secretions. *J. Endocrinol.*, *46*:xx–xxi.

——— and Keverne, E. B. (1968) Pheromones in the sexual status of primates. *Nature (London)*, *218*:746–749.

Chapter 13

Busnel, R. G., ed. (1963) *Acoustic Behaviour of Animals*. Amsterdam/London/New York, Elsevier.

Debetz, G. F. (1961) The social life of early paleolithic man as seen through the work of the Soviet anthropologists. In: *Social Life of Early Man* (ed. S. L. Washburn). Chicago, Aldine, pp. 137–149.

DeBrul, E. L. (1958) *Evolution of the Speech Apparatus*. Springfield, Illinois, Charles C Thomas.

Fossey, D. (1971) More years with mountain gorillas. *Natl. Geographic*, *140*:574–585.

Kelemen, G. (1961) Anatomy of the larynx as a vocal organ: Evolutionary aspects. *Logos*, *4*:46–55.

Lieberman, P. and Crelin, E. S. (1971) On the speech of Neanderthal man. *Ling. Inquiry*, *2*:203–222.

Negus, V. E. (1950) *The Comparative Anatomy and Physiology of the Larynx*. London, Heineman.

Premack, D. (1971) Language in chimpanzee? *Science*, *172*:808–822.

Schultz, A. H. (1961) Some factors influencing the social life of primates in general and of early man in particular. In: *Social Life of Early Man* (ed. S. L. Washburn). Chicago, Aldine, pp. 58–90.

Spencer-Booth, Y. and Hinde, R. A. (1971) Effects of 6 days separation from mother on 18- to 32-week-old rhesus monkeys. *Anim. Behav.*, *19*:174–191.

Theberge, J. B. (1971) "Wolf Music." *Nat. Hist.*, *80* No. 4:37–43.

Washburn, S. L. and DeVore, I. (1961) Social behavior of baboons and early man. In: *Social Life of Early Man*. (ed. S. L. Washburn). Chicago, Aldine, pp. 91–105.

Chapter 14

Carpenter, C. R. (1964) *Naturalistic Behavior of Nonhuman Primates*. University Park, Pennsylvania, Pennsylvania State University Press.

Euler, V. S., von. (1934) Zur Kenntnis der pharmakologischen Wirkungen von Nativsekreten und Extrackten männlicher accessorischer Geschlechtsdrüsen. *Arch. Exp. Pathol. Pharmakol.*, *175*:78–84.

Hinde, R. A. and Spencer-Booth, Y. (1971) Towards understanding individual differences in rhesus mother-infant interaction. *Anim. Behav.*, *19*:165–173.

Johnston, J. W., Jr., Moulton, D. G. and Turk, A. (1970) *Advances in Chemoreception,* Vol. 1. Communication by Chemical Signals. New York, Appleton-Century-Crofts.

Polyak, S. (1957) *The Vertebrate Visual System.* (ed. H. Klüver). Chicago, University of Chicago Press.

Prince, J. H. (1956) *Comparative Anatomy of the Eye.* Springfield, Illinois, Charles C Thomas.

Walls, G. L. (1942) *The Vertebrate Eye and Its Adaptive Radiation.* Bloomfield Hills, Michigan, Cranbrook Institute of Science.

Winkelmann, R. K. (1960) *Nerve Endings in Normal and Pathologic Skin.* Springfield, Illinois, Charles C Thomas.

Chapter 15

Cox, P. R. (1957) *Demography,* 2nd ed. London, Cambridge University Press.

Fiennes, R. (1964) *Man, Nature and Disease.* London, Weidenfield and Nicolson.

Medawar, P. B. (1957) *The Uniqueness of the Individual.* London, Methuen.

———. (1959) *The Future of Man.* New York, Basic Books.

Roberts, J. A. F. (1959) *Introduction to Medical Genetics,* 2nd ed. London, Oxford University Press.

Selye, H. (1956) *The Stress of Life.* New York/Toronto/London, McGraw-Hill.

Chapter 16

Comfort, A. (1956) *The Biology of Senescence.* London, Routledge & Kegan Paul.

Dublin, L. I., Lotka, A. J. and Spiegelman, M. (1949) *Length of Life—A Study of the Life Table.* New York, Ronald Press.

Jones, M. L. (1962) Mammals in captivity—primate longevity. *Lab. Primate Newsletter, 1* No. 3:31–33.

Lansing, A. I., ed. (1952) *Cowdry's Problems of Ageing: Biological and Medical Aspects,* 3rd ed. Baltimore, Williams & Wilkins.

Medawar, P. (1946) Old age and natural death. *Modern Quart.*, 2:30–49.

Schultz, A. H. (1960) Age changes in primates and their modification in man. In: *Human Growth* (ed. J. M. Tanner). London, Pergamon Press, pp. 1–20.

Shock, N. W. (1956) The effects of some of the steroid hormones on the metabolic balances in aged males. In: *Hormones and the Aging Process* (eds. E. T. Engle and G. Pincus). New York, Academic Press, pp. 283–298.

Chapter 17

Bergson, H. L. (1911) *Creative Evolution.* New York, Henry Holt.
Crook, J. H. and Gartlan, J. S. (1966) Evolution of primate societies. *Nature (London)*, *210*:1200–1203.
Darwin, C. R. (1871) *The Descent of Man and Selection in Relation to Sex.* New York, D. Appleton Co.
Lorenz, K. (1966) *On Aggression.* New York, Harcourt, Brace and World.
Sherrington, C. (1963) *Man on His Nature,* 2nd ed. London/New York, Cambridge University Press.
Washburn, S. L. and DeVore, I. (1961). Social behavior in baboons and early man. In: *Social Life of Early Man.* (ed. S. L. Washburn). Chicago, Aldine, pp. 91–105.

INDEX

Abdomen, 159–160
Abdominal wall, 160
Aborigines, Australian, 78
Abortion, 327–328
Acheulian hand-axe industry, 66
Achilles tendon, 151
Achondroplasia, 403
Acne, 191
Acromegaly, 220–221
Adam's apple, 10, 345
Adaptive properties, 71
Addison, T., 227
Addison's disease, 227
Adenohypophysis, 220
Adolescence, 237–238
Adrenal gland,
 cortex, 221
 medulla, 226–229
Adrenaline, 227
Adrenocorticotrophic hormone (ACTH), 221
Aegyptopithecus, 55, 62
Aeolopithecus, 62
Afer niger, 73
Afferent (sensory) nerve cells, 109, 112
Age changes,
 in cells, 409–410
 and growth, 395–396
 Krogman's periods of, 402–407
 of mammals, 33
 of reptiles, 33
 in the skeleton, 402–407
 in soft tissues, 407–409
Aging, 393–413
 definition of, 394
 theories of, 410–413
Ainu, 74
Alae vespertilioni, 241
Albinism, 182
Allantois, 297
Allopatric, 86
Alopecia, 22–23, 188
Alouattinae, 52
Alpines, 74
Ambisexual, 270
Ambrona, 66
American Indian, 73, 76
Amniotic cavity, 327–328
Anatomy,
 ancient, 88–90
 comparative, 104
 Golden Age of, 93–96
 man's, 87–104
 in the Middle Ages, 90–91
 modern, 102–104
 in the Renaissance, 91–93
 transcendental, 98
Androgenic hormones, 235, 256, 286
Anemia, sickle-cell, 81
Angiotensin II, 229
Angiotensinogen, 229
Angwantibos, 31, 45, 62
Ankle,
 dislocation of, 150–151
 joint, 143, 148
Anogenital orifices, 23
Anosmatic, 38, 119
Anteater,
 long-billed, 41

Anteater (*Continued*)
  spiny, 41
Anterior lobe, 220
Anthropoidea, 43–52
Anthropoid families, 5
Antidiuretic principle, 222
Antihormones, 219
Anulus fibrosus, 157
Aortic bodies, 230
Aotinae, 52
Aotus trivirgatus, 368
Apes,
  evolutionary divergence of, 62
  great, 56–61
  lesser, 56–61
Aphasia, 346
Aphonia, 337
Aphrodisiacal, 318
Apidium, 62
Apocrine glands, 192–196
  scent, 195
Appendages, vertebrate, 3
Appendix, vermiform, 59
Aqueduct, 114
Aqueous humor, 369
Aralians, 74
Arboreal life, 37
Arch, 145–148
  branchial, 289
  in dermatoglyphics, 175
  of foot, 145–148
Archeology, 70
Arctics, 76
Arcotcebus calabarensis, 31, 45, 62
Areola, 27, 318
  pigmentation of, during pregnancy, 280
Aristotle, 2, 87, 101, 126
Arm and leg buds,
  horizons XIII and XIV, 291
Arousal reaction, 116
Art, 428
  in Renaissance, 91–93
Arterial vein, 96
Articulate speech, 105
Ascending sensory tracts, 129
Asiaticus luridus, 73
Asphemia, 337
Association areas, 122, 131
  enlargement of, 133
Association fibers, 122, 132
Asthenic, 85
Ateles, 137
Atelinae, 52
Athleticosomatic, 85
Atlanthropus, 65
Atlas, 157
Attitudes, postural, 335
Attraction of sex mate, 309
Auditory,
  area, 123
  meatus,
    external, 347–349
    internal, 350–351
  perception area, 351
  range,
    amphibian, 352
    of bats, 352
    of birds, 352
    cetaceous, 352
    of dogs, 352
    of man, 351
    of nonhuman primates, 352, 354
    of pinnipeds, 352
Aurignacian culture, 163
Australiforms, 77–78
Australopithecines, 64–65, 166
  brain, 64
  dentition, 211
Avicenna of Bokhara, 90
Axilla, 19
  odor, 195
Aye-aye (*Daubentonia madagascariensis*),
    9, 36

Babinski reflex, 296
Baboon (*Papio*), 56
Backbone, man's, 156–160
Baculum, 260
Baldness, 22–23, 188
Baltics, 74
Barbary ape (*Macaca sylvana*), 58, 90
Baroreceptors, 378
  chemoreceptors, 378
Bartholomew, G. A., 311
Basal ganglia, 119–120
Basal nuclei, 119
Basilar artery, 127
Behavior, stereotyped, 117
Bell, J., 26
Benjamin, H., 330
Berengarius of Carpi, 92
Bergson, H., 421
Bernard, C., 216
Betz, V., 121
Bimana, 97
Binovular follicles, 263
Bipedalism, 5, 47, 133, 137
  dynamics of, 141
  evolution of, 165
  gait of, 28
  of man, 139–141
  of primates, 32–33, 137
Birth(s) (parturition),
  canal, 302
  changes during,
    conceptus, 278

environment, 278
requirements, 278
control, 323–324
multiple, 263, 302
primary sex ratio, 270–271
secondary sex ratio, 270–271
stages of, 241
Blastocyst, 240, 275, 285, 326
implantation of, 251, 275
Blood groups,
of animals, 84
in man, 81–84
in skeletal remains, 84
"Blue baby," 296
Body habitus, 84–87
Bolk, L., 35
Bone growth,
ontogeny, 238
Boyle, R., 100
Brachiation, 27, 57, 64, 137, 162
Bradycardia, 127
Brain,
of birds, 117
blood supply of, 126–127
cerebellum, 116, 123–124
cerebrum, 112, 114, 118, 120
development of, horizons XV and XVI, 291
frontal lobes, 12
of great apes, 58
gyrencephalic, 123
human, 105–134
association areas, enlargement of, 133
evolution of, 120
growth of, 133–134
similarity to that of apes, 132
lissencephalic, 123
parts of, 115–127
primate, 38, 127–131
pyramidal,
decussation of, 121
motor system of, 121
motor tracts of, 129, 132
retrocalcarine sulcus, 131
size, 105
evolution of, 106
and spinal cord, 114
stem, 115
vesicles, 112
unequal growth of, 112
Breasts, 8–9, 304, 315
of bats, 9
of elephants, 9
of man, 8–9
of prosimians, 9
of South American monkeys, 9
Breeding season, 253
Broca, P. P., 121, 337
Buccopharyngeal membrane, 287
Buettner-Janusch, J., 44–45, 200–201
Bush baby (*Galago sp.*), 3, 62
Bushmen, 77

Caius, J. 95
Calcarine sulcus, 131
Calcaneus, 142
Calcium, metabolism of, 224
Caldwell, W. H., 41
Callitrichidae, 52
Calvaria, 10
Canal of Schlemm, 369
Canines, 17–18, 38, 201
Capacitation, 265
Caput succedaneum, 302
Caries, 213
Carotid,
body, 378
internal, 126
sinus, 378
Carpenter, C. R., 358
Carrying angle, 28
Castration, effects of, 236
Catarrhines, 52
Caucasian, 73
Cavernous tissue, 260
Cavities,
abdominal, 29
thoracic, 29
Cebidae, 52
Cebuella pygmaea, 34, 52
Cebupithecia, 62
Celebes black apes (*C. niger*), 6, 47, 56, 58, 137, 188, 250, 313
Cell division, 409
Central sulcus, 131
Central Nervous System (CNS), brain and spinal cord, 112–115
Cephalic index, 74
Cercocebus, 56
Cercopithecidae, 52, 54, 56
Cerebellum, 116, 123, 124
Cerebro-cerebellar fibers, 124–125
Cerebro-spinal fluid, 115
Cerebrotonia, 86
Cerebrum, 112, 114, 118, 120
cortex, 121
hemispheres, 112, 114, 118, 120
peduncles, 129
Cervix, 281
dilation of, 300
Cesalpinus, 95
Chambers of eye, 369
Chemoreceptors, 230
Chemosensory messages, 375
Chersiotes, 74
Chimpanzee (*Pan troglodytes*), 56–58, 278

Chloasma, 281, 326
Choanae, 13
Chopping tool industry, 66
Chorio-allantoic placenta, 298
Chorion, 297
  frondosum, 297
Chorionic gonadotrophin, 299
Chorionic sac, 297
Choroid, 367–368
Choukoutien, 66
Chromosomes, 264
  diploid number of, 269–270
  sex, 270
Circle of Willis, 127
Circumvallate papillae (*See Vallate*)
Clark, S., 230
Clark, Sir Wilfred Le Gros, 38, 56
Classification, 2, 5, 33, 39, 41, 73–78, 87, 96–97
  of animals, 87
  of Hominidae, 64
  of Mammalia, 5, 41
  of man, 39, 73–78, 97
Clavicle, 28, 36
Cleavage,
  of fertilized ovum, 272–275
  lines, 26
Clitoris, 317
Cloaca, 259
Coccyx, 156–157
Cochlea, 349–350
  cupola, 358
  modiolus, 350
  organ of Corti, 349–350
    basilar membrane, 349
    hair cells, 349
    tectorial membrane, 349
  Scala vestibuli, 349
Coelom, 287
Coitus, 271, 309, 319–323
  attempts at, by children, 321
  of birds, 322
  face-to-face, 320
  as learned act, 323
  performance of, 320
Colobinae, 56, 62
Color blindness, 81
Color vision, 370
Colostrum, 305
Columbus, R., 296
Columella, 352
Commissural fibers, 122
Communal living, 47–48, 390–391
  and environmental diseases, 391
  stress in, 390
  and venereal diseases, 391
Communication,
  among animals, 42
  internal, 217
  intraspecies, 426
  means of, 333–378
  theory of, 336
Comparative anatomy, 104
Conception, 261–306
Concupiscence, 308, 311
Congolians, 76
Consonants,
  breathed, 344
  voiced, 344
Contraceptive methods, 324–327
Contractions of uterine tubes, 246
Coon, C., 68
Copulation (*See Coitus*)
Copulatory organs, 259
Corner, G. W., 289
Corona radiata, 268
Corpora quadrigemina,
  inferior, 123
  superior, 123–124
Corpus callosum, 122, 129, 132
Corpus luteum, 239, 247–248
Corti, A., 349
  organ of, 349–351
Cotyledons, 297
Courtship, 246, 308, 315
Coxa vara, 142, 387
Craniology, 73
Cranium,
  capacity of, in man, 10
  of gorilla, 9–10
Crelin, E. S., 333–334
Cro-Magnon man, 67, 68
  size, 68
Crown rump length, 289, 293, 295
Cryptorchidism, 259
Culture, 427–428
  evolution of, 419
Cummins, H., 26
Curettage (D and C), 327
Curiosity, 422
Curvatures,
  compensatory, 157
  of the cornea, 367
  primary, 157
  secondary, 157
  spinal, 157
Cusps of teeth, 211
Cushing's syndrome, 227
Cutaneous sensibility, acute, 167–168
Cutaneous sensory system,
  basic mechanism, 362
  in nonhuman primates, 363
Cynopithecus niger, 6, 47, 56, 58, 137, 188, 250, 313
Cytotrophoblast, 275, 299

Darwin, C., 1–2, 10, 15, 98, 101, 419
Deafness, 355

De Beer, Sir Gavin, 14, 98
Defects,
of abdominal wall, 386
hernias, 386
of ankle joint, 388
cancer, 389
cardiovascular diseases, 389–390
fallen arches, 388
of frontal sinuses, 384
of knee, 387–388
of long life, 388–389
of maxillary sinuses, 384
of paranasal sinuses, 384
of vertebral column, 386
de Graaf, R., 246
Della Torre, M., 92
Demographic statistics, 400
Dens serotinus, 213
Denys, J., 81
Dermatoglyphics, 26, 175–177
Dermis, 172, 183–184
Descartes, R., 222
Desire, releasers of, 309
Development, man, 261–306
De Vore, I., 336, 419
De Vries, Hugo,
theories of, 101
Diabetes,
insipidus, 222
mellitus, 226
Diabetogenic hormone (STH), 226
Dialect, 426
Diaphragm, 29, 41, 155, 325
in aquatic mammals, 29
development of, 29
pelvic, 155
in platypus, 41
Diastema, 211
Diencephalon, 112, 119
Dinarics, 74
Dinosaurs, 7
Dislocation,
ankle, 150–151
knee joint, 387
shoulder joint, 385
Dizygotic twins, 262
Dordogne, 334
Dorsal commissure of brain, 129
Dorsal horns,
of spinal cord, 116
Dorsiflexion, 144, 149
Dryopithecus, 63, 211
Du Brul, E. L., 334
Duckbill, 41
Duckworth, W. L. H., 74
Ductus arteriosus, 296
Ductus deferens, 265
Ductus venosus, 296
Dura mater, 127
Dürer, A., 92
Dwarf galagos (*G. demidovii*), 34
Dysarthria, 337
Dyslalia, 346

Ear, 14–15
comparative anatomy, 15
Eustachian tube, 348
external, development, 15
inner, 348–350
middle, 347–348
ossicles, 347
variation in shape, 15
Ectomorphy, 85
Effector cells, 108
Efferent (motor) nerve cells, 109, 112
Ejaculate, 265
Ejaculation, 235, 319, 321
Ejaculatory duct, 260, 265
Elephant,
breasts, 9
skull, 13
Elephant shrew (*Elephantulus*), 253
Embryo,
age, 283
growth, 283
length, crown to rump, 289
Embryogenesis, effects, 218
Embryology, 98
Encephalization, 117
Endocrine,
glands, 216
organs, location of, 219
Endometrium, 239, 252, 275
Endomorphy, 85
Environment,
internal, 229–230
stable, 216
Eoanthropus dawsoni, 200
Eocene period, 33, 45, 61–62
and tarsoid, 55
Ependyma, 115
Epicranius muscle, 10
Epidermis, 172, 177–181
horny layer, 177, 180
as effective barrier, 180
malpighian layer, 179–180
mitosis, 180
stratum lucidum, 180
underside of, 177
Epididymis, 260, 265
Epigamic areas, 169
Erogenous areas, 315
Eroticism, 309
Erythriotes, 74
Estrogenic,
hormones, 239, 304

Estrogenic (*Continued*)
phase, 239
Estrus, 246, 253
cycle, 419
Ethiopian, 73
Ethmoid bone, 12
Eunuchs, 235
Europiforms, 74–75
Eustachius, B., 348
Eutheria, 42
Evolution, 98, 101
divergence of monkeys and apes, 62
events, 32
of races, 70
theory, 87
of tree shrew, 44
Expressions,
"compound," 14
facial, 14, 335
External meatus of ear, 23
Extrapyramidal motor system, 122, 125
Eyeball, 366–367
choroid, 366–367
cornea, 366–367
iris, 366
lens, 367
movement of, 366
pupil, 366
retina, 366
sclera, 366
Eyebrows, 22–23
Eyelashes, 23
Eyelids, 14, 22
Eyes,
evolution, 365
extrinsic muscles, 365
position, 365

Fabrica, 93
Fabricius, H., 95
Face,
expressions of, 14, 335
of man, 10–16
muscles of, 14
Fallopius, G., 241
Family,
human, 416–418
bond of, 330
exogamous, 419
influence of, 417
matriarchal, 418
monogamous, 418–419
patriarchal, 418
polygamous, 418
Fayum region, 54, 62
Feedback mechanisms, 229
Femur, 141
Fernel, J., 104
Ferrier, Sir David, 121
Fertilization, 240, 246, 268–271, 308, 339
prevention of, 324–328
Fertilized ovum, 239
cleavage, 272
Fetal,
abnormality, 283
growth, 283
rotation, 302
Fetalization, 35
Final causes (Galen), 88
Fingerprints, 26
Fire, use of, 66
Flocculo-nodular lobe, 124, 377
Follicle-stimulating hormone (FSH), 221, 243, 246, 325
Foot,
arches of, 145–148, 152
club, 152
eversion of, 151
flat, 152
of gorilla, 152
grasping, 151
Grecian, 145
-hand of lemurs, 163
of horse, 3
human, 3, 37, 145, 152
inversion of, 151
plantar-flexion, 151
of Pliopithecus, 63
-print, 145
of terrestrial primates, 152
Foramen (ovale), 296
Forebrain, 111–112, 116, 118
Fossil remains, 18, 61–68
in America, 62
of teeth, 200
Freedom, 419
Fritsch, R., 121
Frontal bone, 12
Frontal lobes of brain, 12, 123, 131
Furrows, skin, 174

Gait, 136, 140
Galago crassicaudatus, 3, 252
Galagos, 31, 45
Galea aponeurotica, 10
Galenic nature, 89
Galenical system, 89
Galen of Pergamum, 88–90, 126
Gall, F. J., 121
Gastrocnemius muscle, 142
Gelada baboon (*Theropithecus*), 56, 256
Generalized,
organisms, 2
as relative term, 34
structure, 3, 5

General sensory area, 123
Genetic,
  drift, 70
  isolation, 72
  mixture, 72
Genitalia,
  female, 317–318
  male, 359–360
Genius, 425, 428
Geological periods, 33
Gerasimov, M., 9
Germ cells, 242
German measles (*virus rubella*), 300
Gestation, 277–283
  length of, 278
  periods in mammals, 279
  periods in primates, 280
Gibbons, 5, 27, 52, 56
Glucagon, 226
Gluteus maximus muscles, 144
Gluteus medius muscles, 141
Goethe, J. W. von, 9
Golden lion marmoset (*Leontideus rosalia*), 52
Gomphoses, 200
Gonadotrophic hormones, 235
Gonadotrophin, 239, 247
  chorionic, 299
Gorilla (*Gorilla gorilla*), 56, 58
Graafian follicle, 246
Granulosa cell, 243
Granulosa-luteal cells, 247
Great apes, 52, 56–61, 278
  skull of, 10
Greater bush babies (*Galago crassicaudatus*), 252
Grecian foot, 145
Grip,
  hook, 162
  power, 162
  precision, 162
Grooming, 319
Growth hormone, 305
Gubernaculum, 259
Guenon (*Cercopithecus*), 56
Gustatory discrimination, 373–374
Gynandromorphy, 85
Gyrencephalic brains, 123

Haeckel, E., 65
  theory of, 98–99
Hair, 18–23
  on anogenital surfaces, 22, 23
  axillary, 22
  coat, 184–189
  cult of, 22
  on external meatus, 23
  follicles, 186
  growth, 169, 186–187
    control of, 186
  lanugo, 293, 296
  in nostrils, 23
  ornamental, 19
  on scalp, 22
  shedding of, 187
  tactile sensibility of, 185
Hairiness, 19, 25
Halbaffen, 44
Hallux, 6, 137, 139, 145
  valgus, 388
Hamites, 74
Hamstrings, 144
Hand,
  evolutionary trends in development of, 162
  freeing of, 133
  of man, 28–29
  movements, 162–163
  of *Pliopithecus,* 63
Harelip, 14
Harvey, W., 95–96, 230
Haute-Garonne, France, 163
Head-fold, 287
Head injuries, 382–383
Hearing,
  in animals, 351–354
  apparatus, 346–351
  discrimination in, 352, 354
  in man, 351–352
  and ossicles, 351
  and tympanic membrane, 351
  and tympanic ring, 351
Helocene, 33
Hensen's node, 285
Hernias, 386
Herophilus, 88
Hindbrain, 111–112
Hindgut, 287
Hippocrates, 88
Histiotrophe, 275
Histology, 97
Hoel, P. G., 311
Homeostasis, 108
Hominidae, 5, 52, 54, 56, 59
  classification of, 64
Homo erectus, 65–66
  intermediate, 66
  postcranial skeleton of, 66
  and use of fire, 66
Homo habilis, 65, 163
Homo sapiens, 56, 66
Homosexuality, 329
Homunculus, 47, 62
Hooton, E. A., 152
Horizons of fetal development,
  1st week, horizons I-IV, 284
  2nd week, horizons V-VII, 285

Horizons of fetal development (*Continued*)
3rd week, horizons VIII-IX, 285
4th week, horizons X-XII, 287–289
5th week, horizons XIII-XVI, 289–291
6th week, horizons XVII-XX, 291
7th and 8th weeks, horizons XXI-XXIII, 291–292
third month, 292
fourth month, 292
fifth and sixth months, 293–295
seventh to ninth months, 295–296
Hormone(s),
action of, 217
adrenaline, 227
adrenocorticotrophic (ACTH), 221
androgenic, 235, 256
diabetogenic (STH), 226
effects of, 217–218
estrogenic, 304
follicle-stimulating (FSH), 221, 243, 246
glucagon, 226
gonadotrophic, 235
growth, 305
insulin, 226
luteinizing (LH), 221, 243, 246
melatonin, 222
parathormone, 224
placental, 218
renin, 229
somatotrophic (STH), 220
thyrotrophic (TSH), 221, 223
thyroxin, 223
triiodothyronine, 223
vasopression, 305
Hottentots, 77
Howler monkey (*Alouattinae*), 52, 54
Huns, 76
Hunter, J., 85, 96, 99–100
dictum of, 100
Hunter, W., 96, 301
Hunterian museum, 100
Hutchinson, J., 85
Huxley, T. H., 1–2, 57, 101–102, 424
Hylobatidae, 52, 54, 56–57
Hyoid bone, 54
Hypoglossal,
nerve, 116
nucleus, 116
Hypophysis cerebri, 220–222
Hypothalamic nuclei, 222
Hypothalamo-hypophysial tract, 222
Hypothalamus, 119–120, 235, 329

Ilium, 141, 153
Immunity,
cell-mediated, 225
mechanisms of, 225
Implantation, 240, 245, 252, 275–277
of blastocyst, 251, 275
delayed, 248, 252, 277
ectopic, 277
Incisors, 17–18, 38, 201
Incus, 347
Inferior colliculi, 123
Infertility, 271, 320
Information, impartation of, 335
Inguinal canal, 160, 259
Inguinal hernia, 160
Innervation of skin, 359–364
Insectivores, 33
Intercourse, 319
extramarital, 330
premarital, 330
Interstitial cells, 258
Intervertebral discs, 157
Inventiveness, 424
Iris, 367–368
Ischial callosities, 52, 56
Ischium, 153
IUD (intrauterine device), 326

Jackson, J. H., 121, 125
Jacobson organ, 376–377
Japanese macaques, 6, 47, 137
Jones, F. W., 9

Keith, Sir Arthur, 10, 38, 56, 218
Khoisaniforms, 77, 158
Kidney, 229
Kinesthetic sensitivity, 358–359
Kissing, 14, 315, 318–319
Klinefelter, H. F., 270
Knee jerk, 144
Knee joint, 143–144
of man, 152–153
Knuckle,
pads, 58
walkers, 137, 162
Kretschmer, E., 85
Krogman, W. M., 402
Kyphosis, 159, 387

Labor,
spontaneous, 301
stages of, 300–303
Labyrinth,
bony, 348
membranous, 348, 377
endolymph, 377
Labyrinthine placenta, 297
Lactation, 241, 252, 261–306
behavior of uterus during, 305–306
as controlled system, 305
development of, 304

and menstrual cycle, 306
La Ferrassie, 67
Lalling, 346
Lamarck, J., 101
Landsteiner, K., 82–83
Langerhans, P., 225–226
islets in pancreas, 226
Language, 333, 335
olfactory, 374
epithelium, 376
sensory cells, 376
tubercle, 376
Lapps, 74
Larynx, 10, 339–343, 345
of nonhuman primates, 342–343
skeleton of, in man, 340
arytenoid cartilage, 341
cricoid cartilage, 340
thyroid cartilage, 340
Lateral line, 377
organ, 352
Lateral ventricle, 114
Leakey, L. S. B., 65, 163, 166
Le Moustier, 67
Lemuridae, 44
Lemurs, 31, 33, 44, 137
Leonardo da Vinci, 92–93
Leontideus rosalia, 52
Leptosomatic, 85
"Let down," 305
Leukoderms, 25
Leydig cells, 258
Libido, 259
Lie, fetal, 301
Life expectancy, 234
Ligaments,
broad, 241
plantar, 145
terosseous, 145
Limb(s),
buds, 291
generalized structure of, in primates, 36
length, 27
in man, 27–28
upper, freedom of, 160–163
Limbic lobe, 120, 129–130
Linea alba, 160
Linea nigra, 281
Linnaeus, Carolus, 2, 73, 97
Lion-tailed macaque (*M. silenus*), 188
Lips, 13–14
Lissencephalic, 129
brain, 123
Locchia, 241
Locomotion, 135
bipedal evolution, 165
saltatory, 136
Locomotor coordination, 278
Logic, 427
Longevity,
of animals, 398
of man, 393, 398–401
Lordosis, 159
Lorises, 31, 34, 45, 62, 137
Loris tardigradus, 62
Love play, 321
Lower, Richard, 81
Lumbar curve, 158, 160
Lunate sulcus, 123, 131
Luteinizing hormone (LH), 221, 243, 246–247, 325
Luteotrophin (LTH), 247

Macaca fuscata, 6, 47, 137
Macaca mulatta, 253
Macaca nemestrina, 313
Macaca silenus, 188
Macaca sylvana, 58, 90
Macrosmatic, 38, 119, 374, 376
Madagascar, 33
Malayan, 73
Male pattern baldness, 22–23
Malleus, 347
Malocclusion, 212
Malpighian layer, 179
Mammalia, 5, 41, 43
Man, 2, 5
abdomen, 159–160
anatomy, 87–104
backbone, 156–159
bipedalism, 5, 28, 139–141
blood groups, 81–84
brain, 105–134
breasts, 8–9
chin, 15
classification, 39, 73–78, 97
conception, 261–306
defects in structure, 380
definition, 33
degenerative diseases, 380
development, 261–306
errors of metabolism, inborn, 380
face, 10–16
fingerprint patterns, 26
foot, 2, 27, 145–152
forehead, 12
frontal lobes of brain, 12
hair, 18–23
hallux, 6
hand, 28–29
head, 9
height, averages for boys and girls, 7
Hominidae, 52, 54
internal environment, 215–231
jaws, 6
knee joint, 152–153
lactation, 261–306

Man (*Continued*)
limbs, 27–28
lips, 13–14
modern, 67
nakedness, 18–25, 184
nature, 415–428
neck, 9
nose, 13
parturition, 300–304
pelage, 19
pelvis, 6, 153–156
physical development, 7
pollex, 6
position in nature, 39
posture, 159–160
erect, disadvantages, 382–388
pregnancy, 261–300
races, 69–72
relation to primates, 31–39
reproduction,
capacity, 234
cycle, 6
patterns, 233–252
sexual,
behavior, 307–331
cycle, 6
needs, 330–331
shape, 8–10
sinuses, 12–13
size, 6–7
skin, 18–27, 167–197
diseases, 380–382
skull, 9–10
teeth, 16–18, 41
tongue, 14–15
trunk, 29–30
weight, averages for boys and girls, 7
Mandrill (*Mandrillus*), 56
Mangabey (*Cercocebus*), 56
Marmosets, 52
golden lion, 52
pygmy, 34, 52
Marsupials, 41
Masculinity, 308
Maternal care, 278
Mauer man, 65
Medawar, Sir Peter, 411
Medial geniculate body, 124
Medulla oblongata, 116
Meganthropus, 65
Melanin, 181
in human races, 181
Melanoderms, 25
Membrana granulosa, 247
Memory, 120
Meninges,
dura mater, 127
spinal cord, 156
Menopause, 6, 241–243, 252, 259, 311
Menstruation, 239, 247–252, 327
cessation, 280
cycle, 314
discharge (menarche), 235
in primates, 253
purpose, 251
Mesomorphy, 85
Mesopithecus, 62
Metatarsals, 145
Metatheria, 42
Michelangelo, 92
Microsmatic, 38, 119, 318
Midbrain, 111
Midlow, C., 26
Migration, 71
Milieu interne, 216
Minimus muscles, 141
Minkowski, O., 226
Minot's law, 411
Miocene, 33, 47, 62
and apes, 166
strata, 166
Miscegenation, 72
Mitosis, 409
Mittelschmerz, 246
Mivart, St. George, 32
Molars, 18, 38, 201, 211
Monestrous cycles, 253
Monestrous mammals, 252–254
Mongolian, 73
spot, 76
Mongoliforms, 76
Mongolism, 246
Mongoloids, 68
Monkeys,
Celebes black apes, 6, 47, 56, 58, 137, 188, 250, 313
Cercocebus, 56
Cercopithecus, 56
colobus, 56
evolutionary divergence of, 62
gelada baboon, 56, 256
howler, 52, 54
lion-tailed macaque, 188
mandrill, 56
mangabey, 56
marmosets, 34, 52
New and Old World, 52–56
patas, 34, 56
pig-tailed macaque, 313
rhesus, 253
squirrel, 52
spider, 52, 137
stump-tailed macaque, 23
uakaries, 23
woolly, 52
Monoamines, 222
Monogamy, 330
Monotremata, 41, 259

Mons jovis, 235
Mons veneris, 22, 235
Morpheme, 345
Morula, 275
Mossman, H. W., 296
Motor aphasia, 121
Motor area, 121–123
  projection area, 122
  speech area, 337
Mouse lemur (*Microcebus murinus*), 34
Mousterian industry, 66
Müller, J. P., 248
Müllerian ducts, 248
Muscles, 9
  epicranius, 10
  eyeball,
    obliques, 365
    recti, 365
  facial, 14
  gastrocnemius, 142
  gluteus maximus, 141, 144
  gluteus medius, 141
  mimetic, 14
  minimus, 141
  peroneal, 151
  pharyngeal, 348
  popliteus, 153
  quadriceps femoris, 143, 153
  serratus anterior, 162
  soleus, 142
  stapedius, 348
  tensor tympani, 348
  tibialis anterior, 144
  tongue, 15
  trapezius, 161
Musculus, 89–90
Music, 426–428
Myelin sheath, 115
Myxedema, 223

Nails, 296
  of lemurs, 36
Nannopithex, 62
Napier, J. R., 65, 137, 141, 162–163
Nasal hairs, 13
Naturalism,
  in Renaissance, 91–93
Natural selection, 101
Neanderthal,
  man, 66
  period, 67
  tool kit, 67
Neanthropic, 67
Necrolemur, 62
Negriforms, 76–77
Negrillos, 77
Neocerebellum, 124
Neocortical commissure, 132
Neocortical expansion, 117, 119–120, 131
Neonate, average weight of, 295
Nerve,
  cell, 108–109
    afferent, 109
    basic characteristics of, 108
    efferent, 109
  cranial, 116
    abducent, 366
    accessory, 344
    auditory, 350
    hypoglossal, 344
    oculomotor, 365
    trigeminus, 344
    trochlear, 366
    vagus, 344
  impulse, conduction of, 108
  phrenic, 29
Nervous system,
  central, 131–134
  development of, 109–115
  general nature of, 106–109
  peripheral, 114
  visceral, 114
Nesiotes, 76
Neural,
  crest, 112
  folds, 285
  groove, 285
  plate, 109–110, 285
  tube, 110
    embryonic, 114
Neuroendocrine complex, 108
Neurohypophysis, 220–222, 234–255, 283
Neuromuscular spindles, 364
Nilotes, 76
Nipples, 27, 280, 318
Nordics, 74
Nose,
  breathing, 13
  olfactory reception, 13
  racial differences, 13
Notochord, 157, 285
Nucleus pulposus, 157
Nycticebus coucang, 62

Occlusion, 212–213
Occupational marks, 26
Oldowan,
  chopping tools, 64, 163
  pebble industry, 66
Old Stone Age, 67
Olduvai Gorge, 64–65, 163
Olfatory,
  acuity, 314, 374
  bulbs, 129
  placodes, 111

Oligocene, 33, 45, 62
  lower, 54
  prosimians, 62
Oligopithecus, 62
Olympic,
  athletes, 7
  games, 423
Ontogeny, 238
  bone growth, 238
Oocytes, 239
  mature, 243
    size of, 243
  number of, 242
  primary, 241–242
  secondary, 245–246
  viability of, 240
Opossums, 42
Opposition of digits, 36
Optic,
  blind spot, 370
  nerve, 369
  tectum, 116
Orangutan, 56, 58
Organic evolution, 1
Organisms,
  generalized, 2
  specialized, 2
Orgasm, 311, 320–322
Origin of Species, 98, 101
Ornithorhynchus, 41
Orthognathous, 12
Orthograde, 5
Os penis, 260
Ossification centers, 295
Ovarian,
  cycle, 239, 247, 251
  tubes, 272
Ovary, 241–246
Oviparous, 41
"Ovotestis," 270
Ovulation, 239, 244, 246, 251, 271, 283, 324
  in monestrous mammals, 253
  postparturient, 277
Ovum, 41, 262–263
  fertilized,
    cleavage of, 272–275
  life span of, 262
  size of, 262
Oxytocin, 301, 305

"Pacemaker," 230
Pacinian corpuscles, 363
"Packing" of fetus, 301–302
Palaeocene, 33, 61
Palate,
  cleft, 14
  evolution of, 13
Paleoanthropic, 67
Paleolithic epoch, 163–164
Pallium, 118
Pancreas, 225–226
Pan paniscus, 56
Pan troglodytes, 56–58, 278
Papillary layer, 183
Parafollicular cells, 224
Paranthropus, 64, 211
Parapineal, 222
Parapithecus, 55
Parathyroids, 224
Parietal lobe,
  of hemispheres, 131
Parthenogenesis, 268
Parturition, 241, 281, 300–304
  and multiple pregnancies, 303
  in nonhuman primates, 303
  stages of, 241
Patas monkeys (*Erythrocebus patas*), 34
Patella, 144
Pecking order, 312
Peking man, 65
Pelvic,
  bones, 153–154
    iliac, 154
    ilium, 153
    ischium, 153
    pubic, 153
  diaphragm, 155
Pelvis, 6, 153–156
Penfield, W., 130
Penis, 259, 321
Pennant, T., 31
Pericranium, 10
Perineum, 309
Peripheral nervous system (PNS), 114
  somatic, 114
  visceral, 114
Perodicticus potto, 34, 45, 62
Peroneal muscles, 151
Petty, W., 400
Pharyngeal arches, 289
Pharynx, 287
Phenyl-thio-carbamide, 81, 374
  nontasters, 374
  tasters, 374
Pheromones, 312, 314, 318, 374–75
  as chemosensory messages, 375
  as olfactory language, 374
  as phagostimulants, 374
  in sebum, 375
  as sexual attractants, 374
  species-specificity of, 374, 375
Philtrum, 13
Phone, 339
Phoneme, 345
Phonetics, 338–339
Photoreceptor, 222

Phrenic nerve, 29
Phrenology, 121
Phylogeny, 98
Physical maturity, 238
Piebaldism, 182
Pigmentation,
  of man's skin, 23
Pig-tailed macaques (*Macaca nemestrina*), 313
"Pill," the, 239
Pineal body, 222–223
Pinna, 15
Pithecanthropus, 65
Pitheciinae, 52
Pituitary, 220–222, 234–235, 239, 243, 246–247, 283
  anterior lobe, 234–235, 239, 243, 246
  and gonadotrophin (FSH), 243, 247
  posterior lobe, 119, 222
Placenta, 41–42, 281, 296–300
  bidiscoidal, 298
  caducous, 299
  chorio-allantoic, 298
  contradeciduate, 299
  deciduate, 299
  discoidal, 297–298
  early development of, 297
  epitheliochorial, 299
  full-term, 297
  hemochorial, 298
  labyrinthine, 297
  nondeciduate, 299
  primitive, 42
Placental,
  "barrier," 299
  villi, 299
Placentomes, 298
Plantar-flexion, 149–151
Plantar ligaments, 145
Plantigrade,
  fashion, 5
  posture, 159
Platyrrhines, 52, 54
Play, 423
Pleistocene, 33
Plesiadapis, 61
Pliocene, 33, 62
  apes, 166
Pliopithecus, 62–64
  foot and hand, 63
Polar body, 245
Pollex, 6
Pollution, 391–392, 424
Polyestrous mammals, 252–254
Polynesians, 76
Pomum Adami, 10, 345
Pongidae, 52, 54, 56
Pongo pygmaeus, 56
Pons, 124
Pontine nuclei, 124
Popliteus muscle, 153
Population,
  drift, 70
  increase in, 234
  zero, 328
Postnatal immaturity, 133
Postpartum estrus, 306
Postparturient ovulation, 277
Posture,
  erect, 136, 160, 166
  of man, 159–160
  plantigrade, 159
Pott, P., 151
Pott's fracture, 151, 388
Pottos (*Perodicticus potto*), 34, 45, 62, 137
Precoital activity, 317–318
Pregnancy, 240, 247, 261–300
  changes in mother during, 282
  duration of, 240
  growth of uterus during, 279
  hormonal changes during, 282
  mammary tissue and, 305
  and myometrium, 279
  pigmentation of skin during, 281
  tests for, 280
Prehensile tail, 52–53, 137
Premolars, 17, 38, 56, 201
Presbytis, 56
Pressure modalities, 360
Pressure receptors, 230
Primary curvatures, 157
  cervical, 157
  lumbar, 157
  thoracic, 157
Primary sex radio,
  birth of boys and girls, 270–271
Primates, 3, 5, 31–39, 43–61
  arboreal adaptation, 32
    smell, 32
    vision, 32
  brain, 38, 127–131
  definition, 39
  evolution,
    events, 32
    history, 31–39
  limbs, 36
  living, 44
  order, 3, 5
  relationship to man, 31–39
  special senses, 38
  structural adaptation, 32
  teeth, 202
Primigravid, 238, 300
Primitive streak, 285–286
Primordial germ cells, 256
Proboscis monkey (*Nasalis*), 13, 56
Proconsul, 63, 166

Progesterone, 239, 275
Prognathism, 12, 74, 77, 220
Projection fibers, 132
Pronograde, 47
Propithecus, 137
Propliopithecus, 62
Proprioception, 358
Proprioceptive organs, 360, 363–364
Prosimians, 5, 18, 31, 33, 43–45
  earliest primate fossils of, 61–68
    in America, 62
  families of, 5
  galagos, 3, 31, 34, 44–45, 62, 64, 252
  lemurs, 31, 33–34, 44
  lorises, 31, 34, 45, 62, 137
  in Oligocene, 62
  tarsiers, 9, 45, 62
  teeth of, 18, 202
  tree shrews, 31, 33–34, 44, 129
Prostaglandins, 301
Prostate, 265
  of man, 260
  vesicle, 235
Prostatic urethra, 260
Prostitution, 331
Prototheria, 41
Pseudohermaphrodites, 270
Pseudopregnancy, 253
Psychosexual constitution, 330
Puberty, 19, 234–237, 395
  delayed, 236
  in various mammals, 237
Pubic bones, 153
Puerperium, 241
Pulvinar, 132
Pyknic, 85
Pyknosomatic, 85
Pyramidal,
  decussation, 121
  motor system, 121
  tracts, 121
    motor, 129, 132

Quadriceps, 143–144, 153
Quadrumana, 97
Quadruplets, 6, 241
"Quickening," 295
Quintuplets, 6, 263

Race(s), 68–73, 80–81
  definition, 73
  evolution, 70
  nonphysical traits, 80–81
Ramapithecus, 63
Ramsay, E., 300
Ranvier, L., 115
Rathke, M. H., 200
  pouch of, 220
Renaissance, 91–93
Renin, 229
Reproduction,
  female, 236–256
  male, 256–260
Reproductive,
  capacity, man's, 234
  cycle, 6, 324
  patterns,
    in man, 233–252
    in primates and other mammals, 253–257
Retina, 369–370
  blood supply, 371
  cones, 369
  fovea centralis, 369
  macula lutea, 369
  rods, 369
  visual purple, 369
Retrocalcarine sulcus, 131
Rhetians, 74
Rh,
  agglutinogen, 300
  factor, 83
Rhesus monkey, 253
Rhinencephalon, 116
Ribs, 30
Rima glottidis, 345
Running, 136
Rut, 322
Ruysch, F., 100, 230

Sacro-coccygeal curve, 159
Sacrum, 153, 156–157
Salivary glands, 14
Satyr point, 15
Scala vestibuli, 349
Scalp, 10
Schleiden, M. H., 97
Schlemm, F., 369
Schultz, A. H., 337
Schwann, T., 97
Scoliosis, 159, 387
Sebaceous glands, 25, 190–192, 281
Secondary curvatures, 157
Secondary sex ratio,
  birth of boys and girls, 270–271
Secretory phase,
  ovarian cycle, 251
Semen, 265
Semibrachiation, 137
Semicircular canals, 377
  ampulla, 377
  otoliths, 377
  sensory hairs, 377
Semilunar cartilages, 153

Seminal,
  fluid, 260, 265, 268
  vesicles, 235, 260, 265, 301
Seminiferous tubules, 263
Senescence, 252, 394, 400, 411
Sensory,
  projection area, 122–123
  receptors, 112
  systems, 333–378
Septum transversum, 29
Serratus anterior, 162
Sertoli cells, 263
Sex,
  determination, 271
  drive, 311
  lures, 312
  play, 315–319
    in children, 318
    oral exploration during, 315
    visual sensations in, 317–318
  releasers, 318
  skin, 27, 256, 313
  spontaneity in, 325
  stimulants, 309
Sexual,
  attraction, 308–314
  attractiveness, 309
  behavior, 323
    aberrations in, 328–329
    in man, 307–331
    promiscuity in, 319
  courtship, 314
  cycles, 6, 238–239
    in men, 6
    in women, 238
  experiences, 307
  expression, 331
  maturity, 237–238
  needs, 330–331
  skin, 8
  stimulation, 169
Sheldon, W. H., 85
Sherrington, Sir Charles, 104, 121
Siamang (*Symphalangus*), 56–57
Sifaka, 137
Signals,
  chemical, 336
  postural, 335
Simian shelf, 15
Simpson, G. G., 41, 43, 56, 59, 101
Sinanthropus, 65
Singer, C., 90
Sinuses, 12–13
  air, 12
  inflammation of, 13
Skhūl in Mount Carmel, 67
Skin,
  adaptive changes, 18, 181
  aging, 169, 179
  architecture, 170–171
  circulation, 171
  color, 181–183
    in nonhuman primates, 182
  development, 173
  diseases, 380–382
  elasticity, 170
  friction, 26
  glands, 169, 189–190, 196–197
    development of, 197
    eccrine, 196
    function of, 197
    in man, 196
    sebaceous, 190
  innervation, 359–364
  and lymph, 171–172
  of man, 18–27, 167–197
  of nonhuman primates, 26–27
    modifications, 27
  pigmentation, 23, 169
  repair, 172
  and senses, 360–364
  and sex, 27, 257
    attraction, 19
  surface, 173–177
    furrows, 174–175
    lines, 174–175
    wrinkles, 174–175
  texture, 25
  topographic differences in, 25, 169–170
Skull,
  of elephants, 13
  of great apes, 10
  identity, 9
  of man, 9–10
    cranial capacity, 10
  of *Rooneyia*, 62
  simian, 10
Slavehood, Negro, 72
Slipped disc, 157
Smell, 47, 374–376
Snake Creek of Nebraska, 200
Society,
  disintegration of, 418
  human, 416
  and sexual behavior, 418
Soleus muscle, 142
Somatic cells, 272
Somatotonia, 86
Somatotrophic hormone (STH), 220
Somatotyping, 84–87
Somites, 286–287
Speech, 333–334, 338, 342
  in other animals, 334, 338
  articulate, 42
  neural control of, 344
  symbolic, 419
Sperm,
  acrosome, 264

Sperm (*Continued*)
end-piece, 264
head, 264
human, 264
immobilization, 271
production, 256
sensitization, 271
species differences in, 264–265
viability, 240
Spermateliosis, 264
Spermatocytes, 264
Spermatogonia, 263
Spermatozoa, 235, 263–268
formation, 260
movement, 266
in reproductive tract,
entry, 267–268
survival, 267–268
Spermicides, 325
Spider monkeys, 52, 137
Spinal cord, 116, 156
Spinal curvatures, 157
Spinal nerves, 286
Spinocerebellar tracts, 124
Spondylolisthesis, 159
Squirrel monkeys, 52
Stairway of animals, 96–100
Stammering, 346
Stapes, 347
Starling, E. H., 216
Steinheim, 66
Stem cells, 263
Stereoscopic vision, 47
Sterilization, 327
Sternomastoid, 9
Sternum, 30
Stevenson, R. L., 166
Streeter, G. L., 283–284
Stria of Gennari, 119
Stump-tailed macaques (*Macaca speciosa*), 23
Subcutaneous fat, 8
Suckling, 305, 317
Sudanians, 76
Superior corpora quadrigemina, 123–124
Suprarenals, 226–229
Suprasegmental structures, 116
Survival rate,
infant, 234
Suspensory ligament, 369
Swanscombe, 66
Sweat glands, 25, 192–197
apocrine, 192–196
comparative, 196
development, 197
eccrine, 196–197
function, 196–197
in man, 196
in primates, 197
"Swift birth," 301
Swim bladder, 352
Sylvian fissure, 123
Sylvian sulcus, 131
Symphysis pubis, 302
Synaptic contacts, 108–109
Syncytial,
processes (villi), 276
trophoblast, 275–276
Syncytiotrophoblast, 299
Syrinx, 340

Tabūn cave, 66
Tachyglossus, 41
Tactile,
exploration, 315
hairs, 185
sensibility, 167
Target cells, 219
Target tissue, 217
Tarsal bones, 143, 145
Tarsier, 9, 45, 62
Taste, 372–374
buds, 15, 372
gustatory cells, 373
Tasters, 81
Taurics, 74
Taurodontism, 212
Tectum, 123
Teeth, 16–18, 38, 41, 200–213, 295, 384
anatomical specializations of, 56
biting surfaces of, 212
canines, 17–18, 38, 201
caries, 213
cement, 207–208
chemical composition of, 208
comparative, 200
cusps, 211
deciduous, 17,
dentition, 210
in nonhuman primates, 210
dental formula of primates, 202
dentine, 206–207
development, 203, 209–211
enamel, 205–206
eruption, 209–211
evolution, 201
formula, 201
fossils, 200
function, 201
impaction, 211
incisors, 17–18, 38, 201
malocclusion, 212
of man, 16–18, 41, 202, 205
molars, 18, 38, 201, 211
of neandertaloids, 211
occlusion, 212
permanent, 17

periodontal membrane, 207–208
polyphyodont, 209–210
premolars, 17, 38, 56, 201
of prosimians, 18, 202
pulp, 208
replacement, 203–205
types, 200–202
unerupted, 295
versatility in function, 199
wisdom, 384
Temperament, 85–86
Temporal,
association, 129
bone, 347
lobe, 123
Tendo Achillis, 142–143
Teratogens, 283–284, 300
Terosseous ligaments, 145
Tertullian, 88
Testes, 259
function of, 235
Thalamus, 119–120, 129, 132
Thalassaemia, 81
Thalidomide, 284, 300
Theberge, J. B., 334
Theca interna, 246
Thecal gland, 246
Theca-luteal cells, 247
Theropithecus, 56, 256
Thorax, 29–30
Thymus, 225
Thyroid, 223
cartilage, of larynx, 10
evolution, 223
hormone, 223
Thyrotrophic hormone (TSH), 221, 223
Thyroxine, 223
Tibialis anterior, 144
Tinnitus, 348
Tobias, P. V., 65
Tongue,
and articulate speech, 15
in capturing food, 14
of man, 14
muscles, 15
as sensory organ, 14
specialization of, 14
Tools,
in Neanderthal kit, 67
Oldowan, 64, 66
stone, 64
Transplantation, organ, 394
Trapezius muscle, 161
Tree shrews, 31, 33–34, 44
Triiodothyronine, 223
Triplets, 6, 241
Trophoblast, 275–276, 297, 299
Tuatara (*Sphenodon*), 222
Tubercles of Montgomery, 316
Tulp, N., 97
Tupaia javanica, 129
Tupaiidae, 33, 44
Turbinals, 13
Turner, H. H., 270
Turner-Klinefelter syndrome, 246, 270
Twins, 6, 241, 262, 272, 284, 303
dizygotic, 262, 303
fraternal, 303
identical, 284
monozygotic, 262, 272, 303
occurrence of, 6, 241, 303
Tympanic bone, tubular, 56
Tyson, E., 97

Uakaries (*Cacajao*), 23
Urachus, 298
Uralics, 74
Urethra, 259
Uterine,
aspiration, 327
contractions, 300, 303
D and C, 327
glands, 251
"milk," 251
Uterus, 41, 241–246, 278–279
bicornuate, 250, 302
gravid, 301
growth of, 279
of nonpregnant human, 248
prolapse of, 248
unicornuate, 250

Vallate papillae, 372
Vas deferens, 260, 326
Vasectomy, 326
Vasopressin, 305
Veddians, 78
Ventricle,
lateral, 114
third, 114
Vertebral column, 156–157
Vertebrate appendages, 3
Vértesszöllös, 66
Vertex presentation, 301
Vesalius, A., 91, 93, 95
Vestibular perception, 377
equilibration of, 377
Vibrissae, 23, 185
Vicary, T., 95
Villi, 276, 285, 297
placental, 299
Visceral nervous system, 114
Viscerotonia, 86
Vision, 47
Visual,
acuity, 47
apparatus, 365–372

Visual (*Continued*)
area, 123
cortical, 123
center, 123
pathways, 371
geniculate bodies, 371–372
optic chiasma, 371
optic radiation, 372
optic tracts, 371
visual cortex, 372
visuo-sensory area, 272
perception,
in gibbons, 359
in man, 364–365
sense,
in nonhuman primates, 370
stimulation, 315–316
Vitalism theory, 101
Vitelline membrane, 269
Vitiligo, 182
Vitreous body, 369
Viviparous, 41
Vocabulary, silent, 336
Vocal apparatus,
in Cro-Magnon man, 344
folds, 244, 340, 342
in Neanderthal man, 333
sublaryngeal, 333
Vocalization,
in gibbons, 358
Voice, 339
changes with growth, 345–346
in nonhuman primates, 345
production, 343–345
resonators, 345
Vomeronasal organ, 376–377
Von Baer, K., 98
von Frey, M., 360
Von Mering, J., 226
Vulva, 309, 317
secretion of, 313

Walking, 27, 136, 149
in bipedal primates, 32–33, 137
dynamics of, 141
in man, 140–144
Wallace, A. R., 101
Washburn, S. L., 32–34, 336, 419
Weaning, 252
Weberian ossicles, 352
Weight, 7
Weismann, A., 101
Wenden, M., 230
Whales, 8
Whiskers, tactile, 23
Whorl, 175
Wiener, A. S., 83
Wilberforce, Bishop, 1
Willis, T., 127, 226
Window, oval, 348
Wombats, 42
Woolly monkeys (*Atelinae*), 52
Woolner, T., 15
Wren, Sir Christopher, 127
Wrinkles, skin, 174

Xanthoderms, 25

Yolk, sac, 285, 298

Zingians, 76
Zona pellucida, 245, 272, 275
Zygote, 240, 269, 272